NEURAL REGULATION IN THE VERTEBRATE ENDOCRINE SYSTEM

Neuroendocrine Regulation

Edited by

P. D. Prasada Rao
Nagpur University
Nagpur, India

and

Richard E. Peter
University of Alberta
Edmonton, Alberta, Canada

Springer Science+Business Media, LLC

Proceedings of Neural Control of the Endocrine System, held November 26–28, 1997, in Nagpur, India

ISBN 978-1-4613-7177-9 ISBN 978-1-4615-4805-8 (eBook)
DOI 10.1007/978-1-4615-4805-8

Originally published by Kluwer Academic / Plenum Publishers in 1999
Softcover reprint of the hardcover 1st edition 1999

10 9 8 7 6 5 4 3 2 1

A C.I.P. record for this book is available from the Library of Congress

NEURAL REGULATION IN THE VERTEBRATE ENDOCRINE SYSTEM
Neuroendocrine Regulation

PREFACE

The objective of this book is to provide recent information on neural regulation in the endocrine system in vertebrates. Classical studies have revealed that certain neurons synthesize and release chemical messengers into the vascular system. These neurons are endocrine devices that link the brain with the endocrine glands and other target organs. In vertebrates, the hypothalamus is the seat for chemical coordination and integration of environmental and hormonal cues to modulate function of the pituitary gland, and consequently, the functions of other endocrine glands. Exciting information generated during the past few decades has resulted in profound alterations in the conceptual fabric of endocrinology. From the wealth of information that emerged on neuropeptides of the central nervous system, and on the other connectivities of various brain centers, its has become clear that several extra-hypothalamic sites are also involved in regulation of hypophysial hormones.

The brain has assumed a greater importance in the regulation of the endocrine system. However, recent studies have revealed varying degrees of functional autonomy in hypophysial hormone secretion, which may be due to intrapituitary cytokines. Although gonadotropin-releasing hormone (GnRH) is a key regulator of gonadotropin secretion, there exists a GnRH receptor diversity in vertebrates such as the receptor presence in cancer cells. Recent studies have demonstrated the multifactorial nature of the neuroendocrine factors involved in growth hormone regulation in fish. On the other hand, in birds, thyrotropin-releasing hormone plays a major role in growth hormone release. In teleosts, the brain hypophysiotropic neurons regulate anterior pituitary hormone release by direct innervation, while in other vertebrates, peptides released in the median eminence are transported to the anterior pituitary by the hypothalamo–hypophysial blood portal system.

This book deals with neurons contributing to the pituitary innervation in teleosts, median eminence in vertebrates, brain GnRH and regulation of pituitary and ovarian functions in fishes, brain control of growth hormone in fish and birds, brain regulation of mammalian luteinizing hormone, functional autonomy of the adenohypophysis, photoperiodic control of GnRH in birds, changes in avian central steroid receptor expression with respect to reproduction and neural control of other endocrines such as pineal organ, endocrine pancreas, and the immune system. It is hoped that this book will prove useful to researchers, teachers, and students involved in research and teaching in neuroendocrinology and comparative endocrinology.

We would like to acknowledge the cooperation and support received form the authors and the publisher. The review papers in the book arose primarily from presentations made at the Symposium on Neural Control of the Endocrine System, held at Nagpur University, Nagpur, India, November 26–28, 1997. This was a Satellite Symposium of the XIII International Congress of Comparative Endocrinology, held in Yokohama, Japan, November 17–21, 1997.

P. D. Prasada Rao
R. E. Peter

CONTENTS

1

THE MEDIAN EMINENCE

A Mediator in the Regulation of the Pituitary by Brain

Hideshi Kobayashi,[1] Yoshimasa Yamaguchi,[1] and Haruko Uemura[2]

[1]Research Laboratory
Zenyaku Kogyo, Co. Ltd.
2-33-7 Ohizumi-machi
Nerima-ku, Tokyo 178-0062
[2]Biological Laboratory
Kanagawa Dental College
Inaoka-cho, Yokosuka, Kanagawa 238-8580, Japan

1. INTRODUCTION

More than thirty different biologically active peptides have been demonstrated, in addition to several classical neurotransmitters, in the median eminence and, unlike neural information, these messenger molecules can be stored in axon endings. Most of these brain peptides seem to be conveyed from the median eminence to the pars distalis of the adenohypophysis via the hypothalamo-hypophysial portal vessels. They appear to control the activities of a large number of adenohypophysial cells, in contrast to the more limited numbers of cells that are controlled by direct innervation in fishes (Vollrath, 1967). Thus, the median eminence is a site for the collection of the various messenger chemicals that are produced in neurons and, upon demand, they are discharged into the hypophysial portal vessels. The definition of the median eminence will be discussed in the Section 2. Although regulation of the release of the messenger chemicals from the nerve endings is considered to occur predominantly at the level of cell bodies, the possibility of local regulation at the level of the median eminence is supported by extensive data that have been accumulated over the course of the past 40 years. However, the regulatory mechanisms at the level of the median eminence have not yet been fully elucidated and are still speculative.

In this review, we shall enumerate the various brain peptides and neurotransmitters in the median eminence and discuss their significance in relation to their release at the level of the median eminence. The involvement of some other substances such as interleukins, nitric oxide (NO), carbon monoxide (CO), glutamate and gap junctional proteins

Neural Regulation in the Vertebrate Endocrine System, edited by Prasada Rao and Peter, Kluwer Academic / Plenum Publishers, New York, 1999.

in the release of messenger molecules from the median eminence will also be described to some extent. The possible participation of the capillaries of the primary plexus and the portal vessels will be discussed in relation to the transfer of the messenger chemicals from the median eminence to the adenohypophysis. The involvement of tanycytes in the release of the messengers will be discussed. Brain peptides in the adenohypophysis are also considered with respect to the functional relationship between the median eminence and the adenohypophysis.

2. DEFINITION OF THE MEDIAN EMINENCE

The neurohypophysis consists of the median eminence and the pars nervosa (Figure 1). The pars nervosa is a mass that can easily be delineated anatomically, whereas the median eminence is difficult to delineate because it is not a simple lobe. It has not been necessary, for general use, to define the median eminence in detail, and a definition was not completely established until recently. Tilney (1914, 1936) was the first to specify that, in humans, the median eminence is the portion of the tuber cinereum that protuberates from the third ventricle, and that begins to assume prominence immediately behind the optic chiasm, extending backward as far as the cephalic limits of the premammillary area. Diepen (1962) stated that Tilney's median eminence includes only the ventral wall of the neural stalk and that the dorsal wall should also be included (Figures 1, 2a). Green (1951) defined the median eminence as that part of the neurohypophysis that receives its blood supply from the hypophysial portal circulation or that shares vascularization with the adenohypophysis. This definition has physiological implications and is also applicable to the median eminence of lower vertebrates. However, these earlier definitions were based only

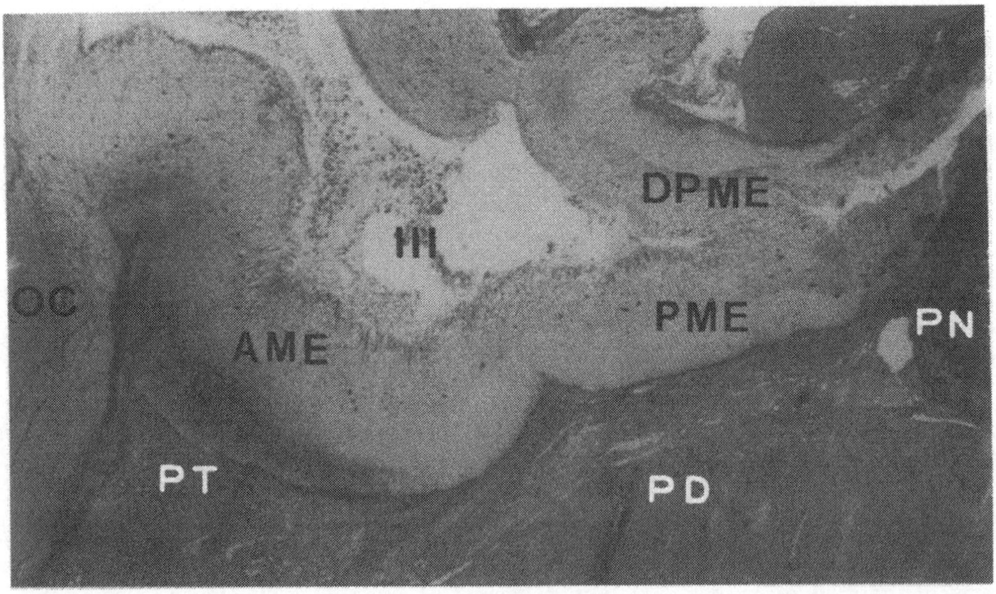

Figure 1. Sagittal section showing the median eminence of the pigeon. AME, Anterior median eminence; DPME, dorsal posterior median eminence; OC, optic chiasma; PD, pars distalis; PME, posterior median eminence; PN, pars nervosa; PT, pars tuberalis; III, third ventricle; Gomori's aldehyde-fuchsin staining.

on observations of the outer surface of the ventral portion of the tuber cinereum and they do not delineate an inner portion in the basal hypothalamus. In the recent usage of "median eminence" by many physiologists, the term seems to include the terminal portions of nerve axons and the processes of tanycytes and glial cells. Details of the structure of the median eminence, including these components, were shown schematically in an earlier review (Kobayashi *et al.*, 1970).

The median eminence is concerned mainly with the release of hypothalamic releasing hormones (RHs) and inhibiting hormones (IHs) from nerve endings into the capillaries of the primary plexus. Since we considered that tanycytes might be involved in the release of RHs and IHs (Matsui, 1966a, b; Kobayashi and Matsui, 1967; Kobayashi *et al.*, 1970; see also Section 6), we defined the median eminence anatomically as follows: exteriorly it is the basal area of the hypothalamus that is vascularized by the capillaries of the primary plexus, which drain the pars distalis of the adenohypophysis; interiorly it is the portion of the basal hypothalamus delineated by the tanycytes and their processes that terminate at the capillaries of the primary plexus (Kobayashi *et al.*, 1970; Figure 2). This definition is applicable to most vertebrates, but in the case of fishes some modifications are necessary (Kobayashi *et al.*, 1970). In general, the median eminence, when defined as above, includes some portion of the arcuate nucleus in mammals (Figure 2) and some portion of the tuberal nucleus or infundibular nucleus in birds, reptiles and certain species of amphibians. The median eminence of fishes does not contain the cells of the nucleus tuberis lateralis. Therefore, the presence of nerve cells is not a limiting factor in defining the median eminence (Kobayashi *et al.*, 1970). Two types of tanycyte, α and β, have been distin-

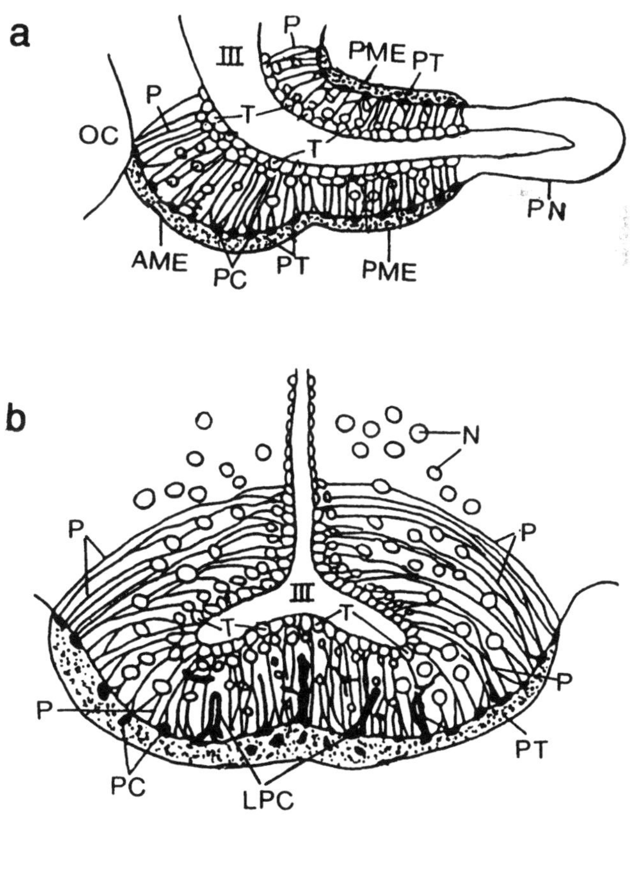

Figure 2. Diagram showing the median eminence, which is clearly delineated by the processes (P) of tanycytes (T). a, Sagittal section of the median eminence of the pigeon; b, frontal section of the median eminence of the rat. AME, Anterior median eminence; LPC, capillary loops of the primary plexus; N, neuronal perikarya of the arcuate nucleus; OC, optic chiasma, PC, capillaries of the primary plexus; PME, posterior median eminence; PN, pars nervosa; PT, pars tuberalis; III, third ventricle (modified from Kobayashi et al., 1970).

guished on the basis of morphological and topographical criteria in rats. The α tanycytes line the inferolateral wall of the third ventricle and the β tanycytes line the floor of the third ventricle (Akmayev *et al.*, 1973; Rodriguez *et al.*, 1979; Flament-Durand and Brion, 1985). These two types of tanycyte are included in the median eminence as defined above. When investigators introduce lesions or Halasz cuts, or when they inject or implant some material in the basal hypothalamic region, the exact site of the intervention must be described with reference to the definition of the median eminence. For example, lesions in the arcuate nucleus include damage to tanycytes in addition to neurons.

3. BIOACTIVE SUBSTANCES IN THE MEDIAN EMINENCE

3.1. Brain Peptides, Neurotransmitters and Their Colocalization

The first detailed electron-microscopic study of the median eminence was performed in the parakeet (Kobayashi *et al.*, 1961). The authors were surprised to see small vesicles that resembled synaptic vesicles and larger granules in the same axon endings. Since then, it has been demonstrated statistically that several populations of granules with significantly different diameters are present in the median eminence of several species of vertebrates. Furthermore, our analysis indicated that at least two populations of granules with different mean diameters coexist in some axon endings in the median eminence of the horse and the hagfish (Kobayashi and Ishii, 1968; Tsuneki *et al.*, 1976). In addition, most endings contain structures that resemble synaptic vesicles. Separation of granules and small vesicles by ultracentrifugation revealed that, in the median eminence of the horse, RHs such as gonadotropin-releasing hormone (GnRH) and corticotropin-releasing hormone (CRH) are included in granules with different diameters and that the small vesicles form a distinct fraction (Kobayashi *et al.*, 1970). It is now known that a large number of brain peptides, including RHs and IHs and several classical neurotransmitters, are present as messenger molecules in the central nervous systems of mammals (see Krieger, 1986; Palkovits, 1988; Hökfelt *et al.*, 1997). Palkovits (1984, 1988) described such messenger molecules in the median eminence in particular. A summary of available information is provided in Table 1.

The coexistence of various neurotransmitters and brain peptides and of various brain peptides is often observed in the central nervous system (Hökfelt *et al.*, 1986, 1987a, b; Crawley, 1990; Hökfelt, 1991; Brown, 1994, p237). Different messengers coexist in the hypothalamus, as indicated in Table 2. Since some hypothalamic nuclei have their nerve endings in the median eminence, colocalization of RHs, IHs and other brain peptides and neurotransmitters must be more extensive than is indicated in Table 2. The various combinations of these different messenger molecules in the median eminence are not always the same as those observed in other brain regions (Table 2). Furthermore, differences in combinations among species have also been reported (Hökfelt *et al.*, 1986). No rules have yet been found that allow us to predict colocalization of specific messenger molecules. In general, it seems that monoamines coexist most frequently with gastrointestinal peptides. The peptides derived from proopiomelanocortin are not colocalized with any of the neurotransmitters or peptides derived from other precursors, but the peptides derived from proenkephalin and prodynorphin are colocalized with classical neurotransmitters and other peptides (see Brown, 1994). Such chemical architecture might be relevant to colocalization in the median eminence.

Subcellular fractionation and centrifugation techniques have revealed that acetylcholine (ACh) and noradrenaline are present in the small and large vesicles, respectively,

Table 1. Brain peptides and transmitters demonstrated mostly immunohistochemically in the median eminence and the adenohypophysis of mammals

Brain peptides: Median eminence	Brain peptides: Adenohypophysis	Transmitters median eminence
Hypothalamic hormones		ACh
GnRH	+	Adrenaline
TRH	+	DA
CRH	+	GABA
GHRH	+	Histamine
Somatostatin	+	5-HT
Vasopressin	+	Noradrenaline
Oxytocin	+	
Pituitary hormones		
TSH	+	
ACTH	+	
PL	+	
Opioid peptides		
Dynorphin	+	
Endorphin	+	
Neo-endorphin	+	
Enkephalin	+	
MSH	+	
Gastrointestinal peptides		
Substance P	+	
Bombesin	+	
VIP	+	
CCK	+	
Glucagon		
PHI		
Motilin	+	
Neuropeptide Y	+	
Galanin	+	
PACAP[a]	+[b]	
Pancreatic polypeptide		
Other peptides		
ANG II	+	
ANP	+	
CGRP	+	
DSIP	+	
Endothelin[c]	+	
FMRFamide		
MCH[d]		
Neurotensin	+	
Urocortin[e]	+[f]	

Original reports of the brain peptides and of the transmitters without superscripts are not cited, but citations can be found in the following reviews: for median eminence, Kobayashi et al., 1970; Palkovits, 1984, 1988; Everitt et al., 1986, and for adenohypophysis, Denef et al., 1986; Schwartz and Cherny, 1992; Houben and Denef, 1994.

[a]Koves et al., 1990; [b]Masuo et al., 1993; [c]Samson et al., 1991; [d]Bittencout et al., 1992; [e]Hara et al., 1997; [f]Iino et al., 1997.

GHRH, growth hormone-releasing hormone; MSH, melanocyte-stimulating hormone; CCK, cholecystokinin; PACAP, pituitary adenylate cyclase-activating polypeptide; ANP, atrial natriuretic peptide; CGRP, calcitonin gene-related peptide; DSIP, delta sleep-inducing peptide; MCH, melanin-concentrating hormone.

Table 2. Colocalization of transmitters and brain peptides and of brain peptides in the arcuate nucleus, PVN, and SON and the median eminence of mammals

Colocalization	References
ANG II and vasopressin	Kilcoyne et al., 1980
CCK and CRH	Cecatelli et al., 1989
CCK and oxytocin	Vanderhaeghen et al., 1981
CCK, vasopressin and CRH	Mezey et al., 1986
CRH and enkephalin[a]	Hisano et al., 1986
CRH and galanin	Cecatelli et al., 1989
CRH and NT	Cecatelli et al., 1989
DA and dynorphin	Everitt et al., 1986
DA and enkephalin	Everitt et al., 1986
DA and galanin	Everitt et al., 1986
DA, galanin, and GHRH	Everitt et al., 1986
DA and GHRH[a]	Meister et al., 1986
DA, GHRH, and NT	Everitt et al., 1986
DA and NT	Everitt et al., 1986
DA and SOM[a]	Fuxe et al., 1986
DSIP and GnRH[a]	Pu et al., 1991
Dynorphin and α-neo-endorphin	Weber et al., 1981
Dynorphin and vasopressin	Watson et al., 1982
Enkephalin and NT	Cecatelli et al., 1989
Enkephalin, NT, and CRH	Cecatelli et al., 1989
Enkephalin and SOM[a]	Beauvillain et al., 1984
GABA and galanin[a]	Melander et al., 1986
Galanin and GHRH	Everitt et al., 1986
Galanin and NT	Cecatelli et al., 1989
Galanin and vasopressin	Gai et al., 1990
GHRH and NPY	Ciofi et al., 1987
GnRH and DSIP[a]	Vallet et al., 1991
Oxytocin and enkephalin	Vanderhaeghen et al., 1983
Oxytocin and vasopressin[a]	Jirikowski et al., 1991
PHI and CRH[a]	Hökfelt et al., 1987a
PHI, CRH, and enkephalin	Hökfelt et al., 1983
Vasopressin and CRH[a]	Whitnall et al., 1985
VIP and CRH[a]	Hökfelt et al., 1987a
VIP and NT	Cecatelli et al., 1989
VIP, NT and CRH	Cecatelli et al., 1989
VIP, PHI, and vasopressin	Okamura et al., 1986

[a]These colocalizations were detected in the median eminence. Since most neurons in the arcuate nucleus, PVN and SON are supposed to extend their axon endings to the median eminence or the pars nervosa, further colocalizations should be found in the median eminence.

and that peptides are exclusively stored in the large granules or vesicles in the submandibular salivary gland and parasympathetic nerves of the rat (Hökfelt *et al.*, 1986). The same is true for the granules and vesicles in the median eminence (Kobayashi *et al.*, 1970). It has been demonstrated that when classical neurotransmitters and neuropeptides coexist in axon endings, the pattern of release of each messenger changes depending on the intensity of neural stimulation. A low frequency of neural stimulation induces release of neurotransmitters from the small or large vesicles and a higher frequency of stimulation or a bursting pattern of stimulation causes the rapid release of neurotransmitters and the release of neuropeptides with delayed onset and long duration from large vesicles or granules (see Hökfelt *et al.*, 1986; Hökfelt, 1991). Thus, differential release occurs between colocalized classical neurotransmitters and neuropeptides, with resultant differential effects on their targets. How can we relate such results, obtained *in vitro*, to the functions of the median eminence? Furthermore, what is the pattern of release of individual neuropeptides from endings that contain two different neuropeptides? More than thirty different brain peptides and several neurotransmitters have been identified in the median eminence (Table 1). No matter whether they are colocalized or not, one type of messenger molecule probably modulates the release of another type of messenger via intracrine, paracrine or autocrine mechanisms. The possible effects of some neuropeptides and neurotransmitters that are released from the nerve endings on the tanycytes, the capillaries of the primary plexus and the portal vessels will be mentioned below.

3.2. Other Bioactive Substances

The findings that the secretion of CRH is increased by interleukin (IL)-1 (Sapolsky *et al.*, 1987) and that immunoreactive IL-1β is localized in the neurons in the human hypothalamus (Breder *et al.*, 1988) stimulated investigators to explore the roles of cytokines in the mediating communication between the endocrine and immune systems. IL-1 is found in the paraventricular nucleus (PVN) and the arcuate nucleus in the hypothalamus; IL-6 is found in the hypothalamus; and tumor necrosis factor (TNF)-α is found in several hypothalamic nuclei, the organum vasculosum of the lamina terminalis (OVLT) and the median eminence (see Wong *et al.*, 1996). Among cytokines, IL-1, IL-6 and TNF-α have been investigated in detail and have been shown to modulate the release of RHs and of pituitary hormones *in vitro* (see Rivier, 1995; Turnbull and Rivier, 1996). It is possible that circulating cytokines enter the median eminence, which is not subject to the blood-brain barrier, and act on nerve endings to modulate the release of messenger molecules. Alternatively, hypothalamic neurons containing IL-1, IL-6 or TNF-α might transport them to the median eminence, thereby modulating the release of messenger molecules at the level of the median eminence. Type 1 receptors for IL-1 have been identified in the median eminence of the mouse (Cunningham *et al.*, 1992) and the rat (Yabuuchi *et al.*, 1994).

In addition to the mediators discussed above, NO functions as a transmitter or neuromodulator in the nervous system. Nitric oxide synthase (NOS) catalyzes the conversion of L-arginine to L-citrulline and NO. The presence of NOS has been demonstrated within the PVN and supraoptic nucleus (SON) of the hypothalamus (Bhat *et al.*, 1995; Turnbull and Rivier, 1996), and the expression of NOS in the magnocellular division of the PVN can be enhanced by experimental treatments that are known to alter the activity of the PVN (see Turnbull and Rivier, 1996). NOS coexists with vasopressin in some neurons in the rostral regions of the PVN and SON (Calka and Block, 1993). Nicotinamide adenine dinucleotide phosphate-diaphorase (NADPH-diaphorase), which is required for the production of NO, coexists with CRH in a subpopulation of cells in the PVN (Siaud *et al.*, 1994). The coexistence of NO and oxytocin, but not vasopressin, has also been demonstrated since immunoreactive NADPH-diaphorase has been found only in oxytocin-producing cells (Miyagawa *et al.*, 1994). However, the release of oxytocin and vasopressin seems to be modulated by NO (Summy-Long *et al.*, 1993: Yasin *et al.*, 1993). NO has also been implicated in the regulation of the release of GnRH, CRH and prolactin (PL) in collaboration with classical neurotransmitters (see Turnbull and Rivier, 1996; Nelson *et al.*, 1997). NADPH-diaphorase is present in the median eminence, pars nervosa and pars distalis, as well as in several hypothalamic nuclei, in *Rana esculenta* (Prasada Rao *et al.*, 1997). The presence of NOS has also been demonstrated in the caudal neurosecretory cells and the urophysis of the teleost *Oreochromis niloticus* (Cioni *et al.*, 1997). Considering these observations, we can speculate that NO might act on nerve cell bodies that produce RHs and IHs and also on the endings of these nerves at the level of the median eminence. Furthermore, NOS is present in the hypophysial portal vasculature and it has been suggested that NO might mediate relaxation of the portal vessels, thereby changing the amount of blood that drains adenohypophysis (Ceccatelli *et al.*, 1992; Nelson *et al.*, 1997).

Heme oxygenase (HO) catalyzes the cleavage of the heme ring into biliverdin and CO, and two isoforms, HO-1 and HO-2, have been recognized. CO seems to be involved indirectly in the secretion of CRH (Pozzoli *et al.*, 1994) and GnRH (Lamar *et al.*, 1996). Since HO-1 has been detected immunocytochemically in neurons of the PVN and SON, in the OVLT and in tanycytes (Ibata *et al.*, 1997; Figure 3), it is possible that CO produced in these nuclei and the tanycytes might affect the release of RHs, IHs and other brain peptides at the level of the median eminence. The mRNA for HO-2 has been detected in the rat hypothalamus, but investigations related to neuroendocrine regulation have been very limited.

It has been demonstrated that ionotropic glutamate receptor subtypes NR1- and KA2-immunoreactivities were colocalized with GnRH-immunoreactivity in nerve terminals in the median eminence. In addition, some glutamate-immunopositive nerve terminals were shown to abut the many GnRH-immunoreactive nerve terminals. Synaptic contacts were not observed on these immunoreactive nerve terminals (Kawakami *et al.*, 1998). From those findings, they suggest that GnRH release may be regulated by excitatory amino acids in a non-synaptic fashion such as the volume transmission at the level of the median eminence.

4. MEDIANOSOMES AND RECEPTORS FOR BIOACTIVE SUBSTANCES

In the external layer of the median eminence, there are numerous nerve endings that contain many different kinds of brain peptide and transmitters as messenger molecules (Ta-

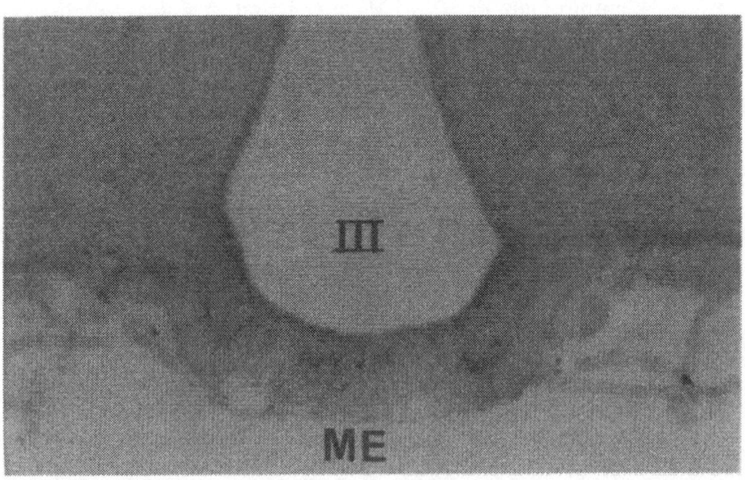

Figure 3. Immunoreactive heme oxygenase-1 (HO-1) in the cell bodies and the processes of tanycytes (black coloration) in the median eminence (ME) of the rat. The distribution of HO-1 suggests the involvement of CO in the release of messenger molecules from the nerve endings in the median eminence. III, Third ventricle; Å~80. (courtesy of Drs. Y. Ibata, N. Iijima and Y. Tamada, Kyoto Prefectural University of Medicine).

ble 2). They are not distributed diffusely over the entire capillary surface of the primary plexus. The nerve endings with identified transmitters form well defined aggregates, namely, chemically identifiable structural compartments, within distinct portions of the median eminence. Nerve endings containing different hypothalamic hormones also form different respective structural compartments in the median eminence, which have been designated "medianosomes" (see Fuxe *et al.*, 1991). Compartments of nerve endings that contain different hypothalamic hormones are considered to be equivalent to different medianosomes in the median eminence. The overlapping zones between medianosomes and other structural compartments are considered to be integrative medianosomes. Terminals of tanycyte processes usually participate in formation of integrative medianosomes. Communication within medianosomes might occur via paracrine mechanisms and/or gap junctions (see Section 6. 3 between the components of medianosomes. For example, GnRH nerve endings are codistributed with dopamine (DA) nerve endings and tanycyte terminals in the lateral palisade zone of the median eminence, forming the so-called integrative GnRH medianosome. DA released from its nerve endings inhibits the release of GnRH from GnRH medianosomes (Fuxe and Hökfelt, 1969). Rats treated locally with DA at the ventricular floor of the median eminence had a disturbed estrous cycle, which suggested inhibition of the release of GnRH by DA (Uemura and Kobayashi, 1971). In such a case, it appears that tanycytes that contain DA receptors might be involved in the inhibition of the release of GnRH (see Section 6. 2). The integrative thyrotropin-releasing hormone (TRH) medianosome is formed by the DA and TRH nerve endings and the terminals of tanycyte processes within the medial palisade zone and, also, within the lateral palisade zone. It seems that TRH neurons have receptors for DA and DA inhibits TRH secretion (see Fuxe *et al.*, 1991). DA neurons contain receptors for thyroid-stimulating hormone (TSH). Thus, a short-loop negative-feedback system exists within the integrative TRH medianosome. Tanycytes that contain DA receptors might be involved in the TRH secretion. It appears, therefore, that the release of hypothalamic hormones might be modulated within medianosomes at the level of the median eminence (see Fuxe *et al.*, 1991). Medianosomes containing RHs or IHs other than GnRH and TRH should be investigated. Further, it is unknown how medianosomes differentiate in the median eminence during the brain development.

Receptors for RHs, such as TRH (Taylor and Burt, 1982) and CRH (De Souza *et al.*, 1985; Wong *et al.*, 1994), have been identified in the external layer of the median eminence (see also Fuxe *et al.*, 1991). The presence of receptors for RHs in the median eminence suggests the existence of an ultra-short-loop feedback system within the medianosomes. Thus, identification of receptors for specific brain peptides and neurotransmitters in the external layer of the median eminence is important for investigations of the local circuitry that regulates the release of messenger molecules from the median eminence. Receptors for the peptides, such as opioids (Mansour *et al.*, 1987, 1994, 1995), galanin (Dutriez *et al.*, 1997), neuropeptide Y (Mikkelsen and Larsen, 1992), PL (Crumeyrolle-Arias *et al.*, 1993), vasopressin (Jurzak *et al.*, 1993), angiotensin (ANG) II (Barnes *et al.*, 1993; Jöhren and Saavedra, 1996), brain natriuretic peptide (Konrad *et al.*, 1990) and endothelin (Yamamoto *et al.*, 1997), have been detected in the median eminence. Furthermore, there are also receptors for neurotransmitters such as ACh (Okuda *et al.*, 1993), γ-aminobutyric acid (GABA) (Inglefield *et al.*, 1994) and 5-hydroxytryptamine (5-HT) (Gonzalez *et al.*, 1997) as well as DA and noradrenalin. Further, receptors for epidermal growth factor (Ma *et al.*, 1994), fibroblast growth factor (Gonzalez *et al.*, 1994) and neurotropin (Moga and Duong, 1997) were found in the median eminence. These receptors for brain peptides and neurotransmitters must have roles in the median eminence but little is known about their physiological significance.

5. ROLE OF CAPILLARIES OF THE PRIMARY PLEXUS OF THE PORTAL VESSELS

Nerve endings in the median eminence must make contact with the capillaries of the primary plexus to release messenger molecules. A question arises, then, whether each medianosome that contains a specific RH or IH releases it into specific nearby capillaries of the primary plexus. In other words, does each portal vessel contain specific RH or IH to drain a portion that contains specific pituitary cells or does each portal vessel contain several RHs or IHs released from different medianosomes? In birds and in some species of reptiles, the median eminence is clearly divisible into anterior and posterior regions, and the primary capillary plexus consists of distinct anterior and posterior capillary plexuses that correspond to the anterior and posterior regions of the median eminence (Figures 1, 4). The two groups of capillary plexuses converge into two groups of portal vessels, anterior and posterior, respectively. The anterior group of portal vessels mainly drains the cephalic lobe of the pars distalis of the adenohypophysis and the posterior group of the portal vessels supplies blood mainly to the caudal lobe of the pars distalis (see Oksche and Farner, 1974; Mikami, 1980). The cephalic lobe contains gonadotropic hormone (GTH), adrenocorticotropic hormone (ACTH), PL and TSH cells and the caudal lobe includes GTH and growth hormone (GH) cells (see Mikami, 1980; Figure 4). These observations suggest the significance of regional differentiation of the median eminence in relation to medianosomes with different RHs or IHs and the relationship between specific medianosomes and specific types of adenohypophysial cell. Now, the question arises as to the presence of different peptides in the anterior and the posterior median eminences and also

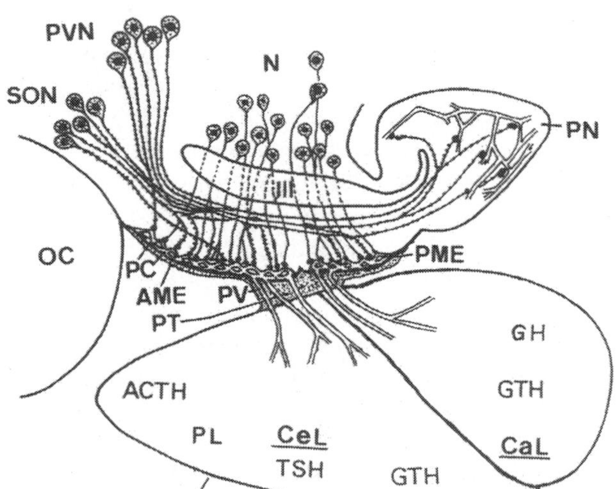

Figure 4. The median eminence consists of anterior (AME) and posterior (PME) divisions in birds. The dorsal portion of the median eminence (see Fig. 1) is not shown in this diagram. The adenohypophysis (AD) is divisible into the cephalic (CeL) and caudal (CaL) regions. Capillaries of the primary plexus (PC) form distinct anterior and posterior capillary plexuses that correspond to the anterior and posterior divisions of the median eminence. These two capillary plexuses converge into two groups of anterior and posterior portal vessels. The anterior group of portal vessels mainly drains the cephalic lobe and the posterior group supplies blood mainly to the caudal lobe of the AD. N, Neurons producing messenger molecules; OC, optic chiasma; PN, pars nervosa; PT, pars tuberalis; PV, portal vessel; PVN, paraventricular nucleus; SON, supraoptic nucleus; III, third ventricle.

in the cephalic and caudal lobes of the pars distalis. As mentioned below, most brain peptides in the median eminence are found in the pars distalis (Table 1, Section 7). The above mentioned relationship, whether loose or strict, seems to be present in vertebrates in general. In vertebrates, cells producing different pituitary hormones aggregate separately, to a greater or lesser extent depending upon species, in the pars distalis. There are many hypothalamo-hypophyseal portal vessels (15–30 vessels in birds), while there is only one hepatic portal vein. These observations suggest that specific medianosomes can release a specific hypothalamic hormone(s) into a specific portal vessel(s), so that each specific hormone reaches cells of a specific type(s). Further detailed studies are needed to validate this hypothesis.

In mammals, the capillaries of the primary plexus invade the median eminence tissues forming long capillary loops within them (Fig. 2). These loops increase the available neurohemal contact area for the endings of neurons that produce hypothalamic hormones. The extent of invasion by capillary loops of the median eminence might differ among species and might change with variations in physiological conditions. In birds, no such loops are observed.

The granules expelled by exocytosis from the nerve endings into the perivascular space of the capillaries of the primary plexus might undergo some chemical modifications in that space (Figures 5, 6). It is also possible that some brain peptides or neurotransmitters might influence endocytosis, exocytosis and production of the cytokines by the endothelial cells of the capillaries of the primary and secondary capillary plexuses. Some of the brain peptides or neurotransmitters might enter the capillary lumen through fenestrations (Fig. 6). Peptides, such as vasoactive intestinal polypeptide (VIP) and peptide histidine isoleucine (PHI), that are released from the median eminence might control the vasodilation of portal vessels (Ceccatelli *et al.*, 1992), thereby changing the amount of blood that drains the adenohypophysis. It might be worthwhile examining the chemical or enzymatic properties of the perivascular space, and chemical changes that might occur in the perivascular space upon sudden release of messenger molecules by high-frequency neural stimulation. The perivascular space is wide, as observed in other endocrine organs in general, and it is possible that messenger molecules might be stored in this space.

6. THE TANYCYTE AND ITS POSSIBLE FUNCTION

6.1. Link between the Cerebrospinal Fluid (CSF) and the Portal Vessels

The ependymal cells lining the ventral portion of the third ventricle have an elongated shape. The apical poles of these ependymal cells make contact with the CSF of the third ventricle and their distal poles abut on the capillaries of the primary plexus of the hypothalamo-hypophysial portal vessels. Horstmann (1954) described these unusual ependymal cells in selachians (elasmobranchs) and he was the first to use the term "tanycyte" in referring to these ependymal cells. Löfgren (1960) proposed that the tanycytes absorb the CSF of the third ventricle and then send absorbed substances to the capillaries of the primary plexus. He suggested a link between CSF and the portal vessels. We demonstrated for the first time that peroxidase injected into the subarachnoid space is absorbed by the tanycytes (Kobayashi *et al.*, 1972) and Nozaki (1975) suggested that the absorption of peroxidase by tanycytes might be regulated by the monoaminergic nerves. Other investigators suggested that, in addition to the absorption of CSF, tanycytes might absorb hormones from the portal blood and secrete them into the third ventricle. Then neurons receive the

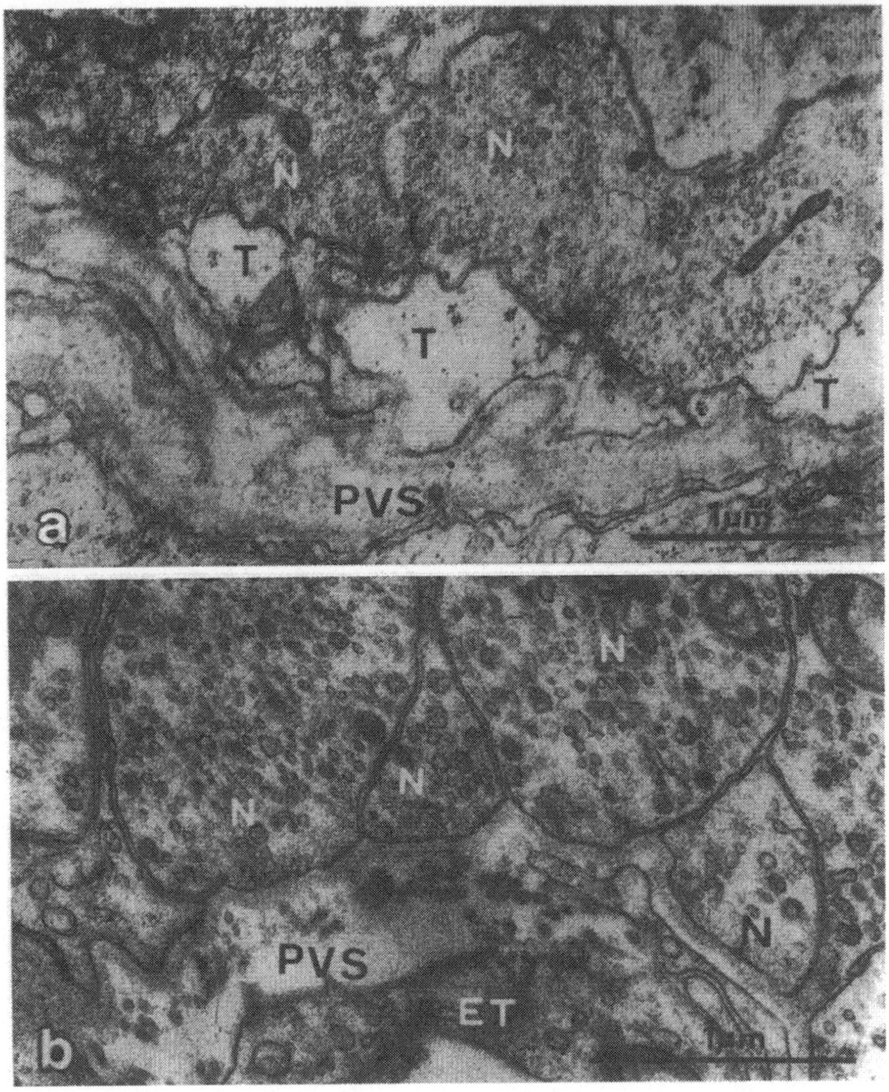

Figure 5. Neurohemal region of the median eminence. a, Posterior median eminence of the lizard, *Lacerta tachydromoides*; b, median eminence of the mouse. ET, endothelial cell; N, nerve endings containing granules and small vesicles; PVS, perivascular space; T, terminals of tanycyte processes. Note several T are interposed between N and capillary wall in the lizard, but in the mouse many N cover the entire surface of the capillary. Bars, 1 μm.

information from the CSF (see Wittkowski, 1967; Kobayashi *et al.*, 1970; Flament-Durand and Brion, 1985). The link between the CSF and the portal vessels, via the tanycytes, might contribute to short-loop negative feedback effects. However, it has been demonstrated that most pituitary hormones are actually synthesized by nerve cells in the brain (see Palkovits, 1988). These pituitary hormones might interact with RH-producing neurons in the hypothalamus. Therefore, a contribution by the tanycytes to short-loop negative feedback effects should be reconsidered. Blood levels of estrogen and corticoids cause morphological changes in the apical portion of tanycytes in several mammalian species (see Kobayashi *et al.*, 1970; Flament-Durand and Brion, 1985). However, the physiological meaning of the morphological changes caused by the steroids is unknown.

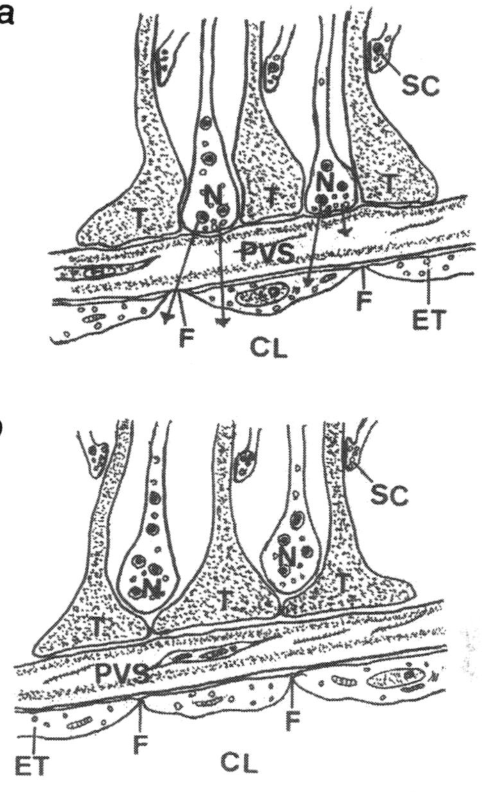

Figure 6. Terminals of tanycyte processes (T) form the gate through which messenger molecules pass to move from nerve endings (N) into the capillaries of the primary plexus. a, The gate is opened by movement of T or a change in the shape of T and, thus, N can make contact with the capillary wall to release messenger molecules into the perivascular space (PVS). The movements of T or a change in the shape of T might be controlled by synaptic contacts (SC) between DA nerves and the T. The left two arrows in **a** show two ways through which the messenger molecules reach the capillary lumen (CL), and the right two arrows show two probable targets, PVS and endothelial cell (ET), of the messenger molecules. b, The gate is closed by T that cover the entire surface of the capillary wall and messenger molecules cannot enter the capillaries. F, fenestration.

It was demonstrated recently that tanycytes probably take up insulin-like growth factor (IGF)-1 from brain parenchyma or CSF and accumulate it since tanycytes do not express mRNA for IGF-1 (Dueñas *et al.*, 1994). The level of IGF-1 in tanycytes fluctuates with changes in blood levels of sex steroids that are associated with the estrous cycle in rats, increasing in the afternoon of proestrus and in the morning of estrus, as compared to levels in the morning of proestrus, diestrus and metestrus. The accumulation of IGF-1 appears to be mediated by IGF-1 receptors that are present in the arcuate neurons and the tanycytes (Fernandez-Galaz *et al.*, 1996). Tanycytes have immunoreactive receptors for estrogen (Langub and Watson, 1992). These observations together suggest that estrogen stimulates the tanycytes to increase the numbers of IGF-1 receptors, with resultant enhanced accumulation of IGF-1 in the tanycytes. The fluctuations in levels of estrogen in the blood are linked to blood levels of gonadotropin, which are regulated by GnRH. Thus, the levels of IGF-1 in tanycytes might be involved in the release of GnRH. Further studies are needed to elucidate the involvement of IGF-1 in the release of hypothalamic hormones.

6.2. Entry of Messenger Molecules into the Capillaries of the Primary Plexus

It is unclear how brain peptides and neurotransmitters enter the capillaries of the primary plexus after their release from nerve endings in the median eminence. In a study by electron microscopy of the external layer of the median eminence of eleven species, some axon endings were found to abut directly on the capillaries while others abutted on the processes of tanycytes that made contact with capillaries. In the latter case, the processes of tanycytes in the median eminence were interposed between the axon endings and the capillary wall of the primary plexus (Oota *et al.*, 1974; Figures 5, 6). In view of these observations, we can speculate that the tanycyte processes form a gate through which messenger molecules, released from nerve endings, reach the capillaries of the primary plexus (Figure 6a). The gate is closed when the terminals of tanycyte processes occupy the entire capillary surface (Figures 5a, 6b) and it is open when the terminals move or change their shape such that they no longer cover the capillary surface (Figures 5b, 6a). The gate can be opened widely or narrowly or it can be closed, depending on the physiological conditions of the animal (Figure 6). When the gate opens more widely, nerve endings make broader contact with the surface of the capillaries and release larger amounts of brain peptides and neurotransmitters into the capillaries (Oota *et al.*, 1974; Kobayashi, 1975). The condition of the gate might be controlled by monoaminergic, perhaps dopaminergic, nerves since synaptic contacts have been observed (Figure 6) between monoaminergic fibers and the cell bodies or processes of the tanycytes in the median eminence in several species of vertebrates (Matsui, 1966a, b; Fuxe and Hökfelt, 1969; Kobayashi and Matsui, 1967; Rodriguez *et al.*, 1979; Krisch, 1986). The proposed scenario is based mainly on observations by electron microscopy. The numbers of terminals of tanycyte processes and nerve endings that abut on the capillaries should each be examined after appropriate experimental manipulations to confirm the scenario. Hökfelt (1973) observed that intraventricularly injected DA increased the extent to which tanycyte processes covered the capillary wall, with a resultant decrease in nerve endings that made contact with the capillary wall.

A similar idea was proposed on the basis of different considerations (see Hökfelt *et al.*, 1986; Fuxe *et al.*, 1991; Everitt *et al.*, 1992). The presence of a DA-and adenosine 3':5'-monophosphate-regulated phosphoprotein (DARPP)-32, with a molecular weight of 32,000, whose phosphorylation is regulated by DA and adenosine 3':5'-monophosphate, was demonstrated mainly in brain areas enriched in DA nerve terminals, suggesting that this protein might be present in DA-sensitive neurons (Hemmings *et al.*, 1984; Ouimet *et al.*, 1984; Walaas and Greengrand, 1984). The tanycytes in the basal hypothalamus contain strongly immunoreactive DARPP-32 (Ouimet *et al.*, 1984). A close anatomical relationship between DARPP-32-containing tanycytes and tyrosine hydroxylase-containing DA nerve endings in the external layer of the median eminence was demonstrated with antibodies against the two macromolecules (Hökfelt *et al.*, 1986). Accordingly, it was suggested that DA might induce tanycytes that contain DARPP-32 to expand along the ventral surface and, in this way, to limit the access of nerve endings that contain GnRH to a secretory position. As a consequence, there would be a decrease in the amount of GnRH released in the capillaries of the primary plexus. Thus, tanycytes seem to be involved in regulation of the release of messenger molecules from nerve terminals into capillaries in a process that is mediated by dopaminergic nerves.

Recently, King and Rubin (1994) suggested that sex steroids may influence the extension or the retraction of tanycyte endfeets surrounding GnRH terminals in rats. Furthermore, they suggested that gonadal steroids may influence the extension of GnRH terminals via changes in the cytoskeleton.

6.3. Gap Junctions in the Median Eminence

It is assumed that many neurons that contain RHs or IHs must release these factors from their terminals into the capillaries of the primary plexus in a coordinated fashion that adapts to changes in the physiological condition of the animal. For example, many GnRH axon endings must be synchronized to release GnRH for the generation of a preovulatory surge of gonadotropin. The mechanisms are unknown, but Hosny and Jennes (1998) observed that all GnRH neurons contain connexin-32, a gap-junctional protein, in their perikarya and axon endings in the median eminence of the female rat. No gap-junctional connections between adjacent GnRH neurons were observed. Hosny and Jennes (1998) suggested, therefore, that gap-junctional coupling of GnRH neurons with neighboring non-GnRH-containing cells might occur and might represent a mechanism whereby GnRH neurons could be synchronized or whereby hormonal or transmitter signals could be conveyed to GnRH neurons (Hosny and Jennes, 1998). It is possible that the neighboring non-GnRH-containing cells to which these authors referred might be the tanycytes, which have estrogen receptors and synaptic contacts with dopaminergic axons, as mentioned above. An intimate topographical relationship between GnRH fibers and tanycyte processes has been demonstrated in the median eminence (Nozaki *et al.*, 1979). The gap-junctional connections between GnRH neurons and tanycytes should be investigated. This "gap-junction hypothesis" might be applicable to the release of other RHs or IHs from the median eminence.

Immunoreactive connexin-43 was found in the tanycytes of the rat median eminence, although the presence of connexin-43 was not specific to tanycytes (Yamamoto *et al.*, 1992). It has been demonstrated that endothelin (Giaume *et al.*, 1992) and DA (Hampson *et al.*, 1992) can modulate the activity of the connexin-43 gap-junctional channels between mouse cultured astrocytes and between mammalian amacrine cells, respectively. Therefore, the activity of connexin-43 gap-junctional channels that might be formed between tanycytes or between tanycytes and nerve fibers might be inhibited by endothelin and DA, which are present in the median eminence (Table 1). However, the localization and the functions of connexin-43 gap junctions in the median eminence are unknown. Connexin-43 immunoreactivity was not detected in the GnRH neurons (Hosny and Jennes, 1998). Further studies are now necessary of the gap junctions and their relationship to the release of neuropeptides from nerve endings in the median eminence.

7. BRAIN PEPTIDES IN THE ADENOHYPOPHYSIS

Most brain peptides, including RHs and IHs, that are found in the median eminence can be detected by immunohistochemical staining or radioimmunoassay in the adenohypophysis (Table 1). Cells in the pars distalis containing certain specific peptides have been identified in many cases (see Houben and Denef, 1990, 1994). Some peptides in the pars distalis might have been transported from the median eminence via the hypophysial portal vessels and been taken up by the cells. However, local synthesis of most of the peptides is suggested by the presence of the appropriate mRNAs (Schwartz and Cherny, 1992). The levels of some peptides in the pituitary do not always change in parallel with those in the hypothalamus, suggesting the local synthesis of these peptides and, also, the specific functional significance of these peptides in the adenohypophysis. Some peptides stimulate and others inhibit the release of adenohypophysial hormones *in vitro* (see McCann, 1991; Schwartz and Cherny, 1992; Denef *et al.*, 1986: Houben and Denef, 1994; Brown, 1994). The extent to which the brain peptides colocalized in the adenohypophysial cells contribute to their physiological functions is unknown. Some classical neurotransmitters are also transported from the median eminence through the portal vessels and regulate the release of hormones from the adenohypophysis (see Fuxe *et al.*, 1980; Denef *et al.*, 1986; McCann, 1991; Houben and Denef, 1994).

8. CONCLUDING REMARKS

A current definition of the median eminence (Kobayashi *et al.*, 1970) includes not only the outside but also the inside portion of the basal hypothalamus. In this definition, the distribution of nerve endings, the terminals of tanycyte processes and the capillaries of the primary plexus are important considerations. More than thirty different brain peptides, several classical neurotransmitters, cytokines and gap-junctional proteins have been identified in the median eminence. An obvious question is, then, why are there so many bioactive substances in the median eminence? There is regional differentiation with respect to different brain peptides and neurotransmitters, and their distribution can be discussed in terms of so-called medianosomes. Receptors in the median eminence for most of the messenger molecules, but not for all of them, have been summarized. In the external layer, most of the messenger molecules might interact to ensure their release in a coordinate fashion through intracrine, autocrine and paracrine activities. The way in which the messenger molecules behave in the perivascular space after they reach the capillaries is unknown. Some of them might affect endocytosis or exocytosis of the endothelial cells and, furthermore, might cause the dilation or constriction of the portal vessels, regulating the blood flow from the median eminence to the pars distalis. The tanycytes seem to regulate the release of hypothalamic hormones and/or other brain peptides from nerve endings by interposing their terminals between the nerve endings and the capillary wall of the primary plexus. This tanycyte-mediated regulation seems to be under the control of dopaminergic nerves. Most of active peptides in the median eminence are found in the pars distalis. Some are synthesized in the pars distalis but others are transported from the median eminence. Some brain peptides seem to stimulate and others seem to inhibit the secretion of pituitary hormones. Cytokines, such as IL-1, IL-6 and TNF-α, seem to be involved in the regulation of the release of messenger molecules from the nerve endings in the median eminence. Involvement of NO, CO, IGF-I, glutamate and gap junctions has also been suggested. Although the release of messenger molecules from the nerve endings is considered to be regulated predominantly at the level of their perikarya, various mechanisms for fine tuning of their release seem to be operative at the level of the median eminence.

ACKNOWLEDGMENTS

The authors are indebted to Mr. Kazuhiro Hashimoto, the President of Zenyaku Kogyo Co. Ltd., for his unwavering encouragement and support.

REFERENCES

Akmayev, I. G., Fidelina, O. V., Kobolova, Z. A., Popov, A. P., and Snitkova, T.A. (1973). Morphological aspects of the hypothalamic-hypophyseal system. IV. Median basal hypothalamus. an experimental morphological study. *Zeit. Zellforsch.* **137**, 493–512.

Barnes, J. M., Steward, L. J., Barber, P. C., and Barnes, N. M. (1993). Identification and characterisation of angiotensin II receptor subtypes in human brain. *Eur. J. Pharmacol.* **230**, 251–258.

Beauvillain, J.-C., Tramu, G., and Garaud, J.-C. (1984). Coexistence of substances related to enkephalin and somatostatin in granules of the guinea- pig median eminence: demonstration by use of colloidal gold immunocytochemical methods. *Brain Res.* **301**, 389–393.

Bhat, G. K., Mahesh, V. B., Lamar, C. A., Ping, L., Aguan, K., and Brann, D. W. (1995). Histochemical localization of nitric oxide neurons in the hypothalamus: association with gonadotropin-releasing hormone neurons and co-localization with N-methyl-D-aspartate receptors. *Neuroendocrinology* **62**, 187–197.

Bittencourt, J. C., Presse, F., Arias, C., Peto, C.,Vaughan, J., Nahon, J. L., Vale, W., and Sawchenko, P. E. (1992). The melanin-concentrating hormone system of the rat brain: an immuno- and hybridization histochemical characterization. *J. Comp. Neurol.* **319**, 218–245.

Breder, C. D., Dinarello, C. A., and Saper, C. B. (1988). Interleukin-1 immunoreactive innervation of the human hypothalamus. *Science* **240**, 321- 324.

Brown, R. E. (1994). "An Introduction to Neuroendocrinology" pp. 221–248. Cambridge University Press, Cambridge.

Calka, J., and Block, C. H. (1993). Relationship of vasopressin with NADPH- diaphorase in the hypothalamo-neurohypophysial system. *Brain Res. Bull.* **32**, 207–210.

Ceccatelli, S., Eriksson, M., and Hökfelt, T. (1989). Distribution and coexistence of corticotropin-releasing factor-, neurotensin-, enkephalin-, cholecystokinin-, galanin- and vasoactive intestinal polypeptide/peptide histidine isoleucine- like peptides in the parvocellular part of the paraventricular nucleus. *Neuroendocrinology* **49**, 309–323.

Ceccatelli, S., Lundberg, J. M., Fahrenkrug, D. S., Bredt, D. S., Snyder, S. H., and Hökfelt, T. (1992). Evidence for involvement of nitric oxide in the regulation of hypothalamic portal blood flow. *Neuroscience* **51**, 769–772.

Ciofi, P., Croix, D., and Tramu, G. (1987). Coexistence of hGHRF and NPY immunoreactivities in neurons of the arcuate nucleus of the rat. *Neuroendocrinology* **45**, 425–428.

Cioni, C., Greco, A., Pepe, A., Devito, L., and Colasanti, M. (1997). Nitric oxide synthase in the caudal neurosecretory system of the teleost *Oreochromis niloticus*. *Neurosci. Lett.* **238**, 57–60.

Crawley, J. N. (1990). Coexistence of neuropeptides and classical transmitters. Functional interactions between galanin and acetylcholine. *Annals N. Y. Acad. Sci.* **579**, 233–245.

Crumeyrolle-Arias, M., Latouche, J., Jammes, H., Djiane, J., Kelly, P. A., Reymond, M. J., and Haour, F. (1993). Prolactin receptors in the rat hypothalamus: autoradiographic localization and characterization. *Neuroendocrinology* **57**, 457–466.

Cunningham, Jr, E. T., Wada, E., Carter D. B., Tracey, D. E., Battey J. F., and De Souza, E. B. (1992). *In situ* histochemical localization of type I interleukin-1 receptor messenger RNA in the central nervous system, pituitary, and adrenal gland of the mouse. *J. Neurosci.* **12**, 1101–1114.

Denef, C., Baes, M., and Schramme, C. (1986). Paracrine interactions in the anterior pituitary: role in the regulation of prolactin and growth hormone secretion. *Front. Neuroendocrinol.* **9**, 115–148.

De Souza, E. B., Insel, T. R., Perrin, M. H., Rivier, J., Vale, W. W., and Kuhar, M. J. (1985). Corticotropin-releasing factor receptors are widely distributed within the rat central nervous system: an autoradiographic study. *J. Neurosci.* **5**, 3189–3203.

Diepen, R. (1962). *In* "Handbuch der Mikroskopischen Anatomie des Menschen"(W. H. W. von Mollendorf, Ed.), Vol. 4, Pt. 7, pp. 152, 186, 368- 372. Springer, Berlin.

Dueñas, M., Luguin, S., Chowen, J. A., Torres-Alemān, I., Naftolin, F., and Garcia-Segura, L. M. (1994). Gonadal hormone regulation of insulin-like growth factor-I-like immunoreactivity in hypothalamic astroglia of developing and adult rats. *Neuroendocrinology* **59**, 528–538.

Dutriez, I., Lagny-Pourmir, I., Spelbaum, J., Beauvillain, J.-C. (1997). Ultrastructural localization of galanin and galanin receptors in the guinea pig median eminence. *Brain Res.* **753**, 36–46.

Everitt, B. J., Meister, B., Hökfelt, T., Melander, T., Terenius, L., Rokaeus, A.,Theodorsson-Norheim, E., Dockray, G., Edwardson, J., Cuello, C., Elde, R.,Goldstein, M., Hemmings, H., Ouimet, C., Ealaas, I., Greengard, P., Vale, W.,Weber, E., Wu, J.-Y., and Chang, K.-J. (1986). The hypothalamic arcuate nucleus-median eminence complex: immunohistochemistry of transmitters, peptides and DARPP-32 with special reference to coexistence in dopamine neurons. *Brain Res. Rev.* **11**, 97–155.

Everitt, B. J., Meister, B., and Hökfelt, T. (1992). The organization of monoaminergic neurons in the hypothalamus in relation to neuroendocrine integration. *In* "Neuroendocrinology" (C. B. Nemeroff, Ed.), pp. 87–128. CRC Press, Boca Raton, Ann Arbor, London.

Fernandez-Galaz, M. C., Torres-Aleman, I., and Garcia-Segura, L. M. (1996). Endocrine-dependent accumulation of IGF-I by hypothalamic glia. *NeuroReport* **8**, 373–377.

Flament-Durand, J., and Brion, J. P. (1985). Tanycytes: morphology and functions: a review. *Intern. Rev. Cytol.* **96**, 121–155.

Fuxe, K., and Hökfelt, T. (1969). Catecholamines in the hypothalamus and pituitary gland. *In* "Frontiers in Neuroendocrinology" (W. F. Ganong and L. Martini, Eds.), pp. 47–96. Oxford Univ. Press, New York.

Fuxe, K., Agnati, L. F., Andersson, K., Locatelli, V., Eneroth, P., Hökfelt, T., Mutt, V., McDonald, T., Etreby, E., Zini, I., and Calza, L. (1980). Concepts in neuroendocrinology with emphasis on neuropeptide-monoamine interactions in neuroendocrine regulation. *In* "Progress in Psychoneuroendocrinology" (F. Brambilla, G. Racagni and D. deWied, Eds.), pp. 47–61. Elsevier/North- Holland Biomedical Press. Amsterdam, New York, Oxford.

Fuxe, K., Andersson, K., Harfstrand, A., Agnati, L. F., Eneroth, P., Janson, A.M., Vale W., Thorner, M., and Goldstein, M. (1986). Medianosomes as integrative units in the external layer of the median eminence. Studies on GRF/catecholamine and somatostatin/catecholamine interactions in the hypothalamus of the male rat. *Neurochem. Int.* **9**, 155–170.

Fuxe, K., Agnati, L. F., Zoli, M., Biagini, G., Cintra, A., and Eneroth, P. (1991). Regulatory peptides in the neuroendocrine system. Aspects of the communicational and computational processes in peptidergic neurons and their steroidal regulation. *In* "Brain Endocrinology" 2nd Ed. (M. Motta, Ed.), pp. 31–69. Raven Press, New York.

Gai, W. P., Geffen, L. B., and Blessing, W. W. (1990). Galanin immunoreactive neurons in the human hypothalamus: colocalization with vasopressin- containing neurons. *J. Comp. Neurol.* **298**, 265–280.

Giaume, C., Cordier, J., and Glowinski, J. (1992).Endothelins inhibit junctional permeability in cultured mouse astrocytes. *Eur. J. Neurosci.* **4**, 877–881.

Gonzalez, A. M., Logan, A., Ying, W. Lappi, D. A., Berry, M., and Baird, A. (1994). Fibroblast growth factor in the hypothalamic-pituitary axis: differential expression of fibroblast growth factor-2 and a high-affinity receptor. *Endocrinology* **134**, 2289–2297.

Gonzalez, M. I., Greengrass, P., Russell, M., Wilson, C. A. (1997). Comparison of serotonin receptor numbers and activity in specific hypothalamic areas of sexually active and inactive female rats. *Neuroendocrinology* **66**, 384–392.

Green, J. D. (1951). The comparative anatomy of the hypophysis, with special reference to its blood supply and innervation. *Am J. Anat.* **88**, 225–311.

Hampson, E. C. G. M., Vaney, D. L., and Weiler, R. (1992). Dopaminergic modulation of gap junction permeability between amacrine cells in mammalian retina. *J. Neurosci.* **12**, 4911–4922.

Hara, Y., Ueta, Y., Isse, T., Kabashima, N., Shibuya, I., Hattori, Y., Yamashita, H. (1997). Increase of urocortin-like immunoreactivity in the rat hypothalamo-neurohypophysial system after salt loading and hypophysectomy. *Neurosci. Lett.* **227**, 127–130.

Hemmings, H. C., Nairn, A. C., Aswad, D. W., and Greengard, P. (1984). DARPP-32, a dopamine - and adenosine 3':5'-monophosphate-regulated phosphoprotein enriched in dopamine-innervated brain regions. II. Purification and characterization of the phosphoprotein from bovine caudate nucleus. *J. Neurosci.* **4**, 99–110

Hisano, S., Daikoku, S., Yanaihara, N., and Shibasaki, T. (1986). Intragranular colocalization of CRF and Met-Enk-8 in nerve terminals in the rat median eminence. *Brain Res.* **370**, 321–326.

Hökfelt, T. (1973). Possible site of action of dopamine in the hypothalamic pituitary control. *Acta Physiol. Scand.* **89**, 606–608.

Hökfelt, T. (1991). Neuropeptides in perspective: the last ten years. *Neuron* **7**, 867–879.

Hökfelt, T., Fahrenkrug, J., Tatemoto, K., Mutt, V., Werner, S., Hulting, A.-L., Terenius, L., and Chang, K. J. (1983). The PHI (PHI-27)/corticotropin- releasing factor/enkephalin immunoreactive hypothalamic neuron: possible morphological basis for integrated control of prolactin, corticotropin, and growth hormone secretion. *Proc. Natl. Acad. Sci. USA* **80**, 895–898.

Hökfelt, T., Everitt, B., Meister, B., Melander, T., Schalling, M., Johansson, O.,Lundberg, J. M., Hulting, A.-L., Werner, S., Cuello, C., Hemmings, H.,Ouimet, C., Walaas, I., Greengard, P., and Goldstein, M., (1986). Neurons with multiple messengers with special reference to neuroendocrine systems. *Recent Prog. Horm. Res.* **42**, 1–70.

Hökfelt, T., Fahrenkrug, J., Ju, G., Ceccatelli, S., Tsuruo, Y., Meister, B., Mutt, V., Rundgren, M., Brodin, E., Terenius, L., Hulting, A.-L., Werner, S.,Björklund H., and Vale W. (1987a). Analysis of peptide histidine-isoleucine/vasoactive intestinal polypeptide-immunoreactive neurons in the central nervous system with special reference to their relation to corticotropin-releasing factor- and enkephalin-like immunoreactivities in the paraventricular hypothalamic nucleus. *Neuroscience* **23**, 827–857.

Hökfelt, T., Millhorn, K., Seroogy, K., Tsuruo, Y., Ceccatelli, S., Lindh, B.,Meister, B., Melander, T., Schalling, M., Bartfai, T., and Terenius, L. (1987b). Coexistence of peptides with classical neurotransmitters. *Experimentia* **43**, 769–780.

Hökfelt, T., Zhang, X., Xu, Z. Q., Ji, R. R., Shi, T., Corness, J., and Kerekes, N. (1997). The ups and downs of neuropeptides. *In* "Neuroendocrinology, Retrospect and Perspectives" (H.-W. Korf, and K.-H. Usadel, Eds.), pp. 5–23. Springer, Heidelberg.

Hosny, S., and Jennes, L. (1998). Identification of gap junctional connexin-32 mRNA and protein in gonadotropin-releasing hormone neurons of the female rat. *Neuroendocrinology* **67**, 101–108.

Horstmann, E. (1954). Die Faserglia des Selachiergehirns. *Zeit. Zellforsch.* **39**, 588–617.

Houben, H., and Denef, C. (1990). Regulatory peptides produced in the anterior pituitary. *Trends in Endocrinol. Metab.* **1**, 398–403.

Houben, H., and Denef, C. (1994). Bioactive peptides in anterior pituitary cells. *Peptides* **15**, 547–582.

Ibata, Y., Tanaka, M., Tamada, Y., Iijima, N., Hayashi, S., and Amaya, F. (1997). Brain and endocrine interaction. *Clinical Neurosci.* **15**, 14–20. (in Japanese).

Iino, K., Sasano, H., Oki, Y., Andoh, N., Shin, R.-W., Kitamoto, T., Totsune, K.,Takahashi, K., Suzuki, H., Nagura, H., and Yoshimi, T. (1997). Urocortin expression in human pituitary gland and pituitary adenoma. *J. Clin.Endocrinol. Metab.* **82**, 3842–3850.

Inglefield, J. R., Sieghart, W., Kellogg, C. K. (1994). Immunohistochemical and neurochemical evidence for GABAA receptor heterogeneity between the hypothalamus and cortex. *J. Chem. Neuroanat.* **7**, 243–252.

Jirikowski, G. F., Ramalho-Ortigai, F. J., and Caldwell, J. D. (1991). Transitory coexistence of oxytocin and vasopressin in the hypothalamo- neurohypophysial system of parturient rats. *Horm. Metab. Res.* **23**, 476–480.

Jöhren, O., and Saavedra, J. M. (1996). Expression of AT_{1A} and AT_{1B} angiotensin II receptor messenger RNA in forebrain of 2-wk-old rats. *Am. J.Physiol.* **271**, E104-E112

Jurzak, M., Fahrenholz, F., and Gerstberger, R. (1993).Vasopressin anti-idiotypic antibody staining in the rat brain: colocalization with $[^{35}S][pGlu^4, Cyt^6]$AVP(4–9) binding sites. *J. Neuroendocrinol.* **5**, 523–531.

Kawakami, S., Hirunagi, K., Ichikawa, M., Tsukamura, H., and Maeda, K. (1998). Evidence for terminal regulation of GnRH release by excitatory amino acids in the median eminence in female rats: a dual immunoelectron microscopic study. *Endocrinology* **139**, 1458–1461.

Kilcoyne, M. M., Hoffman, D. L., and Zimmerman, E. A. (1980). Immunocytochemical localization of angiotensin II and vasopressin in rat hypothalamus: evidence for production in the same neuron. *Clinical Sci.* **59**, 57s-60s.

King, J. C., and Rubin, B. S. (1994). Dynamic changes in LHRH neurovascular terminals with various endocrine conditions in adults. *Horm. Behav.* **28**, 349- 356.

Kobayashi, H. (1975). Absorption of cerebrospinal fluid by ependymal cells of the median eminence. *In* "Brain-Endocrine Interaction" II (K. M. Kniggi, D.E. Scott, H. Kobayashi, and S. Ishii, Eds.), pp. 109–122. Karger AG, Basel.

Kobayashi, H., Bern, H. A., Nishioka, R. S., Hyodo, Y. (1961). The hypothalamo-hypophyseal neurosecretory system of the parakeet, *Melopsittacus undulatus*. *Gen. Comp. Endocrinol.* **1**, 545–564.

Kobayashi, H., and Matsui, T. (1967). Synapses in the rat and pigeon median eminences. *Endocrinol. Japon.* **14**, 279–283.

Kobayashi, H., and Ishii, S. (1968). The median eminence as storage site for releasing factors and other biologically active substances. *Excerpta Medica Found., Intern. Congr. Series No. 184, Progr. in Endocrinol.* 548–554.

Kobayashi, H., Matsui, T., and Ishii, S. (1970). Functional electron microscopy of the hypothalamic median eminence. *Intern. Rev. Cytol.* **29**, 281–381.

Kobayashi, H., Wada, M., Uemura, H., and Ueck, M. (1972). Uptake of peroxidase from the third ventricle by ependymal cells of the median eminence. *Zeit. Zellforsch.* **127**, 545–551.

Konrad, E. M., Thibault, G., Pelletier, S., Genest, J., and Cantin, M. (1990). Brain natriuretic peptide binding sites in rats: *in vitro* autoradiographic study. *Am. J. Physiol.* **259**, E246-E255.

Koves, K., Arimura, A., Somogyvari-Vigh, A., Vigh, S., Miller, J. (1990). Immunohistochemical demonstration of a novel hypothalamic peptide, pituitary adenylate cyclase-activating polypeptide, in the ovine hypothalamus. *Endocrinology* **127**, 264–2671.

Krieger, D. T. (1986). An overview of neuropeptides. *In* "Neuropeptides in Neurologic and Psychiatric Disease" (J. B. Martini and J. D. Barchas, Eds.), pp. 1–33. Raven Press, New York.

Krisch, B. (1986). Ultrastructure of regulatory neuroendocrine neurons and functionally related structures. *In* "Morphology of Hypothalamus and its Connections" (D. Ganten, and D. Pfaff, Eds.), pp. 251–290. Springer-Verlag, Berlin.

Lamar, C. A., Mahesh, V. B., Brann, D. W. (1996). Regulation of gonadotropin- releasing hormone (GnRH) secretion by heme molecules: a regulatory role of carbon monoxide? *Endocrinology* **137**, 790–793.

Langub, M. C., and Watson, R. E. Jr. (1992). Estrogen receptor-immunoreactive glia, endothelia, and ependyma in guinea pig preoptic area and median eminence: electron microscopy. *Endocrinology* **130**, 364–372.

Löfgren, F. (1960). New aspects of the hypothalamic control of the adenohypophysis. *Acta Morph. Neerl.-Scand.* **3**, 55–78.

Ma, Y. J., Hill, D. F., Junier, M.-P., Costa, M. E., Felder, S. E., and Ojeda, S. R. (1994). Expression of epidermal growth factor receptor changes in the hypothalamus during the onset of female puberty. *Mol. Cell. Neurosci.* **5**, 246–262.

Mansour, A., Khachaturian, H., Lewis, M. E., Akil, H., and Watson, S. J. (1987). Autoradiographic differentiation of mu, delta, and kappa opioid receptors in the rat forebrain and midbrain. *J. Neurosci.* **7**, 2445–2464.

Mansour, A., Fox, C. A., Burke, S., Meng, F., Thompson, R. C., Akil, H., and Watson, S. J. (1994). Mu, delta, and kappa opioid receptor mRNA expression in the rat CNS: an *in situ* hybridization study. *J. Comp. Neurol.* **350**, 412–438.

Mansour, A., Fox, C. A., Akil, H., and Watson, S. J. (1995). Opioid-receptor mRNA expression in the rat CNS: anatomical and functional implications. *Trends Neurosci.* **18**, 22–29.

Masuo, Y., Suzuki, N., Matsumoto, H., Tokito, F., Matsumoto, Y., Tsuda, M., and Fujino, M. (1993). Regional distribution of pituitary adenylate cyclase activating polypeptide (PACAP) in the rat central nervous system as determined by sandwich-enzyme immunoassay. *Brain Res.* **602**, 57–63.

Matsui, T. (1966a). Fine structure of the posterior median eminence of the pigeon, *Columba livia domestica*. *J. Fac. Sci., Univ. Tokyo* IV, **11**, 49–70.

Matsui, T. (1966b). Fine structure of the median eminence of the rat. *J. Fac. Sci., Univ. Tokyo* IV, **11**, 71–96.

McCann, S. M. (1991). Neuroregulatory peptides. *In* "Brain Endocrinology" 2nd Ed. (M. Motta, Ed.), pp. 1–30. Raven Press, New York.

Meister, B., Hökfelt, T., Vale, W. W., Sawchenko, P. E., Swanson, L., and Goldstein, M. (1986). Coexistence of tyrosine hydroxylase and growth hormone-releasing factor in a subpopulation of tubero-infundibular neurons of the rat. *Neuroendocrinology* **42**, 237–247.

Melander, T., Hökfelt, T., Rokaeus, A., Cuello, A. C., Oertel, W. H., Verhofstad, A., and Goldstein, M. (1986). Coexistence of galanin-like immunoreactivity with catecholamines, 5-hydroxytryptamine, GABA and neuropeptides in the rat CNS. *J. Neurosci.* **6**, 3640–3654.

Melander, T., Köhler, C., Nilsson, S., Hökfelt, T., Brodin, E., Theodorsson, E., and Bartfai, T. (1988). Autoradiographic quantitation and anatomical mapping of ^{125}I-galanin binding sites in the rat central nervous system. *J. Chem. Neuroanat.* **1**, 213–233.

Mezey, E., Reisine, T. D., Skirboll, L., Beinfeld, M., and Kiss, J. Z. (1986). Role of cholecystokinin in corticotropin release: coexistence with vasopressin and corticotropin-releasing factor in cells of the rat hypothalamic paraventricular nucleus. *Proc. Natl. Acad. Sci. USA* **83**, 3510–3512.

Mikami, S. (1980). Comparative anatomy and evolution of the hypothalamo- hypophysial systems in higher vertebrates. *In* "Hormones, Adaptation and Evolution" (S. Ishii, T. Hirano and M. Wada, Eds.), pp. 57–70. Japan Sci. Soc. Press, Tokyo/Springer-Verlag, Berlin.

Mikkelsen, J. D., and Larsen, P. J. (1992). A high concentration of NPY (Y_1) - receptor mRNA-expressing cells in the rat arcuate nucleus. *Neurosci. Lett.* **148**, 195–198.

Miyagawa, A., Okamura, H., and Ibata, Y. (1994). Coexistence of oxytocin and NADPH-diaphorase in magnocellular neurons of the paraventricular and the supraoptic nuclei of the rat hypothalamus. *Neurosci. Lett.* **171**, 13–16.

Moga, M. M., and Duong, T. (1997). p75 Neurotropin receptor immunoreactivity in the aged human hypothalamus. *Neurosci. Lett.* **231**, 9–12.

Nelson, R. J., Kriegsfeld, L. J., Dawson, V. L., and Dawson, T. M. (1997). Effects of nitric oxide on neuroendocrine function and behavior. *Front. in Neuroendocrinol.* **18**, 463–491.

Nozaki, M. (1975). Tanycyte absorption affected by the hypothalamic deafferentation in the Japanese quail, *Coturnix coturnix japonica*. *Cell Tiss. Res.* **163**, 433–443.

Nozaki, M., Taketani, Y., Minaguchi, H., Kigawa, T., and Kobayashi, H. (1979). Distribution of LHRH in the rat and mouse brain with special reference to the tanycytes. Cell Tiss. Res. 197, 195–212.

Okamura, H., Murakami, S., Fukui, K., Uda, K., Kawamoto, K., Kawashima, S., Yanaihara, N., and Ibata, Y. (1986). Vasoactive intestinal peptide-and peptide histidine isoleucine amide-like immunoreactivity colocalize with vasopressin-like immunoreactivity in the canine hypothalamo- neurohypophysial neuronal system. *Neurosci. Lett.* **69**, 227–232.

Oksche, A., and Farner, D. S. (1974). Neurohistological studies of the hypophyseal system of *Zonotrichia gambelli* (Aves, Passeriformes). *Adv. in Anat. Embryol. and Cell Biol.* 48, pp. 1–136. Springer-Verlag, Berlin.

Okuda, H., Shioda, S., Nakai, Y., Nakayama, H., Okamoto, M., and Nakashima, T. (1993). The presence of corticotropin-releasing factor-like immunoreactive synaptic vesicles in axon terminals with nicotinic acetylcholine receptor-like immunoreactivity in the median eminence of the rat. *Neurosci. Lett.* **161**, 183- 186.

Oota, Y., Kobayashi, H., Nishioka, R. S., and Bern, H. A. (1974). Relationship between neurosecretory axon and ependymal terminals on capillary walls in the median eminence of several vertebrates. *Neuroendocrinology* **16**, 127- 136.

Ouimet, C. C., Miller, P. E., Hemmings, H., Walaas, S. L., and Greengard, P. (1984). DARPP-32, a dopamine- and cyclic adenosine-3':5'-monophosphate- regulated phosphoprotein enriched in dopamine-innervated brain regions. III. Immunohistochemical localization. *J. Neurosci.* **4**, 111–124.

Palkovits, M. (1984). Neuropeptides in the hypothalamo-hypophyseal system: lateral retrochiasmatic area as a common gate for neuronal fibers towards the median eminence. *Peptides* **5** (Suppl. 1), 35–39.

Palkovits, M. (1988). Neuropeptides in the brain. *In* "Frontiers in Neuroendocrinology" 10th Ed. (L. Martini and W. F. Ganong, Eds.), pp. 1–44. Raven Press, New York.

Pozzoli, G., Mancuso, C., Mirtella, A., Preziosi, P., Grossman, A. B., Navarra, P. (1994). Carbon monooxide as a novel neuroendocrine modulator: inhibition of stimulated corticotrophin-releasing hormone release from acute rat hypothalamic explants. *Endocrinology* **135**, 2314–2317.

Prasada Rao, P. D., Sato, T., Ueck, M. (1997). Distribution of NADPH- diaphorase activity in the hypothalamo-hypophysial system of the frog, *Rana esculenta*. *Neurosci. Lett.* **235**, 61–64.

Pu, L.-P., Charnay, Y., Leduque, P., Morel, G., and Dubois, P. M. (1991). Light and electron microscopic immunocytochemical evidence that delta sleep-inducing peptide and gonadotropin-releasing hormone are coexpressed in the same nerve structures in the guinea pig median eminence. *Neuroendocrinology* **53**, 332–338.

Rivier, C. (1995). Influence of immune signals on the hypothalamic pituitary axis of the rodent. *Front. Neuroendocrinol.* **16**, 151–182.

Rodriguez, E. M., Gonzalez, C. B., and Delannoy, L. (1979). Cellular organization of the lateral and post infundibular regions of the median eminence in the rat. *Cell Tissue Res.* **201**, 377–408.

Samson, W. K., Skala, K. D., Alexander, B. D., and Huang, F.-L. S. (1991). Hypothlamic endothelin: Presence and effects related to fluid and electrolyte homeostasis. *J. Cardiovasc. Pharmacol.* **17 (Suppl. 7)**, S346-S349.

Sapolsky, R., Rivier, C., Yamamoto, G., Plotsky, P., and Vale, W. (1987). Interleukin-1 stimulates the secretion of hypothalamic corticotropin-releasing factor. *Science* **238**, 522–524.

Schwartz, J., and Cherny, R. (1992). Intercellular communication within the anterior pituitary influencing the secretion of hypophysial hormones. *Endocrine Rev.* **13**, 453–475.

Siaud, P., Mekaouche, M., Ixart, G., Balmefrezol, M., Givalois, L., Barbanel, G., and Assenmacher, I. (1994). A subpopulation of corticotropin-releasing hormone neurosecretory cells in the paraventricular nucleus of the hypothalamus also contain NADPH-diaphorase. *Neurosci. Lett.* **170**, 51–54.

Summy-Long, J. Y., Bui, V., Mantz, S., Koehler, E., Weisz, J., and Kadekaro, M. (1993). Central inhibition of nitric oxide synthase preferentially augments release of oxytocin during dehydration. *Neurosci. Lett.* **152**, 190–193.

Taylor, R. L., and Burt, D. R. (1982). Species difference in the brain regional distribution of receptor binding of thyrotropin-releasing hormone. *J. Neurochem.* **38**, 1649–1656.

Tilney, F. (1914). An analysis of the juxtaneural epithelial portion of the hypophysis cerebri with an embryological and histological account of a hitherto undescribed part of the organ. *Intern. Monatschr. Anat. Physiol.* Leipzig. **30**, 258–293.

Tilney, F. (1936). The development and constituents of the human hypophysis. *Bull. Neurol. Inst.* N. Y. **5**, 387–436.

Tuneki, K., Adachi, T., Ishii, S., and Oota, Y. (1976). Morphological classification of neurosecretory granules in the neurohypophysis of the hagfish, *Eptatretus burgeri*. *Cell Tissue Res.* **166**, 145–157.

Turnbull, A. V., and Rivier, C. (1996). Cytokine effects on neuroendocrine axes: influence of nitric oxide and carbon monoxide. *In* "Cytokines in the Nervous System" (N. J. Rothwell, Ed.). pp. 93–116. Springer, Heidelberg.

Uemura, H., and Kobayashi, H. (1971). Effects of dopamine implanted in the median eminence on the estrous cycle of the rat. *Endocrinol. Jap.* **18**, 91–100.

Vallet, P. G., Charmay, Y., Boura, C., and Kiss, J. Z. (1991). Colocalization of delta sleep-inducing peptide and luteinizing hormone-releasing hormone in neurosecretory vesicles in rat median eminence. *Neuroendocrinology* **53**, 103- 106.

Vanderhaeghen, J. J., Lotstra, F., Vandesande, F., and Dierickx, K. (1981). Coexistence of cholecystokinin and oxytocin-neurophysin in some magnocellular hypothalamo-hypophyseal neurons. *Cell Tissue Res.* **221**, 227- 231.

Vanderhaeghen, J. J., Lotstra, F., Liston, D. R., and Rossier, J. (1983). Proenkephalin, [Met]enkephalin, and oxytocin immunoreactivities are colocalized in bovine hypothalamic magnocellular neurons. *Proc. Natl. Acad. Sci. USA* **80**, 5139–5143.

Vollrath, L. (1967). Über die neurosekretorische Innervation der Adenohypophyse von Teleostiern, insbesondere von *Hippocampus cuda* und *Tinca tinca*. *Zeit. Zellforsch.* **78**, 234–260.

Walaas, S. I., and Greengrand, P. (1984). DARPP-32, a dopamine- and adenosine-3':5'-monophosphate-regulated phosphoprotein enriched in dopamine-innervated brain regions. I. Regional and cellular distribution in the rat brain. *J. Neurosci.* **4**, 111–124.

Watson, S. J., Akil, H., Fischeli, W., Goldstein, A., Zimmerman, E., Nilaver, G., and van Wimersma Greidanus, T. B. (1982). Dynorphin and vasopressin: common localization in magnocellular neurons. *Science* **216**, 85–87

Weber, E., Roth, K. A., and Barchas, J. D. (1981). Colocalization of a-neo- endorphin and dynorphin immunoreactivity in hypothalamic neurons. *Biochem. Biophys. Res. Commun.* **103**, 951–958.

Whitnall, M. H., Mezey, E., and Gainer, H. (1985). Co-localization of corticotropin-releasing factor and vasopressin in median eminence neurosecretory vesicles. *Nature* **317**, 248–250.

Wittkowski, W. (1967). Zur Ultrastruktur der ependymalen Tanyzyten und Pituizyten sowie ihre synaptischen Verknüpfung in der Neurohypophyse des Meerschweinchens. *Acta Anat.* **67**, 338–360.

Wong, M.-L., Licinio, J., Pasternak, K. I., and Gold, P. W. (1994). Localization of corticotropin-releasing hormone (CRH) receptor mRNA in adult rat brain by *in situ* hybridization histochemistry. *Endocrinology* **135**, 2275–2278.

Wong, M.-L, Al-Shekhlee, A., Gold, P. W., and Licinio, J. (1996). Cytokines in the brain. *In* "Cytokines in the Nervous System" (N. J. Rothwell, Ed.), pp. 3- 20. Springer, Heidelberg, R.G. Landes Co., Austin.

Yabuuchi, K., Minami, M., Katsumata, S., and Satoh, M. (1994). Localization of type I interleukin-1 receptor mRNA in the rat brain. *Mol. Brain Res.* **27**, 27- 36.

Yamamoto, T., Vukelic, J., Hertzberg, E. L., and Magy, J. I. (1992). Differential anatomical and cellular patterns of connexin-43 expression during postnatal development of rat brain. *Dev. Brain Res.* **66**, 165–180.

Yamamoto, T., Suzuki, H., and Uemura, H. (1997). Endothelin B receptor-like immunoreactivity is associated with LHRH-immunoreactive fibers in the rat hypothalamus. *Neurosci. Lett.* **223**, 117–120.

Yasin, S., Costa, A., Trainer, P., Windle, R., Forsling, M. I., and Grossman, A. (1993). Nitric oxide modulates the release of vasopressin from the rat hypothalamic explants. *Endocrinology* **133**, 1466–1469.

2

HYPOPHYSIOTROPIC NEURONS IN THE BRAIN OF TELEOSTS

P. D. Prasada Rao

Department of Zoology
Nagpur University
Nagpur 440 010, India

1. INTRODUCTION

The pituitary gland is primarily under the control of the hypothalamus, although a wealth of recent evidence supports the notion that intrapituitary chemical mediators themselves act as regulators of pituitary function. In teleosts several hypothalamic neuronal processes project into the pituitary forming an intricate association with the adenohypophysis. For several decades, two hypophysiotropic nuclei have been known to control the teleostean adenohypophysial functions. The first was the nucleus preopticus (NPO) and the second was the nucleus lateralis tuberis (NLT). Using classical neurosecretory stains such as aldehyde fuchsin (AF) or chrome alum-hematoxylin-phloxin (CAHP), the NPO neurons were selectively demonstrated in the teleostean brain. On the other hand, the perikarya of the NLT were AF- and CAHP-negative, but stained with the trichrome and tetrachrome stains. Different aspects pertaining to the hypothalamo-hypophysial system have been discussed in several reviews (Peter, 1983, 1986; Peter and Fryer, 1983; Schreibman, 1986).

However, recent immunocytochemical and radioimmunoassay studies on the teleostean brain have revealed the presence of various neuropeptides and hormone-releasing and/or inhibiting factors not only in the hypothalamus, but also in extrahypothalamic sites. For example, gonadotropin hormone-releasing hormone (GnRH)-containing neurons occur in the olfactory bulb and in the midbrain tegmentum; olfactory bulb neurons containing a molluscan cardioexcitatory tetrapeptide, FMRF-amide, project into the pituitary gland. Retrograde neuronal tracing studies showed that hypophysiotropic neurons occur in the midbrain tegmentum and thalamic areas.

Although several studies have demonstrated different immunoreactive brain perikarya, pathways and fibers innervating the hypophysis, the evidence for their hypophysiotropic nature is not authentic. Immunoreactive perikarya and fibers containing a particular neuropeptide may occur in different brain areas and pituitary respectively, but

Neural Regulation in the Vertebrate Endocrine System, edited by Prasada Rao and Peter, Kluwer Academic / Plenum Publishers, New York, 1999.

this does not indicate that all such immunoreactive brain perikarya are hypophysiotropic. In order to obtain more convincing evidence, investigators have conducted lesion- or electron-microscopic studies; even these studies have not yielded conclusive evidence in view of certain shortcomings (see below). Therefore, one of the reliable means for identification of the brain neurons that project to the pituitary gland is to employ retrograde neuronal tracing procedures after applying tracers to the pituitary gland or stalk. Although neuronal tracing techniques have been extensively used for demonstration of the connectivities within the brain or between different sensory organs and the brain, connectivities between the pituitary and the brain have been sparingly studied. This seems to have resulted from the fact that the neuronal tracing techniques have been extensively used by neuroanatomists and neurobiologists, but have been sparingly used by neuroendocrinologists. Consequently, to date the retrograde neuronal tracing method has been applied and a detailed analysis of the hypophysiotropic neurons made in only four species of the 23,000 known forms of extant teleosts.

The origin of pituitary innervation was investigated in two euryhaline fishes, *Gillichthys mirabilis* and *Tilapia mossambica* as early as in 1971 by Bern and colleagues using retrograde transport procedure. However, this aspect remained unexplored for about two decades, but was studied more recently in the electric fish, *Apteronotus leptorhynchus* (Johnston and Maler, 1992), the goldfish, *Carassius auratus* (Anglade *et al.*, 1993), the catfish, *Clarias batrachus* (Prasada Rao *et al.*, 1993b) and the Atlantic salmon, *Salmo salar* (Holmqvist and Ekström, 1995) using different retrograde labeling techniques. However, the data were restricted to the NLT in the euryhaline fishes and to the hypothalamus in the catfish (Prasada Rao *et al.*, 1993b). The present review summarizes the information about the brain neurons that project into the pituitary gland in teleosts, their immunocytochemical nature and possible functional significance.

2. PITUITARY GLAND AND ITS REGULATION

Unlike the tetrapod condition, the teleostean hypophysis shows ramification of neurohypophysial fibers in all the divisions of the adenohypophysis, with maximal interdigitations in the pars intermedia (PI).

2.1. General Morphology of the Pituitary

The teleostean pituitary is divisible into a neural part, the neurohypophysis and a glandular part, the adenohypophysis. The adenohypophysis is distinguishable into the PI characterized by the presence of extensive neurohypophysial ramifications and the pars distalis (PD) predominated by hormone-secreting cells. The PD shows two divisions *viz.*, proximal pars distalis (PPD) and rostral pars distalis (RPD) following the application of staining techniques to reveal the various hormone-secreting cells. Generally, the PPD is composed of gonadotropes, thyrotropes and somatotropes, while the RPD primarily contains adrenocorticotrpic hormone (ACTH) and prolactin (PRL) cells.

2.2. Pituitary Innervation

It has been well-known that the process of synthesis, storage and release of different hypophysial hormones is controlled by hormone-regulating factors or different neuropeptides that originate in the brain and transported via the axons. Several electron microscopic, autoradiographic, Falck-Hillarp fomaldehyde-induced fluorescence and immunocytochemical (see Peter and Fryer, 1983; Peter, 1986) studies have revealed the occurrence of fibers containing a wide array of pituitary-related releasing/inhibiting factors/hormones such as GnRH (Kah *et al.*, 1986a), corticotropin-releasing factor (CRF) (Olivereau *et al.* 1984), thyrotropin-releasing hormone (TRH) (Batten *et al.*, 1990), growth hormone-releasing hormone (GHRH) (Rao *et al.*, 1996b). Apart from the fibers containing classical hypophysiotropic hormones and peptides, the pituitary gland also revealed the presence of fibers containing neuropeptides such as neurotensin (NRT), cholecystokinin (CCK), galanin (GAL) (see Batten *et al.*, 1990), and bombesin (BBS) (Himick and Peter, 1995). In the catfish pituitary gland, using a double immunocytochemical procedure, we observed that GAL-containing fibers are in close association with growth hormone-secreting cells (S.M. Baby and P.D. Prasada Rao, unpublished data). Other neuropeptides such GAL (Prasada Rao *et al.*, 1996) and carassin (CAR) (Rao *et al.*, 1996a) exist in certain adenohypophysial cells and may play a role in autocrine and/or paracrine regulation of the cells.

2.3. Present Status of the Median Eminence (ME)

In tetrapods, the ME plays a pivotal role in brain regulation of the hypophysis; different brain hormones are released into the capillary plexus of the ME, reach their target hormone-secreting cells of the adenohypophysis *via* the portal system and regulate the hypophysial cell activity. In elasmobranchs, primitive actinopterygians and dipnoans, there exists a ME with primary capillary plexus of the portal system as seen in higher vertebrates (see Chiba, 1997 for references). On the other hand, in teleosts there is no convincing evidence favoring the existence of an established ME or a functional portal system similar in structure and function to that present in other vertebrate groups (see Peter and Fryer, 1983; Schreibman, 1986). Although the vascular supply from the loopings of the arterioles in the neurohypophysis at the interface with the pars distalis (Fridberg and Ekengren, 1977) has been compared to the portal system, its homology with the ME awaits further analysis; the ME-like structure described in teleosts should be considered vestigeal (for details see Peter and Fryer, 1983). Furthermore, in many teleosts different peptidergic nerve terminals originating in the brain may establish direct contacts—sometimes like synaptic contacts—on various hormone-secreting cells or may terminate on basement membrane and regulate the adenohypophysial cell functions (Peter *et al.*, 1990).

3. DEMONSTRATION OF HYPOPHYSIOTROPIC NEURONS

A variety of techniques have been used for tracing the origin of pituitary fibers. For convenience, these procedures can be distinguished into indirect and direct ones.

3.1. Indirect Methods

Earlier attempts included electrothermic lesioning of selected hypothalamic nuclei and subsequent identification of degenerating axons and terminals in the pituitary gland; the lesioned nuclei are considered as the sources of pituitary innervation (Fryer *et al.*, 1985). However, lesioning of any given brain area may result in inadvertent destruction of pathways that may originate in other nuclei and pass through the lesioned area and may lead to erroneous conclusions.

The other method adopted was administration of intravenous injections of the neuronal tracer, horseradish peroxidase (HRP) and subsequent identification of retrogradely labeled neurons. Retrograde neuronal labeling results from entry of HRP into the processes where the blood-brain barrier is open. Some neurohypophysial fibers extensively terminate around the capillaries in the PI of the pituitary and the neuronal tracer is picked up and transported by the fibers retrogradely into the brain perikarya. Intravenous injections of HRP resulted in retrograde labeling of the NPO and NLT perikarya, but the other hypophysiotropic neuronal groups of the brain were not labeled (Fryer and Maler, 1981).

3.2. Direct Methods

Two approaches were generally adopted. The first method involved administration of neuronal tracers such as HRP, and cobaltous or cobaltic lysine (CBL) directly into the pituitary gland or severed stalk in live fish and this permitted uptake of the tracer by injured or severed processes and transport into the brain neurons. However, this procedure involved surgical difficulties in exposure of the pituitary or hypophysectomy, and as a result, this approach of direct application of neuronal tracer to the pituitary has been adopted in one species only *i.e.*, the catfish, *Clarias batrachus* (Prasada Rao *et al.*, 1993b).

According to the second method, a lipophylic carbocyanine fluorescent dye, 1-1'-dioctadecyl-3, 3', 3'-tetramethylindocarbocyanine perchlorate (DiI) was applied to the pituitary gland attached to the brain in paraformaldehyde-prefixed tissues; this procedure avoided surgical difficulties. Using this method the origin of pituitary innervation was analyzed in the electric fish, *Apteronotus leptorhynchus* (Johnston and Maler, 1992), the goldfish, *Carassius auratus* (Anglade *et al.*, 1993) and the Atlantic salmon, *Salmo salar* (Holmqvist and Ekström, 1995).

Apart from the above techniques applied for tracing the hypophysiotropic neurons, several more refined neuronal tracers have become available and it would be useful to employ these methods for tracing the sources of origin of the pituitary fibers (see concluding remarks).

4. LOCALIZATION OF HYPOPHYSIOTROPIC NEURONS

Brain neurons that projected into the pituitary gland have been identified in the hypothalamic as well as extrahypothalamic areas such as olfactory bulb, olfactory tract of the goldfish and in the telencephalon, diencephalon, and midbrain tegmentum (Figure 1) in different teleost species.

4.1. Olfactory Bulb and Tracts

Using DiI, small retrogradely labeled perikarya were revealed in the rostral and caudal areas of the olfactory bulbs and isolated cell bodies in the medial olfactory tract (Figure 1) of the goldfish (Anglade *et al.*, 1993), but not in the electric fish (Johnston and Maler, 1992) and Atlantic salmon (Holmqvist and Ekström, 1995). Retrogradely labeled large ganglion cells of the nervus terminalis (NT) which contained GnRH did not contribute to the pituitary innervation (Kobayashi *et al.*, 1992). On the other hand, in a primitive archipterygian fish, *Polypterus senegalus*, electron microscopic studies by Chiba (1997) revealed an overlap of neuropeptide Y- (NPY), GnRH-, and FMRF-amide-like immunoreactivities in the NT cells, fibers and even in the same cytoplasmic granule. In the catfish,

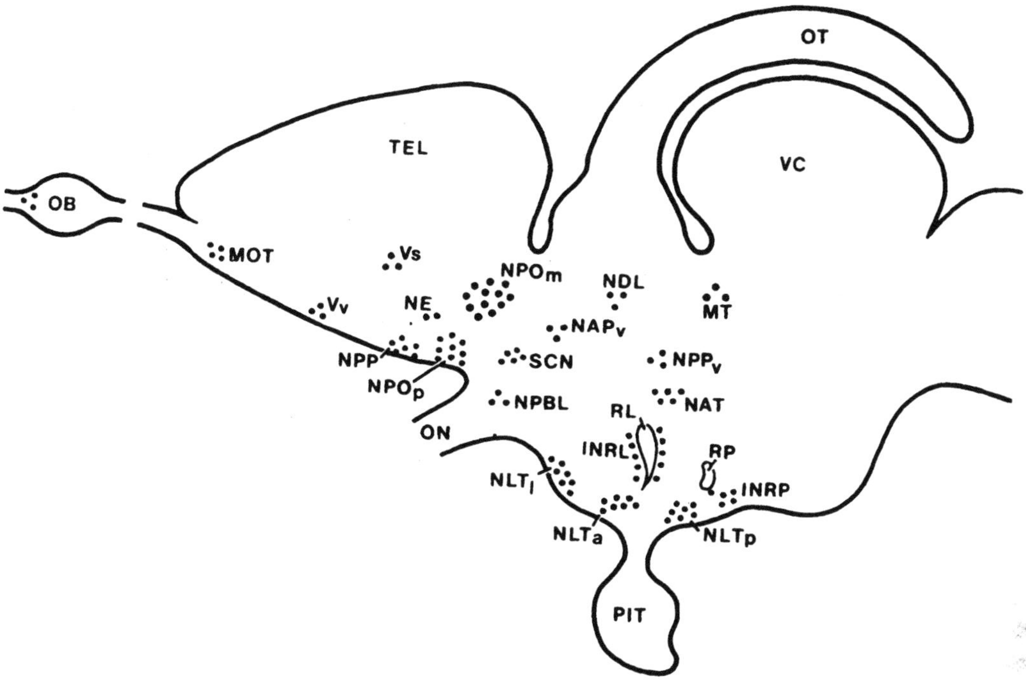

Figure 1. Diagrammatic representation of the sagittal section of the goldfish brain showing the hypophysiotropic nuclei or neurons. INRL, isolated neurons in the vicinity of the nucleus recessus lateralis; INRP, isolated neurons in the vicinity of the nucleus recessus posterioris; MOT, medial olfactory tract; MT, midbrain tegmentum; NAPv, nucleus anterior periventricularis; NAT, nucleus anterior tuberis; NDL, nucleus dorsolateralis thalami; NE, nucleus entopeduncularis; NLTa, nucleus lateralis tuberis, pars anterioris; NLTl, nucleus lateralis tuberis, pars lateralis; NLTp, nucleus lateralis tuberis, pars posterioris; NPBL, nucleus preopticus basalis lateralis; NPOm, nucleus preopticus, pars magnocellularis; NPOp, nucleus preopticus, pars parvocellularis; NPP, nucleus preopticus periventricularis; NPPv, nucleus posterior periventricularis; OB, olfactory bulb; ON, optic nerve; OT, optic tectum; PIT, pituitary gland; RL, recessus lateralis; RP, recessus posterioris; SCN, suprachiasmatic nucleus; TEL, telencephalon; VC, valvula of the cerebellum; Vs, area ventralis telencephali, pars supracommissuralis; Vv, area ventralis telencephali, pars ventralis (modified from Anglade *et al.*, 1993; credit: Springer-Verlag).

Clarias batrachus, olfactory bulb neurons containing FMRF-amide projected into the pituitary gland (Ramakrishna *et al.*, 1992). In various teleosts, the olfactory bulbs or NT neurons showed GnRH (Münz *et al.*, 1981; Stell *et al.*, 1984; Nozaki *et al.*, 1985; Kah *et al.*, 1986a; Batten *et al.*, 1990; Oka and Ichikawa, 1990; Amano *et al.*, 1991, Kim *et al.*, 1995), pancreatic polypeptide (Batten *et al.*, 1990) and NPY (Magliulo-Cepriano and Schreibman, 1993; Peng *et al.*, 1994; Chiba *et al.*, 1996). However, immunoreactive NPY has not been detected by radioimmunoassay in the olfactory bulb of the goldfish (Kah *et al.*, 1989). On the other hand, Fischer *et al.* (1996) have claimed that the goldfish NT contained, at least, two different RF amide-like peptides, of which one was recognized by anti-NPY and anti-A18F amide; it would be useful to investigate the possible existence of these peptides in the NT neurons of other teleosts.

4.2. Telencephalon

Backfilled cell bodies of various sizes were identified in the area ventralis telencephali pars ventralis (Vv) in the electric fish (Johnston and Maler, 1992), goldfish

(Anglade *et al.*, 1993) and Atlantic salmon (Holmqvist and Ekström, 1995). Additionally, in the goldfish, perikarya were revealed in the area ventralis telencephali, pars supracommissuralis (Vs) and in the caudal portion of the nucleus entopeduncularis (NE) (Anglade *et al.*, 1993) (Figure 1). Immunocytochemical studies in different teleosts have shown that Vv is predominated by cell bodies containing salmon GnRH (sGnRH) (Kim *et al.*, 1995, 1997; Amano *et al.*, 1998) and chicken GnRH-II (cGnRH-II) (Kim *et al.*, 1995) and the retrogradely labeled hypophysiotropic neurons might contain one or both types of GnRH. In the Atlantic salmon, the perikarya in this area also showed preproenkephalin (Vecino *et al.*, 1995). Occurrence of backfilled somata in the Vs is of significance since this area was implicated in the control of spawning behavior in the male goldfish (Kyle and Peter, 1982). Although the precise nature of neuropeptide(s) in the retrogradely labeled cells has not been known, certain perikarya in this area contained gamma-aminobutyric acid (GABA) (Martinoli *et al.*, 1990) and GAL (Prasada Rao *et al.*, 1996). The Vs also received a heavy GnRH innervation (Kah *et al.*, 1986a) suggesting that this nucleus participates in reproductive processes of the goldfish.

In the goldfish, small retrogradely labeled cells were visualized in the caudal portion of the nucleus entopeduncularis (Anglade *et al.*, 1993). The nucleus entopeduncularis also revealed NPY- (Pontet *et al.*, 1989; Peng *et al.*,1994), GABA- (Martinoli *et al.*, 1990) and somatostatin- (Kah *et al.*, 1982) containing neurons.

4.3. Diencephalon

In the teleosts investigated so far, several retrogradely labeled nuclei and isolated perikarya were revealed in this brain area (Figures 2 and 3). The somata present in the preoptic area, basal, and caudal hypothalamus and thalamic areas are described below.

4.3.1. Preoptic Region. Extensive neuronal populations were retrogradely labeled in this region in the electric fish, goldfish, catfish and Atlantic salmon. In the electric fish (Johnston and Maler, 1992), the anterior and posterior subdivisions of the nucleus preopticus periventricularis (NPP) were labeled. This nucleus showed several labeled perikarya in the goldfish (Anglade *et al.*, 1993), catfish (Prasada Rao *et al.*, 1993b) and Atlantic salmon (Holmqvist and Ekström, 1995). In the catfish, more NPP cells were labeled with CBL than with HRP. In the goldfish and catfish, several labeled cells were revealed in the ventral wall of the preoptic recess which may represent the extension of the NPP. Immunocytochemical studies revealed a variety of peptides, monoamines and neurotransmitters in the NPP (for review see Peter, 1986). Some NPP neurons contained GnRH (Montero *et al.*, 1994; Kim *et al.*, 1995) which is a potent stimulator of GtH-II, while some showed dopamine (DA) (Kah *et al.*, 1984) or tyrosine hydroxylase (Hornby and Piekut, 1990). DA has been shown to inhibit GtH-II release from the pituitary gland (Omeljaniuk *et al.*, 1989). Electrolytic lesions of this area resulted in disappearance of tyrosine-hydroxylase-positive fibers in the PD of the pituitary and an increase in gonadotropin-II levels (Kah *et al.*, 1987). An increase in GtH-II levels despite possible destruction of GnRH-containing somata following lesioning, indicates that the percentage of hypophysiotropic GnRH-containing cells was low; alternatively, the GnRH-containing neurons of this area may not innervate the pituitary GtH cells. Several GAL-positive cell bodies (Cornbrooks and Parsons, 1991; Power *et al.*, 1996; Prasada Rao *et al.*, 1996) and fibers (Olivereau and Olivereau, 1991; Prasada Rao *et al.*, 1996) were present in the NPP and pituitary respectively. In different teleostean species, cell bodies containing GABA (Martinoli *et al.*, 1990), somatostatin and CRF (Olivereau *et al.*, 1984), thyrotropin-releasing hormone

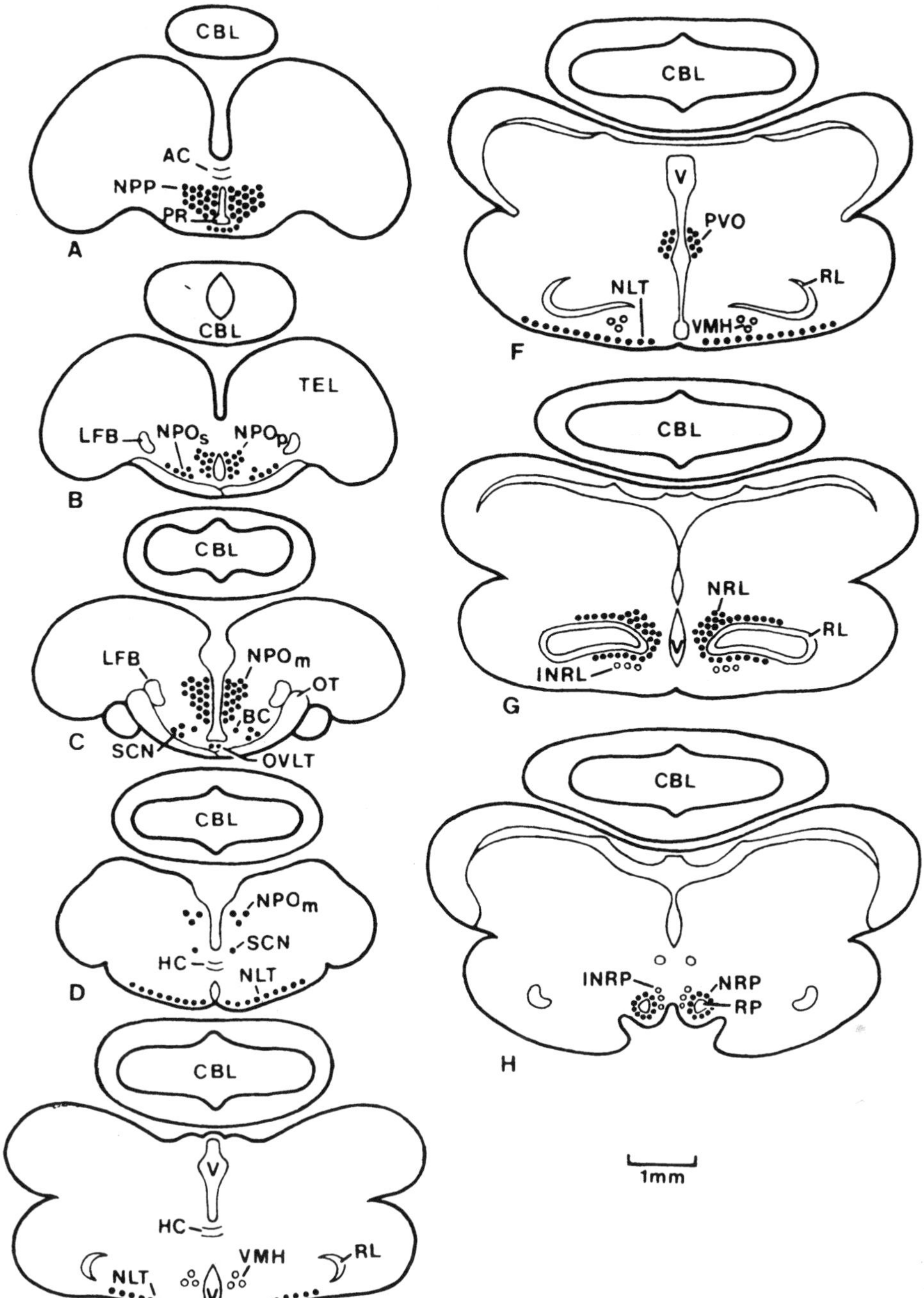

Figure 2. Schematic drawings of representative transverse sections from rostral (A) to caudal (H) hypothalamic area of the catfish, *Clarias batrachus* showing retrogradely labeled nuclei (dark circles) and scattered cell bodies (open circles) following application of cobaltous lysine or HRP to the severed pituitary stalk in hypophysectomized fish. The sections pass through anterior commissure (A), rostral nucleus preopticus (B), nucleus preopticus (C), horizontal commissure (D), rostral lateral recess (E), paraventricular organ (F), lateral recess (G) and posterior recess (H). AC, anterior commissure; BC, bridge cells; CBL, cerebellum; HC, horizontal commissure; LFB, lateral forebrain bundle; NLT, nucleus lateralis tuberis; NPOs, nucleus preopticus, pars supraopticus; NRL, nucleus recessus lateralis; NRP, nucleus recessus posterioris; OVLT, organum vasculosum laminae terminalis; PR, preoptic recess; PVO, paraventricular organ; V, third ventricle; VMH, ventromedial hypothalamic nucleus; arrow indicates the area where neuronal tracer was administered. For explanation of the remaining abbreviations see Figure 1 (modified from Prasada Rao *et al.*, 1993b; credit: Karger, Basel).

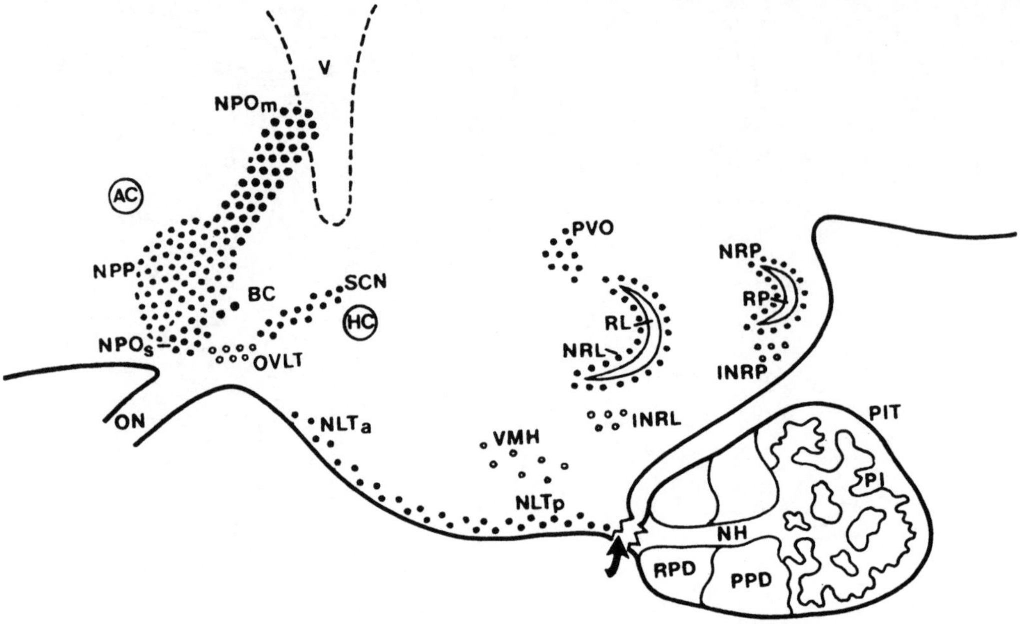

Figure 3. Diagrammatic representation of retrogradely labeled nuclei (dark circles) and scattered perikarya (open circles) in the hypothalamus of the catfish, *Clarias batrachus*, following application of cobaltous lysine/HRP to severed pituitary stalk (arrow), as viewed in superimposed sagittal sections. NH, neurohypophysis; PI, pars intermedia; PPD, proximal pars distalis; RPD, rostral pars distalis. For explanation of the remaining abbreviations see Figures 1 and 2 (modified from Prasada Rao *et al.*, 1993b; credit: Karger, Basel).

(Batten *et al.*, 1990), NPY (Peng *et al.*, 1994), BBS (Himick and Peter, 1995) leu- and met-enkephalin, vasoactive intestinal polypeptide and calcitonin gene-related polypeptide (see Holmqvist and Ekström, 1995) and a few cells containing CAR (Rao *et al.*, 1996a) and GHRH (Rao *et al.*, 1996b) were observed in the NPP. The ventral preoptic region also showed steroid-concentrating cells, high levels of aromatase or androgen receptor immunoreactivity (see Anglade *et al.*, 1993) suggesting its role in control of gonadotropin and/or sexual behavior.

The perikarya of the suprachiasmatic nucleus (SCN) were backfilled in the electric fish, catfish and the Atlantic salmon, suggesting that this nucleus innervated the pituitary gland. In another catfish, *Ictalurus punctatus* the SCN received HRP-labeled fiber terminals from the retina and accumulated radioactive grain after intraocular injection of tritiated proline (Prasada Rao and Sharma, 1982). Similarly, the SCN of apteronoids received retinal input (Sas and Maler, 1986) and showed the presence of tyrosine hydroxylase (Sas *et al.*, 1990), somatostatin (Sas and Maler, 1991) and dopamine (Johnston and Maler, 1992); however, Anglade *et al.* (1993) found no convincing evidence for dopamine as claimed by Johnston and Maler (1992). The SCN of the Atlantic salmon revealed the presence of tyrosine hydroxylase (Holmqvist and Ekström, 1995). In mammals, the SCN serves as a major pace-maker by constituting an integral part of the central circadian oscillators (Moore, 1983). Since the teleostean SCN received retinal output and projected to the pituitary gland, it is suggestive that photic cues may modulate the hypophysial function *via* this pathway and regulate the diurnal rhythmicity of the endocrine system (Holmqvist and Ekström, 1995).

In the preoptic area, hypophysiotropic neurons were identified in the nucleus preopticus basalis, and nucleus anterioris periventricularis (NAPv) of the goldfish (Anglade *et al.*, 1993), NAPv of the Atlantic salmon (Holmqvist and Ekström, 1995) and organum vasculosum laminae terminalis (OVLT) of the catfish (Prasada Rao *et al.*, 1993b). In the goldfish, a few cerebrospinal fluid (CSF)-contacting neurons were visualized among the cells in the ventral wall of the preoptic recess and in the NAPv (Anglade *et al.*, 1993); however, comparable neurons were not discerned in the preoptic region of the remaining fishes investigated. In the goldfish, the OVLT showed densely packed somatostatin-containing cell bodies and varicosities (Pickavance *et al.*, 1992) and prominent CAR-containing fiber terminals (Rao *et al.*, 1996a); however, the role of retrogradely labeled cells in this area with regard to pituitary regulation requires further studies.

In most teleosts, the neurosecretory NPO comprises the major neuronal systems which innervates the pituitary to control various endocrine activities (Peter and Fryer, 1983; Peter, 1986). Accordingly, the NPO represents a very large retrogradely labeled nucleus in the goldfish, catfish and Atlantic salmon and the neurons were strongly labeled. In *Clarias batrachus*, besides the magnocelluar and parvocelluar divisions of the NPO, an aggregation of neurosecretory neurons of the NPO was present dorsal to the optic tract (Figure 2) that is referred to as the NPO pars supraopticus (NPOs). However, in the electric fish, Johnston and Maler (1992) claimed that the NPO was not obvious in cresyl violet-stained preparations of gymnotiform brain; however, it is apparent from their drawings that probably the hypothalamus anterioris (HP) and the nucleus preopticus periventricularis, pars posterior (PPp) which they identified comprise the parvo- and magnocellular divisions of the NPO. Perhaps, application of classical neurosecretory stains such as AF and CAHP or application of immunocytochemical technique using arginine vasotocin (AVT) and isotocin (IST) antibodies would help resolve this issue. In the catfish, most NPO neurons were retrogradely labeled, but a few were unlabeled; these unlabeled neurons may correspond to the perikarya that project into the spinal cord of the catfish (T.C. Job and P.D. Prasada Rao, unpublished data). A few NPO cell bodies were also shown to project into the spinal cord in the goldfish (Demski and Sloan, 1985; Prasada Rao *et al.*, 1987a). Fryer and Maler (1981) also reported that intravenous injections of HRP in the goldfish resulted in retrograde labeling of several NPO cells, but a few were unlabeled. In the goldfish, the neurohypophysial structures and other areas lacking blood-brain barrier appear to have provided access for HRP. The unlabeled neurons in the goldfish may also project to other areas of the brain or may represent interneurons (Fryer and Maler, 1981). In the Atlantic salmon, the processes from the Ha (comparable to the NPO) extended into the pretectal, electrosensory and subelectrosensory nuclei (Johnston and Maler,1992).

The parvocellular and/or magnocellular divisions of the NPO have revealed the presence of several neuropetides, monoamines or neurotransmitters in different teleosts. For example, AVT, IST (Goossens *et al.*, 1977; Batten *et al.*, 1990), AVT, IST and neurophysin (Cumming *et* al., 1982), GnRH (Borg *et al.*, 1982; Goos *et al.*, 1985; Montero *et al.*, 1994), CRF (Olivereau and Olivereau, 1988; Yulis and Lederis, 1987; Olivereau *et al.*, 1988), somatostatin (Olivereau *et al.*, 1984), enkephalin (Reaves and Hayward, 1980; Batten *et al.*, 1990), growth hormone-releasing factor, neurotensin, CCK, and pancreatic polypeptide (Batten *et al.*, 1990), NPY (Danger *et al.*, 1991: Pickavance *et al.*, 1992; Peng *et al.*, 1994), GABA (Martinoli *et al.*, 1990), substance P (Sharma *et al.*, 1989; Moons *et al.*, 1992), vasoactive intestinal polypeptide (Holmqvist *et al.*, 1992), GAL (Batten *et al.*, 1990; Cornbrooks and Parsons, 1991; Prasada Rao *et al.*, 1996), BBS (Himick and Peter, 1995) and aromatase (Gelinas and Callard, 1997). CRF and AVT were colocalized in the NPO neurons (Fryer and Lederis, 1988; Olivereau *et al.*, 1988). It is well known that NPO neurophysins, AVT and IST project into the neurohypophysis (Peter and Fryer, 1983). The hypophysiotropic neurons of the NPO, apart from their role in salt-water balance, and arterial pressure, other possible functions include effects on PRL, ACTH, GtH, growth hormone (GH) and melanophore-stimulating hormone (MSH) secretion in the pituitary.

4.3.2. Basal and Caudal Hypothalamus. A prominent ventral infundibular nucleus that innervated the pituitary in teleosts is the NLT located in the ventral infundibular area (Figures 2 and 3). The NLT was distinguished into several subdivisions *viz.*, pars anterioris, pars inferioris, pars lateralis, pars medialis and pars posterioris in various teleosts. It shows medium- to large-sized neurons extending rostro-caudally. Using iontophoretic transport of cobalt chloride, Bern *et al.* (1971) revealed labeled neurons exclusively in the NLT of *Tilapia mossambica*. In the goldfish, catfish and Atlantic salmon, most cells in the NLT were labeled retrogradely. In the electric fish, the NLT pars anterioris and pars posterioris revealed DiI-labeled cells along the ventricular wall. In the goldfish, electric fish and Atlantic salmon a few DiI-labeled cells were reported to be CSF-contacting in nature.

The NLT is a major hypophysiotropic neuronal assemblage (Fryer and Maler, 1981) and several peptides and neurotransmitters have been identified in this nuclear complex in different teleosts—for example, endorphin (Dubois *et al.*, 1979; Follenius and Dubois, 1979), CCK/gastrin (Notenboom *et al.*, 1981), GnRH (Schreibman *et al.*, 1979; Münz *et al.*, 1981; Stell *et al.*, 1984; Kah *et al.*, 1986a; Batten *et al.*, 1990; Montero *et al.*, 1994), somatostatin (Kah *et al.*, 1982; Olivereau *et al.*, 1984; Power *et al.*, 1996), melanin-concentrating hormone (MCH) (Bird *et al.*, 1989) urotensin I (Fryer, 1989), CRF, NRT and proopiomelanocortins (Batten *et al.*, 1990), GABA (Martinoli *et al.*, 1990), proopiomelanocortin-derived peptides (Olivereau and Olivereau, 1990), serotonin and dopamine (DA) (Corio *et al.*, 1991), tyrosine hydroxylase (Holmqvist and Ekström, 1995), ACTH (Olivereau and Olivereau, 1990), NPY (Pickavance *et al.*, 1992), BBS (Himick and Peter, 1995), GAL (Prasada Rao *et al.*, 1996) and GHRH (Rao *et al.*, 1996b). Some somatostatin-containing perikarya of the NLT project into the CSF and monitor the homeostatic state of the organism by sampling the CSF besides regulating the hypophysial hormones (see Pickavance *et. al.*, 1992). The NLT area is also a steroid-concentrating region (Kim *et al.*, 1978) and represents another major integrative neuroendocrine region which plays an important role in pituitary regulation (Peter, 1986). Fibers immunoreactive to CCK/gastrin have been identified in the PPD of the pituitary gland and their role in GtH-II and GH release demonstrated (Himick *et al.*, 1993).

The data available on the hypophysiotropic nature of the paraventricular organ (PVO), nucleus recessus lateralis (NRL) and nucleus recessus posterioris (NPP) have not been uniform. Fluorescence and immunocytochemical investigations have revealed the presence of catecholaminergic and/or serotoninergic neurons in the hypothalamic PVO, NRL and NRP (Kah and Chambolle, 1983; see Fryer *et al.*, 1985; Prasada Rao *et al.*, 1987b; Meek *et al.*, 1989; Hornby and Piekut, 1990) and fibers in the pituitary (see Corio *et al.*, 1991) of teleosts. It was further suggested that monoaminergic neurons project into the pituitary gland (see Fryer *et al.*, 1985; Sas *et al.*, 1990) and the multiple hypophysiotropic actions of dopamine have been extensively investigated (Peter, 1986; Omeljaniuk *et al.*, 1989).

However, Johnston and Maler (1992) reported that the cells of the PVO, NRL and NRP were never labeled retrogradely with DiI and concluded that the neurons of these nuclei do not project into the pituitary in the electric fish. The situation reported in the goldfish and Atlantic salmon was different. Anglade *et al.* (1993) observed scattered retrogradely labeled cells around the lateral parts of the NRL and around the NRP. However, the CSF-contacting neurons of the NRL and NRP were not fluorescent. They claimed that the dopaminergic innervation of the pituitary in the goldfish originated from the anterior ventral preoptic region and other unidentified nuclei (Kah *et al.*, 1986b). In the Atlantic salmon, Holmqvist and Ekström (1995) have also reported the presence of numerous, but relatively scattered, DiI-labeled neurons in the NRL, NRP and nucleus of the saccus vasculosus (NSV), and that some neurons in these nuclei are CSF-contacting in nature. They have concluded that although there were dopaminergic neurons in the NPO area and throughout the anterior periventricular nucleus (NAPv), the major source of dopaminergic fibers innervating the pituitary is the discrete neurons in the supraoptic/suprachiasmatic position of the NAPv.

In the catfish, *Clarias batrachus*, the perikarya of the PVO, NRL and NRP were not labeled retrogradely with HRP (Ramakrishna and Subhedar, 1989; Prasada Rao *et al.*, 1993b). It has been known that these nuclei contain primarily monoamines and that HRP is not a suitable neuronal tracer for demonstration of cell bodies containing biogenic amines (Ljungdahl *et al.*, 1975). HRP is also not useful as a tracer for revealing diencephalic efferent perikarya (Rusoff and Hapner (1990). On the other hand, the cell bodies of these nuclei were strongly labeled retrogradely with CBL suggesting that these somata projected into the pituitary gland (Prasada Rao *et al.*, 1993b). In view of the discrepancies in the data, it is worthwhile to assess the possible reasons for the reported differences among the catfish and other species investigated. In other catfish species (*Mystus vittatus,* Prasada Rao *et al.*, 1987b; *Clarias gariepinus,* Corio *et al.*, 1991) the majority of monoaminergic perikarya were localized in the periventricular nuclei, PVO, NRL and NRP. In contrast to the condition in the goldfish (Kah *et al.*, 1984), Corio *et al.* (1991) observed that the preoptic region in the African catfish, *Clarias gariepinus* lacked dopaminergic cell bodies and they suggested that the source of hypophysial aminergic fibers may lie in the PVO, NRL and NRP. Alternatively, as pointed above, certain neuronal tracers may be more sensitive than others for retrograde labeling of the monoaminergic cells of the brain. Anglade *et al.* (1993) and Holmqvist and Ekström (1995) found only a few cells in the PVO, NRL and NRP areas. Further, different tracers showed varying rates of transport resulting in either proper backfilling of all the cells, a few cells or only moderate labeling or no labeling at all. Oka *et al.* (1986) reported that cobaltic lysine was more sensitive in retrograde labeling than HRP. Our study on analysis of descending projection neurons into the spinal cord confirms the occurrence of variations in sensitivity of different neuronal tracers. For example, using retrograde HRP study we observed as many as 17 brain nuclei that projected into the spinal cord in the goldfish (Prasada Rao *et al.*, 1987a, 1993a; Sharma *et al.*, 1993). On the other hand, using Dextran Amines as a neuronal tracer, we found several more nuclei and a more extensive fiber system in the goldfish brain (P.D. Prasada Rao and S.C. Sharma, unpublished data). In view of the discrepancies in the data that apparently arose from usage of different tracers, it would be premature to make conclusions on the basis of results generated after using a single tracer. It would be pertinent to examine the hypophysiotropic neurons using more than one retrograde neuronal tracer in order to obtain a more precise information.

Immunocytochemical studies revealed the presence of DA (Meek *et al.*, 1989; Hornby and Piekut, 1990; Sas *et al.*, 1990; Corio *et al.*, 1991), tyrosine hydroxylase (Holmqvist and Ekström, 1995), serotonin (Kah and Chambolle, 1983; Corio *et al.*, 1991; Holmqvist and Ekström, 1995) in the PVO, NRL and NPP.

In the catfish, a few isolated perikarya were retrogradely labeled in the vicinity of the NRL and NRP (Prasada Rao *et al.*, 1993b). The perikarya observed in the goldfish brain near the NRL, NRP (Anglade *et al.*, 1993) seem comparable to those observed in the catfish. The retrogradely labeled cells in the nucleus hypothalamicus ventromedialis of the catfish are comparable to those observed in the NAPv of the Atlantic salmon (Holmqvist and Ekström, 1995) and NAPv, NPPv and NAT in goldfish (Anglade *et al.,* 1993).

4.3.3. Thalamus. Retrogradely labeled cells were reported in the nucleus dorsolateralis thalami (Anglade *et al.*, 1993) and in the nucleus centralis posterioris of the thalamus (Johnston and Maler, 1992) in the brain of the goldfish (Figure 2) and electric fish respectively. However, comparable hypophysiotropic neurons were not identified in the Atlantic salmon.

4.3.4. Mesencephalon. In the goldfish, two neuronal groups of the mesencephalon were reported to project into the pituitary (Anglade *et al.*, 1993). The first neuronal entity included large DiI-labeled cells along the midline caudal to the posterior commissure (Figure 2). The second group of cells was found in the central nucleus of the torus semicircularis. Comparable cells were not identified in the Atlantic salmon or electric fish.

5. CONCLUDING REMARKS

The data available on the brain hypophysiotropic neurons in teleosts have not been uniform perhaps because of species-specificity and sometimes due to terminologies adopted. For example, although the major source of hypophysiotropic afferents is the NPO (Peter, 1986), a typical NPO was not identified in the electric fish (Johnston and Maler,1992). Probably, the other nuclei which they termed differently (PPp and Ha) are comparable to the NPO. But the major cause for the difference in the data stems from the fact that only some of the hypophysiotropic nuclei or neurons become labeled because the sensitivity of different techniques is highly variable. The procedures adopted for administration of neuronal tracers, or duration allowed for transport following administration of the tracer also contribute to such variations since some tracers are transported faster than others; yet others may be more useful for certain tissues or for certain type of cells. For example, CBL was more sensitive (Springer and Gaffney, 1981; Shiga *et al.,* 1985) and HRP was not as useful as CBL for tracing diencephalic efferents (Rusoff and Hapner, 1990). This is also evident from the fact that the catfish PVO, NRL and NRP were not labeled with HRP, but were labeled intensely with CBL (Prasada Rao *et al.*, 1993b).

The fluorescent tracer, DiI has been considered highly sensitive and revealed several brain hypophysiotropic cells. In view of the difficulties in performing hypophysectomies or intrapituitary injections of the tracer in vivo, and its usefulness on prefixed material, apparently DiI has been a convenient tracer. Nevertheless, it may diffuse transsynaptically (Godemont *et al.*, 1987; Von Bartheld *et al.*, 1990). This fluorescent dye used on prefixed material required prolonged duration (3–6 weeks) for complete labeling (Anglade *et al.*, 1993; Holmqvist and Ekström, 1995); sometimes incubation times up to 8 or 12 months are required, although incubation for 1 year resulted in transsynaptic labeling of neurons (Johnston and Maler, 1992).

Such disadvantages have led to development of fluorescent tracers such as 4-acetamido, 4'-isothiocyanostilben-2–2'-disulfonic acid (SITS), Fluoro-Gold (FG), Fluoro-Ruby (FR) which do not diffuse from labeled cells despite wide variations in survival periods (Schmued and Swanson, 1982; Schmued *et al.*, 1990). Recently, two new fluorescent retrograde tracers, Fluoro-Red (FR) and Fluoro-Green (FG) with several unique properties have been introduced (Dong *et al.*, 1996). We have shown that Biotinylated Dextran

Amines of 3kDa molecular weight (BDA 3000) is excellent for labeling cells and fibers not revealed by HRP (P.D. Prasada Rao and S.C. Sharma, unpublished data). Nevertheless, it would be essential to apply more than one tracer for an accurate analysis of the cells and fibers of any system.

The second aspect has been the lack of information on the source of origin of fibers innervating different adenohypophysial components, the RPD, PPD and PI. Johnston and Maler (1992) mentioned that placement of DiI crystals in the RPD resulted in retrograde labeling of Vv and Cp and therefore concluded that these centers exclusively projected to the RPD and/or PPD. However, all the hypophysiotropic brain cells were labeled following placement of DiI crystals either in the PPD or PI (Johnston and Maler, 1992). Although different retrograde labeling patterns resulted from placement of DiI crystals in different zones of the pituitary, they mentioned that it is difficult to interpret the data due to DiI diffusion in the pituitary. To avoid the problem of diffusion of the tracer and erroneous labeling, it may be advantageous to use tracers such as SITS, FG or FR. Employing micro-implants or microinjections of the tracer may also prove advantageous.

The third aspect is the necessity to generate precise knowledge about the nature of the neurons that project into the pituitary gland. Most information available todate is derived through comparing the retrogradely labeled nuclei with other studies conducted on the distribution of different peptides and neurotransmitters in these nuclei. Although such a comparison can be made and the nature of neurons can be analyzed, the information derived cannot be authentic. It is still more difficult to identify the nature of individual isolated hypophysiotropic brain perikarya through such a comparison. Each nucleus may represent an assemblage of different types of neurons and it is known that each neuron may contain more than one neuropeptide. An ideal approach to resolve this issue would be to identify the neuropeptides present in the retrogradely labeled hypophysiotropic neurons after application of immunocytochemical procedures to tissue sections in which hypophysiotropic neurons were demonstrated previously, using retrograde neuronal tracing techniques. However, attempts made to demonstrate neuropeptides immunocytochemically in DiI-labeled sections have not yielded satisfactory results since there was loss of antigenecity during the process of retrograde labeling or the neuronal tracer may have masked some antigenic epitopes or the DiI-labeled cell membranes may become less permeable to antibodies due to fixation. Evolving appropriate techniques to demonstrate neuropeptides in retrogradely labeled perikarya would be useful.

In summary, the pituitary fibers originate also from extrahypothalamic sites; it would be useful to investigate the brain hypophysiotropic neurons in different groups of fishes using more than one neuronal tracer and to identify the brain neurons that innervate each hormone-producing cell type of the pituitary. Considering the available data on the fibers containing hormone-releasing factors and other neuropeptides, it may be suggested that these peptidergic sytems play a role in processing the sensory information and in control or modulation of the hypophysial hormones in conjunction with autonomic functions.

ACKNOWLEDGMENTS

The author wishes to thank Professor R. E. Peter, University of Alberta, Edmonton, Canada, for critical reading of the manuscript and for valuable suggestions, and the Alexander von Humboldt Foundation, Bonn, Germany, for support.

REFERENCES

Amano, M., Oka, Y., Aida, K., Okumoto, N., Kawashima, S., and Hasegawa, Y. (1991). Immunocytochemical demonstration of salmon GnRH and chicken GnRH-II in the brain of masu salmon, *Oncorhynchus masou*. *J. Comp. Neurol.* **314**, 587–597.

Amano, M., Ashihara, M.O., Yoshiyura, Y., Kitamura, S., Ikuta, K., and Aida, K. (1998). Two different salmon GnRH precursor mRNAs are co-expressed in the brain of sockeye salmon (*Oncorhynchus nerka*). *Cell Tissue Res.* **292**, 267–273.

Anglade, I., Zandbergen, T., and Kah, O. (1993). Origin of the pituitary innervation in the goldfish. *Cell Tissue Res.* **273**, 345–355.

Batten, T.F.C., Cambre, M.L., Moons, L., and Vandesande, F. (1990). Comparative distribution of neuropeptide-immunoreactive systems in the brain of the green molly, *Poecilia latipinna*. *J. Comp. Neurol.* **302**, 893–919.

Bern, H.A., Zambrano, D., and Nishioka, R.S. (1971). Comparison of innervation of the pituitary of two euryhaline teleost fishes, *Gillichthys mirabilis* and *Tilapia mossambica* with special reference to the origin and nature of type 'B' fibres. *Mem. Soc. Endocrinol.* **19**, 817–822.

Bird, D.J., Baker, B.I., and Kawauchi, H. (1989). Immunocytochemical demonstration of melanin-concentrating hormone and proopiomelanocortin-like products in the brain of the trout and carp. *Gen. Comp. Endocrinol.* **74**, 442–450.

Borg, B., Goos, H.J.Th., and Terlou, M. (1982). LH-RH immunoreactive cells in the brain of the three-spined stickleback *Gasterosteus aculeatus* L. (Gasterosteidae). *Cell Tissue Res.* **226**, 695–699.

Chiba, A. (1997). Distribution of neuropeptide Y-like immunoreactivity in the brain of the bichir, *Polypterus senegalus*, with special regard to the terminal nerve. *Cell Tissue Res.* **289**, 275–284.

Chiba, A., Sohn, Y.C., and Honma, Y. (1996). Distribution of neuropeptide Y and gonadotropin-releasing hormone immunoreactivities in the brain and hypophysis of the ayu, *Plecoglossus altivelis* (Teleostei). *Arch. Histol. Cytol.* **59**, 137–148.

Corio, M., Peute, J., and. Steinbusch, H.W.M. (1991). Distribution of serotonin- and dopamine-immunoreactivity in the brain of the teleost *Clarias gariepinus*. *J. Chem. Neuroanat.* **4**, 79–95.

Cornbrooks, E.B., and Parsons, R.L. (1991). Sexually dimorphic distribution of a galanin-like peptide in the central nervous system of the teleost fish *Poecilia latipinna*. *J. Comp. Neurol.* **304**, 639–657.

Cumming, R., Reaves, T.A., and Hayward, J.N. (1982). Ultrastructural immunocytochemical characterization of isotocin, vasotocin and neurophysin neurons in the magnocellular preoptic nucleus of the goldfish. *Cell Tissue Res.* **223**, 685–694.

Danger, J.-M., Breton, B., Vallarino, M., Fournier, A., Pelletier, G., and Vaudry, H. (1991). Neuropeptide Y in the trout brain and pituitary: Localization, characterization and action on gonadotropin release. *Endocrinology* **238**, 2360–2368.

Demski, L.S., and Sloan, H.E. (1985). A direct magnocellular-preopticospinal pathway in goldfish: Implications for control of sex behavior. *Neurosci. Lett.* **55**, 283–288.

Dong, K., Qu, T., Ahmed, F.A.K.M., Zhang, L., Yamada, K., Guison, N.G., Miller, M., and Yamadori, T. (1996). Fluoro-Green and Fluoro-Red: Two new fluorescent retrograde tracers with a number of unique properties. *Brain Res.* **736**, 61–67.

Dubois, M.P., Billard, B., Breton, B., and Peter, R.E. (1979). Comparative distribution of somatostatin, LH-RH, neurophysin, and α-endorphin in the rainbow trout: An immunocytochemical study. *Gen. Comp. Endocrinol.* **37**, 220–232.

Fischer, A.J., Reisch, H.M., Kyle, A.L., and Stell, W.K. (1996). Characterization of the RF amide-like neuropeptides in the nervus terminalis of the goldfish (*Carassius autarus*). *Regul. Pept.* **62**, 73–87.

Follénius, E., and Dubois, M.P. (1979). Localisation des sites immunoréactifs avec un antiserum contre la met-enkephaline et contre l'α-endorphine dans le cerveau de la Carpe. *C. R. Acad. Sci. Paris, Ser. D* **288**, 903–906.

Fridberg, G., and Ekengren, B. (1977). The vascularization and the neuroendocrine pathways of the pituitary gland in the Atlantic salmon, *Salmo salar. Can. J. Zool.* **55**, 1284–1296.

Fryer, J.N. (1989). Neuropeptides regulating the activity of goldfish corticotrophes and melanotrophes. *Fish Physiol. Biochem.* **7**, 21–27.

Fryer, J.N., Bourdreault-Chateauvert, C., and Kirby, R. (1985). Pituitary afferents originating in the paraventricular organ (PVO) of the goldfish hypothalamus. *J. Comp. Neurol.* **242**, 474–484.

Fryer, J.N. and Lederis, K. (1988). Comparison of actions of posterior pituitary hormones in cortocotropin secretion in mammals and teleosts. *In* "Recent Progress in Posterior Pituitary Hormones" (M. Yoshida, and L. Share, Eds.), pp. 337–344. Elsevier Press, Amsterdam.

Fryer, J., and Maler, L. (1981). Hypophysiotropic neurons in the goldfish pituitary demonstrated by the retrograde transport of horseradish peroxidase. *Cell Tissue Res.* **218**, 93–102.

Gelinas, D., and Callard, G. V. (1997). Immunolocalization of aromatase- and androgen receptor-positive neurons in the goldfish brain. *Gen. Comp. Endocrinol.* **106**, 155–168.

Godemont, P., Vaneslow, J., Thanos, S., Bonhoeffer, F. (1987). A study in developing visual system with a new method of staining neurons and their processes in fixed tissue. *Development* **101**, 697–713.

Goos, H. J. Th., de Leeuw, R. D., de Zoeten-Kamp, Peute, J., and Blähser, S. (1985). Gonadotropin hormone-releasing hormone-immunoreactive neuronal structures in the brain and pituitary of the catfish, *Clarias batrachus* (Burchel). *Cell Tissue Res.* **241**, 593–596.

Goossens, N., Dierickx, K., and Vandesande, F. (1977). Immunocytochemical localization of vasotocin and isotocin in the preopticohypophysial neurosecretory system of teleosts. *Gen. Comp. Endocrinol.* **32**, 371–375.

Himick, B.A., Golosinski, A.A., Jonsson, A.-C., and Peter, R.E. (1993). CCK/Gastrin-like immunoreactivity in the goldfish pituitary: Regulation of pituitary hormone secretion by CCK-like peptides *in vitro*. *Gen. Comp. Endocrinol.* **92**, 88–103.

Himick, B.A., and Peter, R.E. (1995). Bombesin-like immunoreactivity in the forebrain and pituitary and regulation of anterior pituitary hormone release by bombesin in goldfish. *Neuroendocrinology* **61**, 365–376.

Holmqvist, B.I., and Ekström, P. (1995). Hypophysiotrophic systems in the brain of the Atlantic salmon: Neuronal innervation of the pituitary and the origin of pituitary dopamine and nonapeptides identified by means of combined carbocyanine tract and imunocytochemistry. *J. Chem. Neuroanat.* **145**, 125–145.

Holmqvist, B.I., Östholm, T., and Ekström, P (1992). Retinohypothalamic projections and the suprachiasmatic nucleus of the teleostean brain. *In* "Rhythms in fishes" (M.A. Ali, Ed.), pp. 394–418. Plenum Press, New York.

Hornby, P.J., and Piekut, D.T (1990). Distribution of catecholamine-synthesizing enzymes in goldfish brain and pituitary: Presumptive dopamine and norepinephrine neuronal organization. *Brain Behav. Evol.* **35**, 49–64.

Johnston, S.A., and Maler, L. (1992). Anatomical organization of the hypophysiotrophic system in the electric fish, *Apteronotus leptorhynchus*. *J. Comp. Neurol.* **317**, 421–437.

Kah, O., Breton, B., Dulka, J.G., Nunez-Rodriguez, J., Peter, R.E., Corigan, A., Rivier, J.J., and Vale, W.W. (1986a). A reinvestigation of the Gn-RH (gonadotropin-releasing hormone) systems in the goldfish brain using antibodies to salmon Gn-RH. *Cell Tissue Res.* **244**, 327–337.

Kah, O., and Chambolle, P. (1983). Serotonin in the brain of the goldfish, *Carassius auratus*: An immunocytochemical study. *Cell Tissue Res.* **234**, 319–333.

Kah, O., Chambolle, P., Dubourg, P., and Dubois, M.P. (1982). Localisation immunocytochemique de la somatostatine dans le cerveau antérieur de l'hypophyse de deux téléostéens , le Cyprin (*Carassius auratus*) et *Gambusia* sp., *C. R Acad. Sci. (III)* **294**, 519–524.

Kah, O., Chambolle, P., Thibault, J., and Geffard, M. (1984). Existence of dopaminergic neurons in the preoptic region of the goldfish. *Neurosci. Lett.* **48**, 293–298.

Kah, O., Dubourg, P., Onteniente, B., Geffard, M., and Calas, A. (1986b). The dopaminergic innervation of the goldfish pituitary: An electron microscope immunocytochemical study using antibodies to dopamine. *Cell Tissue Res.* **244**, 577–582.

Kah, O., Dulka, J.G., Dubourg, P., Thibault, J., and Peter, R.E. (1987). Neuroanatomical substrate for the inhibition of gonadotropin secretion in goldfish: Existence of a dopaminergic preoptico-hypophyseal pathway. *Neuroendocrinology* **45**, 451–458.

Kah, O., Pontet, A., Danger, J.M., Dubourg, P., Pelletier, G., Vaudry, H., and Calas, A. (1989). Characterization, cerebral distribution and gonadotropin release activity of neuropeptide Y (NPY) in the goldfish. *Fish Physiol. Biochem.* **7**, 69–76

Kim,Y.S., Stumf, W.E., and Sar, M. (1978). Topography of estrogen target cells in the forebrain of the goldfish, *Carassius auratus*. *J. Comp. Neurol.* **182**, 611–620.

Kim, M.-H., Amano, M., Suetake, H., Kobayashi, M., and Aida, K. (1997). GnRH neurons and gonadal maturation in masu salmon and goldfish. *In* "GnRH Neurons: Gene to Behavior" (I.S. Parhar and Y. Sakuma, Eds.), pp. 313–324, Brain Shuppan, Tokyo.

Kim, M.-H., Oka, Y., Amano, M., Kobayashi, M., Okuzawa, K., Hasegawa, Y., Kawashima, S., Suzuki, Y., and Aida, K. (1995). Immunocytochemical localization of sGnRH and cGnRH-II in the brain of goldfish, *Carassius auratus*. *J. Comp. Neurol.* **356**, 72–82.

Kobayashi, M., Amano, M., Hasegawa, Y., and Aida, K. (1992). Effects of olfactory tract section on brain GnRH distribution and gonadal development in goldfish: *In* "Proceedings of the 2nd International Symposium on Fish Endocrinology, St. Malo," June 1992 P. 87.

Kyle, A.L., and Peter, R.E. (1982). Effects of forebrain lesions on spawning behavior in the male goldfish. *Physiol. Behav.* **28**, 1103–1109.

Ljungdahl, A., Hökfelt, T., Goldstein, M., and Park, D. (1975). Retrograde peroxidase tracing of neurons combined with transmitter biochemistry. *Brain Res.* **84**, 313–319.

Magliulo-Cepriano, L., and Schreibman, M.P. (1993). The distribution of neuropeptide Y and dynorphin immunoreactivity in the brain and pituitary gland of the platyfish, *Xiphophorus maculatus* from birth to sexual maturity. *Cell Tissue Res.* **271**, 87–92.

Martinoli, M.G., Dubourg, P., Geffard, M., Calas, A., and Kah, O. (1990). Distribution of GABA-immunoreactive neurons in the forebrain of goldfish. *Cell Tissue Res.* **260**, 77–84.

Meek, H., Joosten, H.W.J., and Steinbusch, H.W.M. (1989). The distribution of dopamine-immunoreactivity in the brain of the Mormyrid teleost *Gnathonemus petersii*. *J. Comp. Neurol.* **281**, 362–383.

Montero, M., Vidal, B., King, J.A., Tramu, G., Vandesande, F., Dufour, S., and Kah, O. (1994). Immunocytochemical localization of mammalian GnRH (gonadotropin-releasing hormone) and chicken GnRH-II in the brain of the European silver eel (*Anguilla anguillla* L.). *J. Chem. Neuroanat.* **7**, 227–241.

Moons, L., Batten, T.F.C., and Vandesande, F. (1992). Comparative distribution of substance P (SP) and cholecystokinin (CCK) binding sites and immunoreactivity in the brain of the sea bass (*Dicentrarchus labrax*). *Peptides* **13**, 37–46.

Moore, R.Y. (1983). Organization and function of a central nervous system circadian oscillator: The suprachiasmatic hypothalamic nucleus. *Fed. Proc.* **43**, 2783–2789.

Münz, H., Stumf, W.E., Jennes, L. (1981). LH-RH systems in the brain of platyfish. *Brain Res.* **221**, 1–13.

Notenboom, C.D., Garaud, J.C., Doerr-Schott, J., and Terlou, M. (1981). Localization by immunofluorescence of a gastrin-like substance in the brain of the rainbow trout, *Salmo gairdneri*. *Cell Tissue Res.* **214**, 247–255.

Nozaki, M., Fujita, I., Saito, N., Tsukahara, T., Kobayashi, H., Ueda, K., and Oshima, K. (1985). Distribution of LH-RH-like immunoreactivity in the brain of the Japanese eel (*Anguilla japonica*) with special reference to the nervus terminalis. *Zool. Sci.* **2**, 537–547.

Oka, Y., and Ichikawa, M. (1990). Gonadotropin-releasing hormone (GnRH) immunoreactive system in the brain of the dwarf gorami (*Colisa lalia*) as revealed by light microscopic immunocytochemistry using a monoclonal antibody to common aminoacid sequence of GnRH. *J. Comp. Neurol.* **300**, 511–522.

Oka, Y., Satou, M., and Ueda, K. (1986). Ascending pathways from the spinal cord in the Himé salmon (landlocked red salmon, *Oncorhynchus nerka*). *J. Comp. Neurol.* **254**, 104–112.

Olivereau, M., Moons, L., Olivereau, J., and Vandesande, F. (1988). Coexistence of corticotropin-releasing factor-like immunoreactivity and vasotocin in perikarya of the preoptic nucleus in the eel. *Gen. Comp. Endocrinol.* **70**, 41–48.

Olivereau, M., and Olivereau, J.M. (1988). Localization of CRF-like immunoreactivity in the brain and pituitary of teleost fish. *Peptides* **9**, 13–21.

Olivereau, M., and Olivereau, J.M. (1990). Corticotropin-like immunoreactivity in the brain and pituitary of three teleost species (goldfish, trout and eel). *Cell Tissue Res.* **262**, 115–123.

Olivereau, M., and Olivereau, J.M. (1991). Immunocytochemical localization of a galanin-like peptidergic system in the brain and pituitary of some teleost fish. *Histochemistry* **96**, 343–354.

Olivereau, M., Ollevier, F., Vandesande, F., and Verdonck, W. (1984). Immunocytochemical identification of CRF-like and SRIF-like peptides in the brain and the pituitary of cyprinid fish. *Cell Tissue Res.* **237**,379–382.

Omeljaniuk, R.J., Tonon, M.C., and Peter, R.E. (1989). Dopamine inhibition of gonadotropin and alpha-melanocyte stimulating hormone release in vitro from the pituitary of the goldfish (*Carassius auratus*). *Gen. Comp. Endocrinol.* **74**, 451–467.

Peng, C., Gallin, W., Peter, R.E., Blomqvist, A.G., and Larhammar, D. (1994). Neuropeptide-Y gene expression in the goldfish brain: Distribution and regulation by ovarian steroids. *Endocrinology* **134**, 1095–1103.

Peter, R.E. (1983). The brain and neurohormones in teleost reproduction. *In* "Fish Physiology, Reproduction, Part A, Endocrine Tissues and Hormones" (W.S. Hoar, D.J..Randall and E.M. Donaldson, Eds.) Vol. IX A, pp. 97–135. Academic Press, New York.

Peter, R.E. (1986). Vertebrate neurohormonal systems. *In* "Vertebrate Endocrinology: Fundamentals and Biomedical Implications" (P.K.T. Pang and M.P. Schreibman, Eds.) Vol.1, pp. 57–104. Academic Press, New York.

Peter, R.E. and Fryer, J.N. (1983). Endocrine functions of the hypothalamus of actinopterygians. *In* "Fish Neurobiology, Higher Brain Areas and Functions" (R.E. Davis and R.G. Northcutt, Eds.) Vol. 2, pp.165–201. Univ. of Michigan Press, Ann Arbor.

Peter, R.E., Yu, K.L., Marchant, T.A. and Rosenblum, P.M. (1990). Direct neural regulation of the teleost adenohypophysis. *J. Exp. Zool.* (Suppl. 4), 84–89.

Pickavance, C.L., Staines, W.A., Fryer, J.N. (1992) Distributions and co-localization of neuropeptide Y and somatostatin in the goldfish brain. *J. Chem. Neuroanat.* **5**, 221–233.

Pontet, A., Danger, J.M., Dubourg, P., Pelletier, G., Vaudry, H., Calas, A., and Kah, O. (1989). Distribution and characterization of neuropeptide Y-like immunoreactivity in the brain and pituitary of the goldfish. *Cell Tissue Res.* **255**, 529–538.

Power, D.M., Canario,A.V.M., and Ingleton, P.M. (1996). Somatotropin-release-inhibiting factor and galanin innervation in the hypothalamus and pituitary of seabream (*Sparus aurata). Gen. Comp. Endocrinol.* **101**, 264–274.

Prasada Rao, P.D., A.G. Jadhao, and Sharma, S.C. (1987a). Descending projection neurons to the spinal cord of the goldfish, *Carassius auratus. J. Comp. Neurol.* **265**, 96–108.

Prasada Rao, P.D., Jadhao, A. and S.C. Sharma (1993a). Topographic organization of descending projection neurons to the spinal cord of the goldfish, *Carassius auratus. Brain Res.* **620**, 211–220.

Prasada Rao, P.D., Job, T.C. and Schreibman, M. (1993b). Hypophysiotropic neurons in the hypothalamus of the catfish, *Clarias batrachus*: A cobaltous lysine and HRP study. *Brain Behav. Evol.* **42**, 24–38.

Prasada Rao, P.D., Murthy, C.K., Cook, H., and Peter, R.E. (1996). Sexual dimorphism of galanin-like immunoreactivity in the brain and pituitary of the goldfish, *Carassius auratus. J. Chem. Neuroanat.* **10**, 119–135.

Prasada Rao, P.D., Rajendra Babu, P., and Jadhao, A.G. (1987b). Distribution of serotonin-like immunoreactivity in the hypothalamus and pineal organ of the catfish, *Mystus vittatus* (Bl.). *In* "Proc. First Congr. Asia and Oceania Soc. Comp. Endocrinol." (E. Ohnishi, Y. Nagahama and H. Ishizaki, Eds.), pp.115–116. Nagoya University Corp., Nagoya, Japan.

Prasada Rao, P.D., and Sharma, S.C. (1982). Retinofugal pathways in juvenile and adult channel catfish, *Ictalurus punctatus*: An HRP and autoradiographic study. *J. Comp. Neurol.* **210**, 37–48.

Ramakrishna, N.S., and Subhedar, N. (1989). Hypothalamic innervation of the pituitary in the catfish, *Clarias batrachus* (L.): A retrograde horseradish peroxidase study. *Neurosci. Lett.* **107**, 39–45.

Ramakrishna, N.S., Subhedar, N., and Schreibman, M.P. (1992). FMRF-amide-like immunoreactive nervus terminalis innervation of the pituitary in the catfish, *Clarias batrachus* (Linn.): demonstration by lesion and immunocytochemical techniques. *Gen. Comp. Endocrinol.* **85**, 111–117.

Rao, S.D., Prasada Rao, P.D., and Peter, R.E. (1996a). Brain and pituitary immunocytochemistry of carassin in the goldfish, *Carassius auratus*: A new neurohormone peptide? *Zool. Sci.* **13**, 415–420.

Rao, S.D., Prasada Rao, P.D., and Peter, R.E. (1996b). Growth hormone releasing hormone immunoreactivity in the brain, pituitary and pineal of the goldfish, *Carassius auratus. Gen. Comp. Endocrinol.* **102**, 210–220.

Reaves, T.A., and Hayward, J.N. (1980). Functional and morphological studies of peptide-containing neuroendocrine cells in the goldfish hypothalamus. *J. Comp. Neurol.* **193**, 777–788.

Rusoff, A.C., and Hapner, S.J. (1990). Organization of retinopetal axons in the optic nerve of the cichlid fish, *Herotilapia multispinosa. J. Comp. Neurol.* **237**, 100–115.

Sas, E., and Maler, L. (1986). Retinofugal projections in a weakly electric gymnotiform fish (*Apteronotus leptorhynchus*). *Neuroscience* **18**, 247–259.

Sas, E., and Maler, L. (1991). Somatostatin-like immunoreactivity in the brain of an electric fish (*Apteronotus leptorhynchus*) identified with monoclonal antibodies. *J. Chem. Neuroanat.* **4**, 155–186.

Sas, E., Maler, L., and Turner, B. (1990). Catecholaminergic system in the brain of a gymnotiform teleost fish: An immunocytochemical study. *J. Comp. Neurol.* **292**, 127–162.

Schmued, L.C., Kyriakidis, K., and Heimer, L., (1990). In vivo anterograde and retrograde axonal transport of fluorescent rhodamine-dextran-amine, fluoro-ruby, within the CNS. *Brain Res.* **526**, 127–134.

Schmued, L.C., and Swanson, L.W. (1982). SITS: a covalently bound fluorescent retrograde tracer that does not appear to be taken up by fibers-of-passage. *Brain Res.* **249**, 137–141.

Schreibman, M.P., Halpern, L.R., Goos, H.J.Th., and Margolis-Kazan, H. (1979). Identification of luteinizing hormone-releasing hormone (LH-RH) in the brain and pituitary gland of a fish by immunocytochemistry. *J. Exp. Zool.* **210**, 153–160.

Schreibman, M.P. (1986). Pituitary gland. *In* "Vertebrate Endocrinology: Fundamentals and Biomedical Implications" (P.K.T. Pang and M.P. Schreibman, Eds.) Vol. I, pp.11–55. Academic press, New York.

Sharma, S.C., Berthoud, V.M., and Breckwoldt, R. (1989). Distribution of substance P-like immunoreactivity in the goldfish brain. *J. Comp. Neurol.* **279**, 104–116.

Sharma, S.C., Prasada Rao, P.D. and Jadhao, A.G. (1993). Regeneration of supraspinal projection neurons in the adult goldfish. *Brain Res.* **620**, 221–228.

Shiga, T., Oka, Y., Satou, M., Okumoto, N., and Ueda, K. (1985). An HRP study of afferent connections of the supracommissural ventral telencephalon and the medial preoptic area in himé salmon (landlocked red salmon, *Oncorhynchus nerka*). *Brain Res.* **361**, 162–177.

Springer, A.D., and Gaffney, J.S. (1981). Retinal projections in the goldfish: A study using cobaltous-lysine. *J. Comp. Neurol.* **203**, 401–424.

Stell, W.K., Walker, S.E., Chohan, K.S., and Ball, A.K. (1984). The goldfish nervus terminalis: A luteinizing hormone-releasing hormone and molluscan cardioexcitatory peptide immunoreactive olfactoretinal pathway. *Proc. Natl. Acad. Sci., USA* **81**, 940–944.

Vecino, E., Perez, M.T., and Ekström, P. (1995). Localization of enkephalinergic neurons in the central nervous system of the salmon (*Salmo salar* L.) by in situ hybridization and immunocytochemistry. *J. Chem. Neuroanat.* **9**, 81–97.

Von Bartheld, C.S., Cunningham, D.E., and Rubel, E.W. (1990). Neuronal tracing with DiI: Decalcification, cryosectioning, and photoconversion for light and electron microscopic analysis. *J. Histochem. Cytochem.* **38**, 725–733.

Yulis, C.R., and Lederis, K. (1987). Co-localization of immunoreactivities of corticotropin-releasing factor and arginine vasotocin in the brain and pituitary system of the teleost *Catostomus commersoni*. *Peptides* **7**, 79–86.

3

FUNCTIONAL AUTONOMY OF THE ADENOHYPOPHYSIS

D. R. Naik*

Department of Zoology
Utkal University, Vani Vihar
Bhubaneswar, 751 004, India

1. INTRODUCTION

The role of hypothalamus in the regulation of adenohypophysial functions is well established (see Holmes and Ball, 1974). The functional integrity of the pituitary gland is largely dependent upon its intact anatomical connection with the brain. Generally, ectopic transplantation of the pituitary gland results in reversible functional impairments, which are restored to normal by its retransplantation beneath the median eminence (ME). On the other hand, it has also been shown that the ectopically transplanted pituitary is capable of secreting some of the tropic hormones, thereby displaying a variable degree of functional autonomy. Although considerable literature has accumulated regarding ectopically transplanted pituitaries in mammals, and to a lesser extent in fishes and amphibians, little information is available in respect of reptiles and birds. Moreover, immunohistochemical studies on the ectopically transplanted pituitary have been undertaken mostly in mammals. Therefore, we examined, immunohistochemically, the ectopically transplanted pituitaries in three representative nonmammalian species (toad, lizard, and quail), which forms the background of this discussion. Ever since the publication of the excellent monograph on the pituitary gland by Holmes and Ball (1974), in which earlier works on ectopic transplantation of the pituitary were reviewed, no attempt seems to have been made to compile the results of subsequent investigations on the subject. In this review, it has been endeavoured to knit a comprehensive account on ectopically transplanted pituitary gland in different vertebrate groups.

* Present Address: Vice-Chancellor, Sambalpur University, Jyoti Vihar, Sambalpur-768 019, India

Neural Regulation in the Vertebrate Endocrine System, edited by Prasada Rao and Peter, Kluwer Academic / Plenum Publishers, New York, 1999.

2. REVIEW OF LITERATURE

2.1. Ectopic Pituitary Transplantation in Mammals

Early reports indicated that when the pituitary gland was transplanted to an ectopic site, it developed adequate vascular connections but exhibited little functional activity (Westman and Jacobson, 1940). Some investigators did, however, report active hormone secretion by pituitary transplants, although doubts were invariably raised regarding possible incomplete removal of the gland from the sella turcica. By removing pituitaries of rat from the hypothalamic site and transplanting them either under the temporal lobe of the brain or beneath the ME, Harris and Jacobson (1952) concluded that anatomical connection of the pituitary with the hypothalamus was essential for its functional integrity. Dependence of the pituitary gland on hypothalamus for normal functioning was also demonstrated by Nikitovitch-Winer and Everett (1957, 1958). When the pituitary of female rats was autotransplanted in the kidney capsule they became well vascularised, but the rats became acyclic and the ovaries were regressed. However, when the pituitaries of these rats were retransplanted in their normal hypothalamic site, i.e., beneath the ME, the gonadotropic activity was restored; the rats became cyclic and some of them even became pregnant after mating. Also, the pioneering studies of Desclin (1950), Everett (1954, 1956) and Sanders and Rennels (1959) suggested that the pituitary grafted in the kidney capsule was capable of secreting prolactin (PRL). From studies of the hypothalamo-hypophysial relationship by various experimental designs (e.g., stalk sectioning, ectopic transplantation, *in vitro* culture) it became clear that, when the pituitary gland was deprived of its hypothalamic connection, secretion of most of the tropic hormones was considerably reduced, although it was invariably accompanied by hyperprolactinemia (see Adler, 1986). Conversely, just as retransplantation of the ectopically grafted pituitary beneath the ME restored normal secretion of all tropic hormones, so also, addition of hypothalamic extracts to cultured pituitary tissue *in vitro* resulted in a decrease in the amount of PRL released into the culture medium. These findings prompted the development of the concept that the hypothalamus, through its regulatory factors, stimulates the secretion of most of the pituitary tropic hormones, but inhibits the secretion of PRL (see Holmes and Ball, 1974). Notwithstanding this generalization, several investigators reported that hypophysectomized animals bearing pituitary grafts did secrete other tropic hormones as well, *albeit* in low concentrations. For example, intraocular hypophysial grafts of fetal pituitaries in hypophysectomized male rats were believed to secrete growth hormone (GH), thyrotropin (TSH), corticotropin (ACTH), and probably also follicle stimulating hormone (FSH) and luteinizing hormone (LH) (Goldberg and Knobil, 1957). It was also shown that pituitary tissue grafted in the renal capsule of hypophysectomized rat secreted small amounts of ACTH (Hertz, 1959). While studying the function of pituitary ocular grafts in the rat, Knigge (1961) reported low levels of TSH production by the grafts. Similarly homotransplantation of juvenile rat pituitaries in the anterior eye chamber of hypophysectomized hosts reversed the regressive changes due to hypophysectomy indicating the restoration of pituitary activity following their ocular transplantation (Martinovitch and Pavic, 1960).

Early histological studies correlated one category of acidophilic (erythrosinophilic/carminophilic) cells of the pituitary with PRL secretion (see Purves, 1966). The preponderance of such acidophilic cells in the grafted pituitaries indicated that they were the source of PRL (Everett, 1956; Nikitovitch-Winer and Everett, 1959; Sanders and Rennels, 1959; Kovacs, 1961). Siperstein and Greer (1956) studied the morphology and histochemistry of the mouse pituitaries implanted in the anterior eye chamber. The ultrastructural characteristics of the PRL secreting cells of the pituitary grafts were reported by several authors (Rennels, 1962; Virag *et al.*, 1967; Santolaya and Rodriguez, 1977; Aguado *et al.*, 1979). These ultrastructural studies revealed that, in addition to the PRL cells, other types of cells were also present in the ectopic pituitary grafts. Subsequently, the appearance of a few GH and gonadotropic (GTH) cells, along with PRL cells, in ectopically transplanted pituitary primordia of rat, was shown by Schechter *et al.*(1985). While putting forth the anterior pituitary grafted rat as a valid model of chronic hyperprolactinemia, Adler (1986) also reported the occurrence of GH, GTH, TSH and ACTH cells in the allografts of adult pituitary tissue.

Like the rat, in the hamster also the ectopically transplanted pituitary tissue produced hyperprolactinemia (Bex *et al.*, 1978). It was further revealed that GH cells could be recognized in the pituitary tissue transplanted in the cheek pouch of hamsters for two weeks (Campbell *et al.*, 1979). In addition, pituitary tissue of neonatal hamsters transplanted beneath the kidney capsule of adult animals for 19 days contained LH cells as well (Gregerson and Campbell, 1982). Thus, in the ectopic pituitary grafts of both rat and hamster, clear evidences regarding the presence of cell types other than PRL cells were provided. In recent years, the most extensive studies on the ectopic pituitary grafts were undertaken by G.T. Campbell and his collaborators (see Campbell *et al.*, 1988, 1996). Neonatal hamster pituitaries were grafted in the kidney capsule of hypophysectomized adult females. The serum levels of PRL, GH, and LH were measured by radioimmunoassay after different durations. The allografts were examined immunohistochemically and the PRL, GH, LH, FSH, and ACTH cells were identified. Their results suggested that: "(a) development of normal adenohypophysial morphology can occur in an ectopic position; (b) intracellular hormones are present in an ectopic site; (c) development and retention of intracellular FSH is more dependent on occupation of normal position of the adenohypophysis than is retention of intracellular LH; and (d) release of PRL occurs from atypical cells in allografts of adult adenohypophysial tissue" (Campbell *et al.*, 1988). Using anterior pituitary allografts in hypophysectomized and orchidectomized hamsters, they also showed the "trophic influences of luteinizing hormone-releasing hormone (LH-RH) on gonadotrophs, growth hormone-releasing hormone (GH-RH) on somatotrophs, and corticotrophin-releasing hormone (CRH) on corticotrophs" (see Campbell *et al.*, 1996).

That the hypophysiotropic neurohormones of the hypothalamus regulated the synthesis and release of pituitary tropic hormones was well established, but their precise role in the functional differentiation of the fetal pituitary was unclear and divergent views were expressed (Bégeot *et al.*, 1983; Watanabe, 1987; Nemeskéri and Halász, 1989). Employing the 'ectopic pituitary graft' model, Nemeskéri *et al.* (1990) grafted fetal rat pituitary in the kidney capsule of ME-removed adult rats and showed that "the undifferentiated fetal pituitary does not require hypothalamic hypophysiotropic neurohormones for proliferation and cytodifferentiation".

Several diverse aspects of ectopically transplanted pituitaries were studied by other investigators. Vuillez *et al.* (1989) stereotaxically implanted neurointermediate lobes from 10-day-old or adult rats into the third ventricle of adult male rats. Immunocytochemistry and ultrastructure of the implants as well as the nature of their innervation revealed that in the early stages of transplantation the MSH cells exhibited intense secretory activity but in later stages (2½ months after transplantation), when reinnervation was established, the overall effect was inhibitory (Vuillez *et al.*, 1989). The effect of 2-bromo-α-ergocryptine, a dopamine agonist drug, on the *in situ* pituitary as well as on the pituitaries transplanted in the renal capsule of hypophysectomized rats was reported by McComb *et al.* (1979).

Their results showed that bromocryptine decreased prolactin content of the grafted anterior pituitary, indicating that even in the absence of hypothalamic connection, the PRL cells were responsive to the effects of bromocryptine. Employing the protein A-gold method of ultrastructural immunocytochemistry, Ishibashi and Shiino (1988) studied the cytological changes in the PRL cells of ocular homografts of rat pituitary after administration of dopamine. They showed that "injection of dopamine into the common carotid artery of the graft-bearing rat rapidly suppressed the granule extrusion and then gradually induced a remarkable morphological atrophy in the prolactin cells".

The appearance of specialized structures in the ectopic pituitary grafts and the possible role of the folliculo-stellate (FS) cells in the process also attracted the attention of investigators. Gon *et al.* (1987) reported the appearance of cyst-like, and ductule-like structures in the rat pituitary autografts and recognized the close association of the FS cells with them. They suggested that "FS cells can play an important role in the restoration of degenerated pituitary glandular tissue during the early stages of transplantation". Inoue *et al.* (1987) noticed the differentiation of striated muscle fibres in the pituitary gland grafts transplanted beneath the kidney capsule of rats. These muscle fibres generally appeared in the central region of the grafts where glandular cells became necrotic, but the FS cells were well preserved. The authors concluded that "this close association between heterotopically differentiated muscle fibres and the normally occurring folliculo-stellate cells strongly suggested a close relationship in differentiation and function between these different cells."

2.2. Sites for Ectopic Transplantation

Ectopic transplantation of the pituitary gland has been done in widely different locations of the body (see Naik *et al.*, 1997a). The kidney capsule being the most convenient site in mammals, in most studies ectopic transplantation has been done in the kidney capsule (see Adler, 1986; Campbell *et al.*, 1996). However, some investigators have chosen other sites, such as the anterior eye chamber (Siperstein and Greer, 1956; Kovács, 1961; Virag *et al.*, 1967; Ishibashi and Shino, 1988), the cheek pouch of hamster (Campbell *et al.*, 1979), the third cerebral ventricle (Vuillez *et al.*, 1989), the so-called hypophysiotropic area of the brain (Halász *et al.*, 1962, 1965) the liver of toad (Naik, 1978) and the tail of tadpole larva of toad (Kikuyama *et al.*, 1993; Kawamura and Kikuyama, 1995). Unlike mammals, most nonmammalian vertebrates do not offer such a convenient site as the kidney capsule for ectopic transplantation. When pituitaries were transplanted in other sites, eg. liver (Naik, 1978), retrieval of the grafts posed problems. Therefore, we developed a novel and simple technique for ectopic transplantation of the pituitary gland. In this, pituitaries are placed in Silastic tubes which are then implanted in the body cavity or liver. The technique is particularly suited for nonmammalian species (Naik *et al.*, 1997a).

2.3. Ectopic Pituitary Transplantation in Fishes

As in mammals, in fishes also the hypothalamus was believed to exert an inhibitory influence on PRL secretion, although the existence of stimulatory factors were also reported. In the ectopic grafts of *Poecilia* pituitary the PRL cells were reported to be active (Ball *et al.*, 1972; Peter, 1973; Chester-Jones *et al.*, 1974). It was also found that when *Poecilia latipina* entered into freshwater from sea water the PRL cells of the *in situ* pituitary as well as of the ectopically transplanted pituitary were rapidly activated, which suggested that hypothalamic connection was not essential for activation of these cells.

However, there appeared to exist considerable species difference, and it was suggested that there might be "a large degree of autonomy" in PRL secretion in teleosts (Holmes and Ball, 1974). It is now well established that a variety of aminergic (e.g. dopamine) and peptidergic (somatostatin, thyrotropin-releasing hormone, vasoactive intestinal polypeptide, peptide histidine isoleucine) agents as well as certain extra-hypothalamic factors regulate PRL secretion in teleosts (Nishioka *et al.*, 1988).

Although results of ectopic transplantation of the pituitary gland *vis a vis* regulation of GH secretion in fishes varied with species, at least in some species like *Anguilla* GH secretion was said to be "strongly autonomous" (Holmes and Ball, 1974). It is now well established that the hypothalamus has both stimulatory and inhibitory control on GH secretion. The presence of GRF-like peptide (Pan *et al.*, 1985; Olivereau *et al.*, 1990) and somatostatin-like peptide (Cook and Peter, 1984; Marchant *et al.*, 1987) in the brain and pituitary of certain fishes has been demonstrated.

The observation that ectopically transplanted fish pituitary secreted abundant TSH, which was also reflected in its histology, clearly indicated that TSH secretion in teleosts could occur even in the absence of hypothalamic connection (Holmes and Ball, 1974). It was shown that administration of reserpine to intact *Poecilia latipina* significantly increased TSH secretion. This was in conformity with the view that the hypothalamus exerted an inhibitory control on TSH secretion in fish. It has also been shown that in goldfish somatostatin has thyrotropin release-inhibiting factor (TIF)-like activity (Peter and McKeown, 1975). It is noteworthy that the presence of receptors for thyrotropin releasing hormone (TRH) in the teleost pituitary (Burt and Ajah, 1984) and of immunoreactive TRH in the brain and pituitary of *Dicentrarchus labrax* (Batten *et al.*, 1990) has been recently demonstrated. However, it has been pointed out that in teleost fishes TRH may be related to pars intermedia (PI) function, rather than to TSH- or PRL-releasing function (Batten *et al.*, 1990).

Ectopic transplantation of fish pituitary resulted in the regression of GTH cells in the grafts indicating the stimulatory nature of the hypothalamic control (see Holmes and Ball, 1974). It is now well established that GTH secretion in teleosts is regulated by the stimulatory actions of gonadotropin-releasing hormone (GnRH) as well as the inhibitory effects of dopamine (Peter *et al.*, 1986; van Oordt, 1987). *In vitro* culture experiments and ectopic transplantation of pituitaries showed that in teleosts, ACTH secretion was stimulated by the hypothalamus (Ball *et al.*, 1972; Peter, 1973). The ACTH cells in the pituitary transplants were "hypoactive" and the plasma cortisol level in the graft-bearing fishes was comparable to that of the hypophysectomized fishes. However, when the graft-bearing fishes were subjected to stress, plasma cortisol levels were elevated to normal, indicating that ACTH secretion from the transplants was elevated. It was also suggested that the stress-induced cortisol secretion was mediated through hypothalamic CRF (Holmes and Ball, 1974). Subsequently Fryer and Peter (1977), through a series of elegant experiments, clearly demonstrated the presence of CRF activity in the goldfish brain and also identified the area involved in the control of ACTH secretion, which might be the source of CRF.

It was shown that, unlike ACTH, MSH secretion in fishes was under dual hypothalamic control (Holmes and Ball, 1974; Olivereau, 1975). When *Poecilia* pituitary was transplanted ectopically or maintained in culture, the PI cells exhibited decreased secretory activity or became completely atrophied, which suggested that stimulatory factor from the hypothalamus was necessary for maintaining the PI cells in a functional state. The presence of MSH-releasing factor (MRF) and MSH release-inhibiting factor (MIF) in certain species of teleost fishes has been demonstrated. It has also been reported that the dopaminergic systems inhibit MSH release (Olivereau *et al.*, 1987; Omeljaniuk *et al.*,

1989), while the serotonergic system seemed to have a stimulatory effect (Olivereau and Olivereau, 1979). Direct innervation of the MSH cells of goldfish pituitary by dopamine fibres has also been demonstrated (Kah *et al.*, 1986).

2.4. Ectopic Pituitary Transplantation in Amphibians

Results of several earlier studies indicated that the amphibian hypothalamus exerted an inhibitory influence on pituitary PRL secretion. Ectopically transplanted pituitaries in newts secreted enough PRL to induce precocious water drive response (Masur, 1969; Peyrot *et al.*,1969) a characteristic function of PRL in some urodeles. Autonomous secretion of PRL *in vitro* was also reported in *Necturus* and *Bufo* (Nicoll, 1971). McKeown (1972) measured the plasma PRL levels in *Bufo bufo* bearing ectopic pituitary transplants, which confirmed increased PRL secretion by the transplants. It was, therefore, argued that in both anurans and urodeles the hypothalamus probably secretes a PRL-inhibiting factor (Holmes and Ball, 1974). Recently in *Bufo japonicus*, Kikuyama *et al.* (1993) autografted pituitary primordia removed from tail-bud embryos (stages 25–26) into the tail primordium. Immunohistochemical examination revealed that normal PRL cells were developed in the ectopically transplanted pituitary primordia. In another study, Kawamura and Kikuyama (1995) surgically removed the central part of the neural plate (NP) of *Bufo japonicus* embryos at the open neurula stage such that in these NP-ectomized embryos the infundibulum did not develop, but the pituitary developed away from the brain. These animals developed and metamorphosed almost normally. In the pituitary of these NP-ectomized animals, which had no morphological contact with the brain, normal PRL cells were present, but were relatively fewer in number.

Initial studies in some urodeles revealed that animals bearing autotransplanted pituitaries exhibited more rapid growth (Jorgenson, 1968; Pasteels, 1960). However, this did not provide conclusive evidence regarding the possible autonomy in GH secretion, because acidophilic cells that persisted in the transplants were correlated with PRL secretion (Holmes and Ball, 1974). McKeown (1972) measured the plasma GH levels in *Bufo bufo* bearing ectopic pituitary graft and found that very little GH was secreted by the transplant, which was believed to suggest a predominantly stimulatory control by the hypothalamus on GH secretion. In a recent study, however, Kawamura and Kikuyama (1995) observed that in the pituitary of NP-ectomized toads immunoreactive GH cells were present which might suggest a certain degree of ontogenic as well as functional autonomy of these cells.

Several studies indicated that TSH secretion in urodeles was much less dependent on hypothalamic connections, and animals bearing ectopically transplanted pituitary exhibited near normal thyroid function (see Holmes and Ball, 1974). The TSH cells in ectopic transplants were also active (Masur, 1969) and, following thiouracil treatment, they exhibited conspicuous hypertrophy and even transformed into the so-called ‘thyroidectomy cells’ (Mazzi, 1970; Compher and Dent, 1970). Thus, TSH secretion in urodeles was said to be ‘highly (but not completely) autonomous’. Unlike the urodeles, in anurans TSH secretion was dependent on hypothalamic TRF stimulation, although in some a TSH inhibitory mechanism seemed to exist (Rosenkilde, 1972). It may also be mentioned that in *Bufo japonicus*, Kawamura and Kikuyama (1995) examined the pituitaries of NP-ectomized animals immunohistochemically and found that TSH cells constituted the predominant (46%) cell population, even though the pituitaries had no contact with the brain.

In urodeles normal gonadotropic function of the pituitary was dependent on its hypothalamic connection, and ectopic autotransplantation of the pituitary induced testicular atrophy (Mazzi, 1970). Similarly, ectopic transplantation of the pituitary pars distalis in anurans resulted in decreased GTH secretion, although the GTH cells seemed to have a certain degree of autonomy (see Holmes and Ball,1974). It is noteworthy that in the pituitaries of NP-ectomized *Bufo japonicus*, devoid of hypothalamic connection, a sizable population (36%) of cells were gonadotropes (Kawamura and Kikuyama, 1995).

It was shown that ectopically transplanted urodele pituitary secreted normal amounts of ACTH, which was evidently due to a considerable degree of autonomy. On the other hand, in anurans, ectopic transplantation of the pituitary resulted in a failure of ACTH secretion which varied with the distance of the graft from the hypothalamus (see Holmes and Ball, 1974). Using plasma corticosterone as a parameter, in *Bufo bufo*, it was shown that pituitary grafts placed on an eye muscle secreted significantly less ACTH (Büchmann *et al.*, 1972). Recently Kawamura and Kikuyama (1995) demonstrated the inductive influence of the posterior hypothalamus (infundibulum) on the development and differentiation of pituitary proopiomelanocortin (ACTH / MSH) cells in *Bufo japonicus*. In their NP-ectomized animals, ACTH and MSH cells did not appear, indicating the dependence of these cells on the hypothalamus for their differentiation.

Numerous earlier studies showed that the amphibian hypothalamus exerted a predominantly inhibitory control over MSH secretion, although the existence of stimulatory factor was also postulated. The PI is poorly vascularized receiving little blood from the ME, but is richly innervated. Sectioning of these nerves, without severing the portal vessels, resulted in hypersecretion of MSH indicating the predominant role of direct innervation in hypothalamic inhibition of MSH secretion (see Holmes and Ball, 1974). In a recent study in *Bufo japonicus*, pituitary primordia taken from tail-bud stage embryos, when transplanted into the tail primordia of intact embryos, developed normal MSH cells for which hypothalamic connection was not necessary (Kikuyama *et al.*, 1993).

2.5. Ectopic Pituitary Transplantation in Reptiles and Birds

A review of earlier literature (see Holmes and Ball,1974) shows that no attempt was made to reveal the nature of hypothalamic regulation in reptiles and birds by employing ectopic pituitary transplantation. Therefore, we studied, in detail, the immunohistochemistry of ectopically transplanted pituitaries in a reptile (monitor lizard) and a bird (Japanese quail), which are briefly presented below along with some information that accrued from our pilot study on the ectopically transplanted toad pituitary.

3. IMMUNOHISTOCHEMISTRY OF ECTOPIC PITUITARY GRAFTS IN THE TOAD, LIZARD, AND QUAIL

In the Indian toad (*Bufo melanostictus*), monitor lizard (*Varanus flavescens*) and the Japanese quail (*Coturnix coturnix japonica*) pituitaries of homologous donors, killed by decapitation, were transplanted either between the coin-like lobes of the kidney (in the lizard), or in the abdominal cavity (in the toad and quail) by placing them in Silastic tubes (Naik *et al.*, 1997a). The transplanted pituitaries were retrieved by killing the host animals after different durations (4/10/20 days for toad; 4/7/14 days for quail; 10/15/35 days, and 5 months for lizard), and paraffin sections were stained immunohistochemically (PAP method) using heterologous antisera: anti-human (h) TSHβ, anti-rat (r) TSH, anti-rPRL, anti-hPRL (gift from National Hormone and Pituitary Program, Baltimore, USA), anti-porcine (p) LHβ, anti-bovine (b) GH (gift from the late Prof. M.P. Dubois, INRA, Nouzilly, France), anti-ACTH$^{1-24}$, anti-αMSH (gift from Prof. J.A. Edwardson, MRC

Neurochemical Pathology Unit, New Castle upon Tyne, UK), and anti-chicken (c) FSH (gift from Prof. S. Ishii, Waseda University, Tokyo, Japan) also see Naik *et al.*, 1997b).

In the 4/10/20-day-old pituitary transplants of the toad normal PRL (Figure 1) and GTH (Figure 2) cells as well as a few diminutive ACTH cells were present. However, GH and TSH cells were not detected. In the monitor lizard all the adenohypophysial cell types, viz. GH (Figures 3, 4), ACTH (Figure 5), GTH (Figures 6, 7), TSH, and PRL, were found

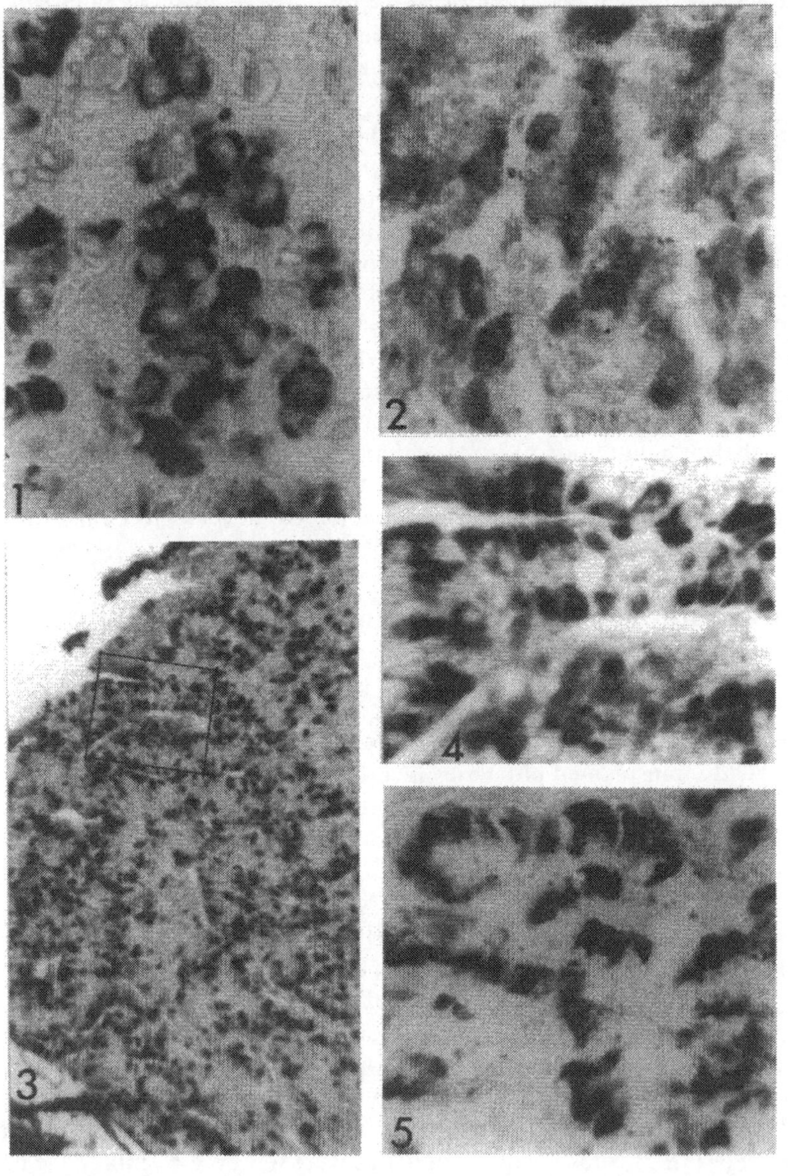

Figure 1-5. (1) Immunoreactive PRL cells in 10-day-old pituitary homograft (Silastic tube implant) of toad. anti-hPRL. X 660. (2) Immunoreactive GTH cells in 20-day-old pituitary homograft (Silastic tube implant) of toad. anti-hLH. X 660. (3) Immunoreactive GH cells in 5-month-old renal pituitary homograft of monitor lizard. anti-bGH. X 180. (4) Marked portion in Fig. 3 enlarged to show GH cells. X 660. (5) Immunoreactive ACTH cells in 15-day-old renal pituitary homograft of monitor lizard. anti-ACTH1–24. X 660.

in all stages of the transplants, including the 5-month-old ones, although there was variation in their numbers and cytological features. In the quail viable PRL (Figure 8), GH (Figure 9), GTH, and ACTH cells were present in the 4/7/14-day-old transplants, but TSH cells were detected only in the 4-day-old grafts. The persistence of various adenohypophysial cell types in the ectopically transplanted pituitaries of the three nonmammalian species possibly indicates the existence of varying degree of functional autonomy in the secretion of pituitary tropic hormones.

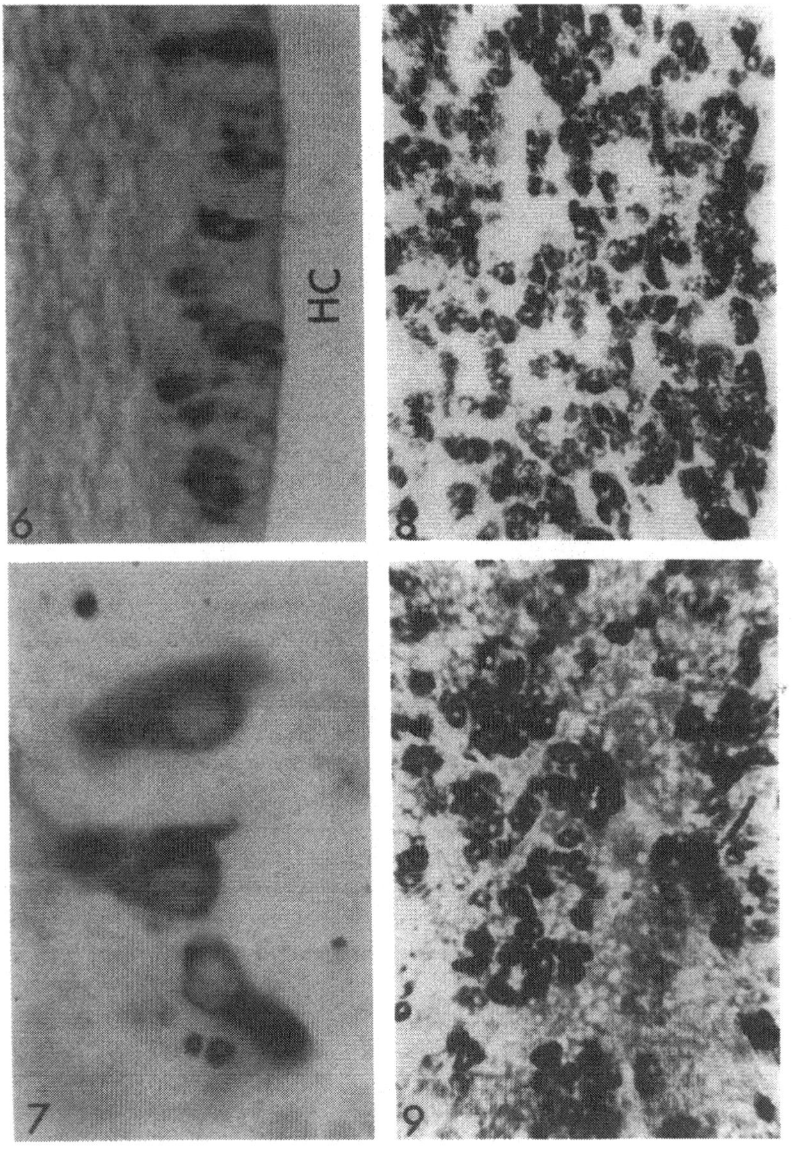

Figure 6-9. (**6 and 7**) Immunoreactive GTH cells in 5-month-old renal pituitary homograft of monitor lizard. Normal cells are usually found in the vicinity of the hypophysial cleft (HC) (Fig. 6) and in the peripheral region (Fig. 7). anti-pLHβ. X 660. (**8**) Immunoteactive PRL cells in 7-day-old pituitary homograft (Silastic tube implant) of Japanese quail. anti-rPRL. X 660. (**9**) Immunoreactive GH cells in 14-day-old pituitary homograft (Silastic tube implant) of Japanese quail. anti-bGH. X 660.

4. CONCLUSIONS

When the pituitary gland is transplanted at an ectopic site, the neurovascular route of hypothalamic control does not operate. However, hypophysiotropic substances have been shown to be present in the general circulation (Brodish and Long, 1962; Saito *et al.*, 1997; Arimura and Schally, 1979). These regulatory factors can reach the ectopically transplanted pituitary and modulate its secretory activity (Nemeskéri *et al.*, 1990).

On the basis of the presence of various adenohypophysial cell types in the fetal rat pituitary transplanted in the kidney capsule of ME-removed isogenic hosts for 28 days, it has been concluded that the "undifferentiated fetal pituitary does not require hypothalamic hypophysiotropic neurohormones for proliferation and cytodifferentiation"(Nemeskéri *et al.*, 1990). However, it has been assumed that "in the absence of the ME, the hypophysiotropic neurohormone cannot be released into the circulation". This hypothesis needs to be verified.

By transplanting neonatal female hamster pituitaries in the kidney capsule of adult hypophysectomized female hosts, it has been concluded that "development of normal adenohypophysial morphology can occur in an ectopic position" (Campbell *et al.*, 1988).

In addition to facilitating the elucidation of the concept of hypothalamic and neurohormonal control of adenohypophysis, the hypophysial grafts beneath the renal capsule has also proved to be a valuable model for other endocrine control systems: (a) Very early studies using pituitary isografts helped in implicating PRL with the growth of mammary tumours; (b) Pituitary allografts can be used to provide PRL replacement with accompanying minimal amounts of other pituitary hormones in hypophysectomized hosts, or to produce hyperprolactinemia in almost any type of host. (c) Pituitary allografts can be used to elucidate the possible physiological roles of PRL in immune response, calcium regulation, and cardiovascular and renal functions (Campbell *et al.*, 1996).

The newly developed novel and simple technique for ectopic transplantation of the pituitary gland using Silastic tube is expected to be useful and may "find wider application because it can also be employed for transplanting other organs/tissues into a wide range of animals, including invertebrates" (Naik *et al.*, 1997a).

ACKNOWLEDGMENT

The work was supported by grants under the Special Assistance Program (DRS) from the University Grants Commission of India. Thanks are due to Dr. K.C. Mohanty, Miss Littli Patnaik, and Ms. H. Samantaray for providing the photomicrographs, and to Prof. S. Das for his valuable assistance during the preparation of the manuscript.

REFERENCES

Adler, R.A. (1986). The anterior pituitary-grafted rat: a valid model of chronic hyperprolactinemia. Endocr. Rev. 7, 302–313.

Aguado, L.I., Rodriguez, S., Bilbao, A., and Rodriguez, E.M. (1979). Cytological changes in the pars distalis of the female rat hypophysis grafted under the kidney capsule. Cell Tissue Res. **199**, 539–543.

Arimura, A., and Schally, A.V. (1979). Hypothalamic releasing and inhibiting hormones and factors. *In* "Hormones in Blood", Vol.I (C.H. Gray and V.H.T. James, Eds.), pp. 1–53. Academic Press, London.

Ball, J.N., Baker, B.I., Olivereau, M., and Peter, R.E. (1972). Investigations on hypothalamic control of adenohypophysial functions in teleost fishes. Gen. Comp. Endocrinol. **Suppl. 3**, 11–21.

Batten, T.F.C., Moons, L., Cambre, M.L., Vandesande, F., Seki, T., and Suzuki, M. (1990). Thyrotropin-releasing hormone-immunoreactive system in the brain and pituitary gland of the sea bass (*Dicentrarchus labrax*, Teleostei). Gen. Comp. Endocrinol. **79**, 385–392.

Bégeot M., Hemming, F.J., Martinet, N., Dubois, M.P., and Dubois, P.M. (1983). Gonadotropin releasing hormone (GnRH) stimulates immunoreactive lactotrope differentiation. Endocrinology **112**, 2224–2226.

Bex, F., Bartke, A., Goldman, B.D., and Dalterio, S. (1978). Prolactin, growth hormone, luteinizing hormone receptors and seasonal changes in testicular activity in the golden hamster. Endocrinology **103**, 2069–2080.

Brodish, A., and Long, C.N.H. (1962). ACTH-releasing hypothalamic neurohumor in peripheral blood. Endocrinology **71**, 298–306.

Burt, D.R., and Ajah, M.L. (1984). TRH receptors in fish. Gen. Comp. Endocrinol. **53**, 135–142.

Büchmann, N.B., Spies, I., and Vijayakumar, S. (1972). Hypophysial corticotropic function and its hypothalamic control in *Bufo bufo* (L) evaluated by the plasma concentration of corticosterone. Gen. Comp. Endocrinol. **18**, 306–314.

Campbell, G.T., Wagoner, J., Gregerson, K.A., and Joyner, W.L. (1979). Immunohistochemical visualizations of prolactin, growth hormone, and a substance resembling placental lactogen in the in situ and ectopic pituitary in the hamster. Endocrinology **105**, 905–910.

Campbell, G.T., Wagoner, J., Colosi, P., Soares, M.J., and Talamantes, F. (1988). Development and retention of phenotypically specialized cells in pituitary allografts in the hamster (*Mesocricetus auratus*). Cell Tissue Res. **251**, 215–220.

Campbell, G.T., Steele, M.A., and Blake, C.A. (1996). Hypophysial grafts beneath the renal capsule: A model to study endocrine control system. *In* "Endocrine Methods" (J.A. Thomas, Ed.) pp. 101–114. Academic Press, New York.

Chester-Jones, I., Ball, J.N., Henderson, I.W., Sandor, T., and Baker, B.I. (1974). The endocrinology of fish. *In* "Chemical Zoology", Vol. 8. (M. Florkin and B.T. Scheer, Eds.) pp. 523–593. Academic Press, New York.

Cook, A.F., and Peter, R.E. (1984). The effects of somatostatin on serum growth hormone levels in the goldfish, *Carassius auratus*. Gen. Comp. Endocrinol. **54**, 109–113.

Compher, M.K., and Dent, J.N. (1970). Responses to thiourea and to surgical thyroidectomy by the autotransplanted pituitary gland in the red spotted newt. Gen. Comp. Endocrinol. **14**, 141–147.

Desclin, L. (1950). A propos du mechanisms d'action des oestrogens sur le lobe anterieur de l'hypophyse chez le rat. Ann. Endocrinol. (Paris) **11**, 656–659.

Everett, J.W. (1954). Luteotrophic function of autografts of the rat hypophysis. Endocrinology **54**, 658–690.

Everett, J.W. (1956). Functional corpora lutea maintained for months by autografts of rat hypophysis. Endocrinology **58**, 786–796.

Fryer, J.N., and Peter, R.E. (1977). Hypothalamic control of ACTH secretion in goldfish. I. Corticotrophin-releasing factor activity in teleost brain tissue extracts. II. Hypothalamic lesioning studies. III. Hypothalamic cortisol implant studies. Gen. Comp. Endocrinol. **33**, 196–225.

Goldberg, R.C., and Knobil, E. (1957). Structure and function of intraocular hypophyseal grafts in the hypophysectomised male rat. Endocrinology **61**, 742–752.

Gon, G., Shirasawa, N., and Ishikawa, H. (1987). Appearance of cyst- and ductule-like structures and their role in the restoration of the pituitary autograft. Anat. Rec. **217**, 371–378.

Gregerson, K.A., and Campbell, G.T. (1982). Influences of luteinizing hormone releasing hormone, hypophysectomy and orchidectomy on the differentiation of luteinizing hormone and follicle-stimulating hormone cells in an ectopic pituitary in the hamster. Biol. Reprod. **27**, 169–181.

Halász, B., Pupp, L., and Uhlarik, S. (1962). Hypophysiotrophic area in the hypothalamus. J. Endocr. **25**, 147–154.

Halász, B., Pupp, L., Uhlarik, S., and Tima, L. (1965). Further studies on the hormone secretion of the anterior pituitary transplanted into hypophysiotropic area of the rat hypothalamus. Endocrinology **77**, 343–355.

Harris, G.W., and Jacobson, D. (1952). Functional graft of the anterior pituitary gland. Proc. R. Soc. Ser. B. **139**, 263–276.

Hertz, R. (1959). Growth in the hypophysectomised rat sustained by pituitary grafts. Endocrinology **65**, 926–931.

Holmes, R.L., and Ball, J.N. (1974). "The Pituitary Gland. A Comparative Account". Cambridge Univ. Press, London.

Inoue, K., Taniguchi, Y., and Kurosumi, K. (1987). Differentiation of striated muscle fibres in pituitary gland grafts transplanted beneath the kidney capsule. Arch. Hist. Jpn. **50**, 567–578.

Ishibashi, T., and Shiino, M. (1988). Ultrastructural and immunocytochemical changes of prolactin cells of grafted pituitary after the injection of dopamine in the albino rat. Acta. Anat. **131**, 66–72.

Jorgensen, C.B. (1968). Central nervous control of adenohypophysial functions. *In* "Perspectives in Endocrinology" (E.J.W. Barrington and C.B. Jorgensen, Eds.), pp. 469–541. Academic Press, New York.

Kah, O., Dubourg, P., Onteniente, B., Geffard, M., and Calas, A., (1986). The dopaminergic innervation of the goldfish pituitary. Cell Tissue Res. **244**, 577–582.

Kawamura, K., and Kikuyama, S. (1995). Induction from posterior hypothalamus is essential for the development of the pituitary proopiomelanocortin (POMC) cells of the toad (*Bufo japonicus*). Cell Tissue Res. **279**, 233–239.

Kikuyama, S., Inaco, H., Jenks, B.G., and Kawamura, K. (1993). Development of ectopically transplanted primordium of epithelial hypophysis in *Bufo japonicus*. J. Exp. Zool. **266**, 216–220.

Knigge, K.M. (1961). Pituitary ocular graft function in the rat, with a comparison of the qualitative nature of TSH from normal pituitaries and ocular grafts. Endocrinology **68**, 101–114.

Kovács, K. (1961). Histological alterations in the rat pituitary transplanted to the eye. J. Endocr. **23**, 109–117.

Marchant, T.A., Fraser, R.A., Andrews, P.C., and Peter, R.E. (1987). The influence of mammalian and teleost somatostatins on the secretion of growth hormone from goldfish (*Carassius auratus* L.) pituitary fragments *in vitro*. Regul. Pep. **17**, 41–52.

Martinovitch, P.N., and Pavić, B. (1960). Functional pituitary transplants in rats. Nature, Lond. **185**, 155–156.

Masur, S.K. (1969). Fine structure of the autotransplanted pituitary in the red eft, *Notophthalmus viridescens*. Gen. Comp. Endocrinol. **12**, 12–32.

Mazzi, V., (1970). The hypothalamus as a thermodependent neuroendocrine center in urodeles. *In* "The Hypothalamus" (L. Martini, M. Motta and F. Frashchini, Eds.), pp. 663–676. Academic Press, New York.

McComb, D.J., Ryan, N., Horvath, E., Kovács, K., Domokos, I., and Laszolo, F.A. (1979). An immunohistochemical and ultrastructural comparison of the effects of 2-bromo-α-ergocryptine on intrasellar and transplanted rat pituitaries. Experientia **35**, 1409–1410.

McKeown, B.A. (1972). Prolactin and growth hormone concentrations in the plasma of the toad *Bufo bufo* (L.) following ectopic transplantation of the pars distalis. Gen. Comp. Endocrinol. **19**, 167–174.

Naik, D.R. (1978). The nature of hypothalamic influences on prolactin secretion in the yellow monitor *Varanus flavescens* (Gray) and the common Indian toad *Bufo melanostictus* Schneider. *In* "Proceedings, 3rd All India Symposium on Comparative Endocrinology, Bhubaneswar", p.18 (Abstract).

Naik, D.R., Das, S., Patnaik, L., and Samantaray, H. (1997a). A novel and simple technique for ectopic transplantation of the pituitary gland. Gen. Comp. Endocrinol. **107**, 304–310.

Naik, D.R., Das, S., Mohanty, K.C., Patnaik, L., Samantaray, H. and Ishikawa, H. (1997b). Immunohistochemistry of ectopically transplanted pituitary in nonmammalian vertebrates. *In* "Advances in Comparative Endocrinology" Vol. I (S. Kawashima and S. Kikuyama, Eds.), pp. 767–770. Monduzzi Editore, Bologna (Italy).

Nemeskéri, A., and Halász, B. (1989). Cultured fetal rat pituitaries kept in synthetic medium are able to initiate synthesis of trophic hormones. Cell Tissue Res. **255**, 645–650.

Nemeskéri, A., Sétáló, G., Kacsóh, B., and Halász, B. (1990). Fetal pituitary graft is capable of initiating hormone synthesis in median eminence removed adult rat. Endocrinologia Experimentalis 24, 283–292.

Nicoll, C.S. (1971). Aspects of the neural control of prolcatin secretion. *In* "Frontiers in Neuroendocrinology" (L. Martini and W.F. Ganong, Eds.). pp. 291–330. Oxford University Press, London.

Nikitovitch-Winer, M., and Everett, J.W. (1957). Resumption of gonadotrophic function in pituitary grafts following retransplantation from kidney to median eminence. Nature, (London). **180**, 1434–1435.

Nikitovitch-Winer, M., and Everett, J.W. (1958). Functional restitution of pituitary grafts re-transplanted from kidney to median eminence. Endocrinology **63**, 916–930.

Nikitovitch-Winer, M., and Everett, J.W. (1959). Histological changes in grafts of rat pituitary on the kidney and upon re-transplantation under the diencephalon. Endocrinology **65**, 357–368.

Nishioka, R.S., Killey, K.M., and Bern, H.A. (1988). Control of prolactin and growth hormone secretion in teleost fishes. Zoological Science **5**, 267–280.

Olivereau, M. (1975). Hypothalamic control of MSH secretion in lower vertebrates. Folia. Anat. Iugosl. **4**, 15–21.

Olivereau, M., and Olivereau, J. (1979). Effect of serotonin on prolactin and MSH-secreting cells in the eel. Comparison with the effect of 5-hydroxytryptophan. Cell Tissue Res. **196**, 397–408.

Olivereau, M., Olivereau, J., and Lambert, J. (1987). *In vivo* effect of dopamine and dopamine antagonists on melanocyte-stimulating hormone cells of the goldfish (*Carassius auratus* L.) pituitary. Gen. Comp. Endocrinol. **68**, 12–18.

Olivereau, M., Olivereau, J., and Vandesande, F. (1990). Localization of growth hormone-releasing factor-like immunoreactivity in the hypothalamo-hypophysial system of some teleost species. Cell. Tissue Res. **259**, 73–80.

Omeljaniuk, R.J., Tonon, M.C., and Peter, R.E. (1989). Dopamine inhibition of gonadotropin and α-melanocyte-stimulating hormone release *in vitro* from the pituitary of the goldfish (*Carassius auratus*). Gen. Comp. Endocrinol. **74**, 451–467.

Pasteels, J.L. (1960). Étude expérimentale des différentes catégories d'éléments chromophiles de l'hypophyse adulte de *Pleurodeles waltlli*, de leur fonction et de leur contrôle par l'hypothamus. Archs Biol., Liége **71**, 409–471.

Pan, J.X., Lechan, R.M., Lin, H.D., and Jackson, I.M.D. (1985). Immunoreactive neuronal pathways of growth hormone-releasing hormone (GRH) in the brain and pituitary of the teleost *Gadus morhua*. Cell Tissue Res. **241**, 487–493.

Peter, R.E. (1973). Neuroendocrinology of teleosts. Am. Zool. **13**, 743–755.

Peter, R.E., and McKeown, B.A. (1975). Hypothalamic control of prolactin and thyrotropin secretion in teleosts, with special reference to recent studies on the goldfish. Gen. Comp. Endocrinol. **25**, 153–165.

Peter, R.E., Chang, J.P., Nahorniak, C.S. Omeljaniuk, R.J., Sokolowska, M., Shih, S.H., and Billard, R. (1986). Interactions of catecholamines and GnRH in regulation of gonadotropin secretion in teleost fish. Recent Prog. Horm. Res. **42**, 513–548.

Peyrot, A., Mazzi, V., Vellano, C., and Lodi, G. (1969). Prolactin activity of short-term and long-term autografted pituitaries of the hypophysectomized crested newt. J. Endocr. **45**, 515–530.

Purves, H.D. (1966). Cytology of the adenohypophysis. *In* "The Pituitary Gland", Vol. I (G.W. Harris and B.T. Donovan, Eds.) pp. 147–232. Butterworths, London.

Rennels, E.G. (1962). An electron microscope study of pituitary autograft cells in the rat. Endocrinology **71**, 713–722.

Rosenkilde, P. (1972). Hypothalamic control of thyroid function in Amphibia. Gen. Comp. Endocrinol. Suppl. **3**, 32–40.

Saito, T., Sawano, S., Arimura, A., and Schally, A.V. (1967). Follicle-stimulating hormone releasing activity in peripheral blood. Endocrinology **81**, 1226–1230.

Sanders, A.E., and Rennels, E.G. (1959). Evidence on the cellular source of luteotrophin derived from a study of rat pituitary autografts. Z. Zellforsch. **49**, 263–274.

Santolaya, R.C., and Rodriguez, E.M. (1977). Ultrastructure of the male rat hypophysis chronically grafted under the kidney capsule. Cell Tissue Res. **179**, 271–284.

Schechter, J., Gash, D., and Ahmad, M. (1985). Mesenchymal influences on the development of the adenohypophysis in the rat. Cell Tissue Res. **241**, 67–76.

Siperstein, E.R., and Greer, M.A. (1956). Observations on the morphology and histochemistry of the mouse pituitary implanted in the anterior eye chamber. J. Natl. Cancer Inst. **17**, 569–600.

van Oordt, P.G.W.J. (1987). Modern trends in reproductive endocrinology of teleosts. *In* "Proc. Vth Cong. Europ. Ichthyol." (S.O. Kullandes and B. Fernholm, Eds.) pp. 247–268. Swedish Museum of Natural History, Stockholm.

Virágh, S., Tiboldi, T., Kovács, K., Julesz, M., and Törö, I. (1967). Electron microscopic structure of anterior pituitaries transplanted into anterior chamber of the eye in rats. Acta. Biol. Hung. **18**, 77–89.

Vuillez, P., Moss, F., and Stoeckel, M.E. (1989). Immunocytochemical and ultrastructural studies on allografts of the pituitary neurointermediate lobe in the third cerebral ventricle of the rat. Cell Tissue Res. **255**, 393–404.

Watanabe, Y.G. (1987). Failure of luteinizing hormone-releasing hormone (LHRH) to affect the differentiation of LH cells in the rat hypophyseal primordium in serum-free culture. Cell Tissue Res. **250**, 35–42.

Westman, A., and Jacobson, D. (1940). Experimentelle untersuchung über hypophysentransplantate bei der ratte. Acta Path. Microbiol. Scand. **17**, 328–347.

4

BRAIN REGULATION OF GROWTH HORMONE SECRETION AND FOOD INTAKE IN FISH

Richard E. Peter and John P. Chang

Department of Biological Sciences
University of Alberta
Edmonton, Alberta T6G 2E9, Canada

1. INTRODUCTION

Growth in fish is regulated by the brain neuroendocrine—growth hormone (GH)—insulin-like growth factor axis (Peter and Marchant, 1995; Peng and Peter, 1997). However, growth cannot be realized without adequate food intake. A question to be answered in fish is whether there is any connection between the regulation of food intake and the brain neuroendocrine – GH – insulin-like growth factor axis. An additional question of relevance to aquaculture is whether the brain neuroendocrine – GH – insulin-like growth factor axis and food intake can be manipulated to stimulate faster growth rates of farmed fish?

2. REGULATION OF FOOD INTAKE

Brain electrical stimulation and lesioning studies on fish have demonstrated that feeding behaviour is regulated by the posterior ventromedial–ventrolateral hypothalamus (for review: Peter, 1979; Himick and Peter, 1995b). Given the availability of an adequate food supply, food intake is regulated to maintain growth in an orderly fashion, including variations in growth that may occur on a seasonal basis. Treatment of fish with GH, by repeated injection, intubation, pellet implantation or other means, increases growth rates (McLean and Donaldson, 1993). Treatment with GH for two days, or a more extended period also causes increased feeding and food conversion efficiency in teleosts (Donaldson *et al.*, 1979; Peter and Marchant, 1995; Peng and Peter, 1997). This indicates that GH not only regulates growth, but that it also has influences on food intake.

Is there a relationship between food intake and GH secretion in fish? We have recently found that goldfish, fed a normal ration at a regular time, have an acute elevation in

Neural Regulation in the Vertebrate Endocrine System, edited by Prasada Rao and Peter, Kluwer Academic / Plenum Publishers, New York, 1999.

serum GH levels at 30 minutes following feeding (Himick and Peter, 1995b). This initial rise in serum GH levels, which occurs independent of body weight, is followed by a sharp decrease and then a more gradual decrease to serum GH levels significantly lower than in unfed control fish. This characteristic pattern in post-prandial serum GH levels does not occur in goldfish fed on a maintenance diet. This indicates that there is a relationship between normal feeding and GH secretion in goldfish, and presumably other fish as well.

2.1. Neuropeptide and Neurotransmitter Regulation of Feeding in Fish

The brain and gut peptide bombesin (BBS) suppresses food intake within 45 minutes following intraperitoneal or brain intraventricular injection (Himick and Peter, 1994a). Likewise, intraperitoneal or brain intraventricular injection of the eight amino acid sulfated form of cholecystokinin (CCK) also acutely suppresses food intake in goldfish (Himick and Peter, 1994b). At 30 minutes following suppression of food intake by either BBS or CCK, serum GH levels increase (Himick and Peter, 1994b, 1995a), similar to the situation that occurs in fish fed a normal ration at a regular time (Himick and Peter, 1995b). Notably, immunocytochemistry studies demonstrate that BBS and CCK are present in the brain hypothalamic feeding area of goldfish (Himick and Peter, 1994b, 1995a) and other teleosts (Himick and Peter, 1995b). Receptor radioautography studies indicate the presence of BBS (Himick *et al.*, 1995) and CCK (Himick *et al.*, 1996) receptors in the brain hypothalamic feeding area of goldfish. The complementary DNA (cDNA) encoding prepro-CCK in goldfish has been isolated and sequenced, and the presence of prepro-CCK mRNA demonstrated throughout the brain, with highest levels of expression occurring in the hypothalamus and olfactory bulbs (Peyon *et al.*, 1998). *In situ* hybridization studies demonstrated prepro-CCK mRNA in the posterior ventrolateral hypothalamus, the hypothalamic feeding area (Peyon *et al.*, 1998). Shortly following feeding, an acute elevation of prepro-CCK mRNA occurs in goldfish brain (Peyon and Peter, 1997). Together these data provide strong evidence that BBS and, particularly, CCK are involved in satiation in goldfish, and presumably also in other fish.

Corticotropin-releasing factor (CRF) has been reported to suppress food intake in goldfish following either intraperitoneal or brain intracerebroventricular injection (De Pedro *et al.*, 1993). Immunocytochemical studies demonstrated CRF in the goldfish in the posterior ventral and posterior ventrolateral hypothalamus (Lederis *et al.*, 1994), appropriate for involvement in regulation of food intake. De Pedro *et al.* (1998) also demonstrated that brain intracerebroventricular injection of serotonin (5-HT) is inhibitory to feeding in goldfish, and is in part mediated by CRF. Interestingly, subordinate rainbow trout, which have suppressed food intake and growth rates relative to dominants, have elevated brain serotonergic activity (Winberg *et al.*, 1993).

Neuropeptide Y (NPY), which plays a prominent role in stimulating food intake in mammals, is present in the goldfish ventromedial-posterior hypothalamus and hypothalamic inferior lobes (Pontet *et al.*, 1989), the appropriate brain areas for involvement in food intake. *In situ* hybridization studies in goldfish demonstrate that mRNA encoding NPY is expressed in these same brain areas (Peng *et al.*, 1994). Further, NPY expression in the goldfish preoptic area is stimulated by ovarian steroids (Peng *et al.*, 1994), and NPY expression in the preoptic area is stimulated by food deprivation in chinook and coho salmon (Silverstein *et al.*, 1998). While more direct studies on neuropeptides that stimulate food intake in fish would be helpful, these studies strongly implicate NPY in feeding behaviour in fish.

Stimulatory effects of galanin (De Pedro *et al.*, 1995) and ß-endorphin (De Pedro *et al.*, 1995) on food in goldfish have also been demonstrated. These studies involve brain intracerebroventricular injection of peptides or peptide analogs, indicating that central mechanisms underlie the effects observed. The effects of galanin were shown to be mediated, in part, by α_2-adrenergic mechanisms (De Pedro *et al.*, 1995). In anatomical studies, only isolated galanin-immunoreactive perikarya were found in the posterior hypothalamic region of goldfish (Prasada Rao *et al.*, 1996). The lack of localization of galanin in the hypothalamic feeding area suggests that it may have only indirect actions on food intake. Actions of GH-releasing factor (GRF) on food intake have not been tested, to our knowledge; however, GRF-immunoreactive perikarya have been demonstrated in parts of the posterior hypothalamic region involved in food intake regulation in the goldfish (Rao *et al.*, 1996).

3. NEUROENDOCRINE REGULATION OF GH SECRETION IN FISH

The neuroendocrine regulation of GH secretion in goldfish and other fish is multifactorial, with a balance of stimulatory and inhibitory neurohormones acting on the somatotrophs (Peter and Marchant, 1995; Peng and Peter, 1997). Somatostatin (SRIF) is the primary inhibitor of basal and stimulated GH secretion; norepinephrine (NE), serotonin (5HT) and glutamate also have inhibitory actions on GH release. GH secretion is stimulated by GRF, pituitary adenylate cyclase-activating polypeptide (PACAP), gonadotropin-releasing hormone (GnRH), dopamine (DA), NPY, thyrotropin-releasing hormone (TRH), CCK, BBS, and activin.

3.1. Stimulation of GH Secretion

GRF and PACAP are members of the secretin / glucagon / vasoactive intestinal peptide superfamily of peptides. Stimulation of GH release by GRF in teleosts is well established (Peng and Peter, 1997). GRF in common carp has been isolated and sequenced (Vaughan *et al.*, 1992), and the structure of GRF deduced from cDNAs cloned from salmon (Parker *et al.*, 1993) and catfish (McRory *et al.*, 1995). Immunoreactive perikarya and fibers were demonstrated in the preoptic and ventrobasal hypothalamic hypophysiotropic regions in goldfish brain using antisera specific to common carp GRF (Rao *et al.*, 1996). Fish GRFs share only 35%–40% identities with human GRF. Notably, in salmon (Parker *et al.*, 1993, 1997) and catfish (McRory *et al.*, 1995) GRF and PACAP are encoded on the same gene. Fish PACAP shares greater than ~90% primary sequence homology with mammalian PACAP. PACAP-like peptides or mRNA have also been detected in Thai catfish (McRory *et al.*, 1995), Atlantic, chinook, coho and sockeye salmon, and rainbow trout (Parker *et al.*, 1997). Tissue distribution studies of PACAP mRNA suggest that PACAP is produced in the brain but not the pituitary in salmonids (Parker *et al.*, 1997). Immunohistochemical studies demonstrate PACAP in nerve fibers in the goldfish pituitary, in particular in the proximal pars distalis in proximity to the somatotrophs (Wong *et al.*, 1998). PACAP also acts as a GH regulator in teleosts. Salmon PACAP stimulates GH release from cultured coho salmon pituitary cells (Parker *et al.*, 1997). Several mammalian and non-mammalian PACAPs were demonstrated to stimulate GH release from cultured goldfish pituitary cells; these actions of PACAP are inhibited by SRIF, 5HT and NE, but are additive to the responses to GnRH (Leung *et al.*, 1997; Wong *et al.*, 1998). Goldfish type 1 PACAP receptors have been cloned and expression of PACAP receptor mRNA demonstrated in brain and pituitary by Northern blot analysis, and RT-PCR and Southern blots (Wong *et al.*, 1998). Together these studies indicate that PACAP is a physiological regulator of GH secretion in goldfish.

GnRH stimulation of GH release was first demonstrated in goldfish (Marchant *et al.*, 1989a; for review: Peter and Marchant, 1995; Peng and Peter, 1997). Studies on the biological activity of GnRH agonists (Habibi *et al.*, 1992) and GnRH antagonists (Murthy *et al.*, 1993, 1994a) indicate that GnRH receptors on somatotrophs are functionally distinct from those on gonadotrophs. Moreover, circulating serum GH levels in sexually mature goldfish decrease following treatment with the GnRH antagonist [Ac-Δ^3-Pro^1, 4FD-Phe^2, D-Trp^{3,6}]-mGnRH (Murthy *et al.*, 1994b), indicating that GnRH has a role in the physiological regulation of basal GH levels. *In vivo* treatment of goldfish with GnRH has a potent stimulatory action on the expression of GH mRNA (Mahmoud *et al.*, 1996). GnRH also stimulates GH release in common carp (Lin *et al.*, 1993a). While GnRH stimulates GH release from tilapia pituitary cells *in vitro*, there is no influence on gene expression (Melamed *et al.*, 1996), indicating that stimulation of GH release can be dissociated from regulation of GH gene expression. In rainbow trout, initial studies indicated a lack of effect of GnRH on GH release (Blaise *et al.*, 1995). However, more recently it has been discovered that insulin-like growth factor I (IGF-I) pretreated somatotrophs are responsive to GnRH (Blaise *et al.*, 1997). This permissive action of IGF-I on GnRH stimulated GH release was observable in pituitary cells taken from sexually immature or early gametogenic trout, indicating some unknown interaction with sex steroids. In catfish no effects of GnRH on GH release were found (Bosma *et al.*, 1997).

Native goldfish NPY, as well as human NPY are highly potent in stimulating GH release from the goldfish pituitary *in vitro* (Peng *et al.* 1990, 1993b). The stimulation of GH release by NPY has at least 2 components, a direct action on the pituitary cells and an indirect action through the release of GnRH from neurosecretory terminals in the pituitary (Peng *et al.* 1993a). NPY stimulates GH release from mixed populations of enzymatically dispersed goldfish pituitary cells, as well as enriched somatotrophs (Peng *et al.* 1993a), indicating that NPY exerts its stimulation on GH secretion by acting directly on the somatotrophs. On the other hand, NPY stimulates GnRH release from the neurosecretory nerve terminals remaining in goldfish pituitary fragments and the stimulatory effects of NPY on GH release from pituitary fragments can be partially blocked by a GnRH antagonist (Peng *et al.* 1993a). Therefore, the actions of NPY on GH release appear to be in part mediated by GnRH.

DA stimulation of GH release is directly at the level of somatotrophs and is mediated by D1 receptors (Chang *et al.*, 1990; Wong *et al.*, 1992; and, for review: Peng and Peter, 1997). DA D1 receptors have been characterized in goldfish pituitary cells using enzymatically dispersed mixed cell populations as well as enriched somatotrophs (Wong *et al.* 1993a), the first demonstration of D1 receptors in a vertebrate pituitary. Autoradiographic studies reveal that the D1 specific binding sites are localized in the proximal pars distalis of the goldfish pituitary where the somatotrophs are located (Wong *et al.* 1993a). Results from fluorescent-labeled D1 ligand-binding studies confirmed that D1 receptors are selectively located on identified somatotrophs, but not gonadotrophs and lactotrophs, in goldfish (D. Taylor and J.P. Chang, unpublished results). DA and D1 agonists also stimulate GH release in common carp (Lin *et al.*, 1993a), tilapia (Melamed *et al.*, 1996), and catfish (Bosma *et al.*, 1997).

GH release is stimulated by CCK both *in vitro* (Himick *et al.*, 1993) and *in vivo* (Himick and Peter, 1994b) in goldfish. The presence of a high density of CCK-like immunoreactive fibers in the goldfish proximal pars distalis (Himick *et al.*, 1993), the region of the pituitary containing somatotrophs and gonadotrophs, supports that CCK is involved in regulation of GH release. The presence of CCK-like immunoreactive perikarya and fibers (Himick and Peter, 1994b) and CCK receptor binding sites (Himick at al., 1996) in preoptic and ventromedial hypothalamic hypophysiotrophic areas of the brain in goldfish further supports the neuroendocrine function of CCK.

In goldfish, BBS stimulates GH release from perifused goldfish pituitary fragments *in vitro* (Himick and Peter 1995a). Intraperitoneal injection or brain intraventricular injection of BBS also stimulates serum GH levels (Himick and Peter, 1994a). The demonstration of BBS binding sites in goldfish pituitary (Himick *et al.* 1995) further supports the idea of involvement of BBS in the regulation of GH secretion. However, the relative importance of BBS in regulation of GH secretion is not clear because the distribution of BBS/gastrin-releasing peptide IR fibers in the pituitary is primarily in the neurointermediate lobe, or along the border of the neurointermediate lobe and the proximal pars distalis, with relatively few fibers directly in the proximal pars distalis (Himick and Peter 1995a). Likewise, autoradiographic localization of BBS/gastrin-releasing peptide binding sites demonstrates a high density in the neurointermediate lobe, and a relatively low density in the proximal pars distalis (Himick *et al.* 1995).

In goldfish, TRH increases serum GH levels following intraperitoneal injection (Cook and Peter 1984) and stimulates GH release from perifused pituitary fragments (Trudeau *et al.* 1992). The actions of TRH on GH have been extended to common carp (Lin *et al.*, 1993b, Kagabu *et al.*, 1998) and tilapia (Melamed *et al.*, 1995), although in the latter case TRH was effective only *in vivo* but not *in vitro*.

Immunocytochemical staining of goldfish pituitary indicates that the activin ßA subunit is predominantly produced in the pituitary in somatotrophs, whereas staining for ßB in somatotrophs is relatively weak and staining for the α-subunit of inhibin is localized to fibers in the neurointermediate lobe (Ge and Peter 1994). These results suggest that goldfish somatotrophs produce activin. Recent studies confirm the expression of mRNA encoding activin ß subunits in the pituitary of goldfish (Ge *et al.* 1997). Porcine inhibin and activin both stimulate GH release (Ge and Peter 1994) from perifused goldfish pituitary fragments, suggesting that activin produced by somatotrophs may have autocrine and paracrine actions, respectively, in the pituitary.

3.2. Inhibition of GH Secretion

SRIF-14 has been demonstrated to inhibit basal and stimulated GH release in a number of species of teleost fish both *in vivo* and *in vitro* (for review: Peter and Marchant, 1995; Peng and Peter, 1997). As in mammals, SRIF-14 is a potent inhibitor of basal as well as stimulated GH secretion in teleost fishes; SRIF inhibits GH release stimulated by all of the neuroendocrine factors indicated above (for review: Peter and Marchant, 1995; Peng and Peter, 1997). Interestingly, SRIF-14 does not suppress expression of GH mRNA in rainbow trout (Yada and Hirano, 1992) or tilapia (Melamed *et al.*, 1996). Catfish SRIF-22 and salmon SRIF-25 are not effective in suppressing GH release in goldfish (Marchant *et al.*, 1987, Marchant and Peter, 1989).

Forebrain concentrations of SRIF-14 are highest in autumn in sexually regressed goldfish and are lowest in sexually mature (i.e., prespawning) goldfish in spring; whereas, serum GH levels are highest in sexually mature goldfish in spring and early postspawning goldfish in late spring or early summer, while lowest serum GH levels are in sexually regressed goldfish in autumn (Marchant and Peter, 1986, Marchant *et al.*, 1989b). This inverse relationship between seasonal changes in forebrain concentrations of SRIF, and serum GH levels and reproductive cycles suggests that seasonal variations in GH reflect changes in the intensity of the SRIF-14 inhibitory tone.

5HT, NE and glutamate are each inhibitory to GH secretion (for review: Peter and Marchant, 1995; Peng and Peter, 1997). 5HT and NE inhibit GH release *in vivo* and *in vitro* (for review: Peter and Marchant, 1995; Peng and Peter, 1997). The inhibitory actions of glutamate agonist N-methyl-D,L-aspartic acid (NMA) on GH secretion have been tested *in vivo* in goldfish (Trudeau *et al.*, 1996). Estradiol enhances the inhibitory effects of NMA on GH secretion (Trudeau *et al.*, 1996). Similar to SRIF, NE and 5HT can completely suppress the stimulatory actions of GnRH, DA, and PACAP on GH release *in vitro* (Wong, 1993; Leung *et al.*, 1997). Additional studies are needed to understand these inhibitory mechanisms.

3.3. Actions of Sex Steroids on GH Secretion

Serum GH levels are highest in sexually mature goldfish in spring and early post-spawning goldfish in late spring or early summer, while lowest serum GH levels are in sexually regressed goldfish in autumn (Marchant and Peter, 1986, Marchant *et al.*, 1989b). Sex steroids, in particular estradiol, influence the basal secretion rate of GH. Estradiol treatment of goldfish, at any stage of the annual sexual cycle, causes an increase serum GH levels; however, testosterone treatment is ineffective (Trudeau *et al.*, 1992). Pituitary GH content is increased by estradiol treatment, but no changes in GH mRNA are found (Zou *et al.*, 1997), suggesting that while translation of GH mRNA is enhanced by estradiol, GH gene expression is not. On the other hand, testosterone and 11ß-hydroxyandrosterone treatment of goldfish, either *in vivo* or *in vitro*, causes increased GH gene expression (Huggard *et al.*, 1996).

The responsiveness of the somatotrophs to neuroendocrine factors is influenced by sex steroids. The stimulatory action of DA on GH release varies with the seasonal reproductive cycle, being highest in sexually regressed fish, intermediate in recrudescent fish, and lowest in sexually mature fish (Wong *et al.* 1993b, c). Castration enhances the GH *in vitro* responses to DA (Wong *et al.*, 1993d). Pituitary cells taken from sexually regressed goldfish are much more responsive to cGRF than those from sexually recrudescent fish (C. Peng and R.E. Peter, unpublished results). Likewise, perifused goldfish pituitary fragments from sexually regressed fish are more responsive to CCK than those from sexually mature fish, suggesting that gonadal steroids may decrease the responsiveness of somatotrophs to these neuroendocrine factors.

Contrary to this, the GH release-response to sGnRH, both *in vivo* and *in vitro*, is greatest in sexually mature goldfish and lowest in sexually regressed fish (Trudeau *et al.*, 1992). Pretreatment of sexually regressed and gonadal recrudescing fish with estradiol *in vivo* enhances the *in vitro* GH release-response of pituitaries to sGnRH (Trudeau *et al.*, 1992). The magnitude of the *in vitro* GH response to NPY changes with the seasonal reproductive cycles, with the greatest responses observed in sexually mature goldfish (Peng *et al.*, 1993c); implantation of testosterone into sexually regressed goldfish significantly enhanced the *in vitro* GH response to NPY (Peng *et al.*, 1993c). This may in part be due to actions of the sex steroid on the GnRH system, as implantation of testosterone in sexually regressed goldfish enhances the GnRH release response to NPY (Peng *et al.*, 1993c). Similarly, pituitaries from sexually mature fish have a greater sensitivity to TRH than pituitaries from sexually regressed fish (Trudeau *et al.*, 1992), and the GH response to TRH in sexually regressed fish can be increased by pretreatment with estradiol (Trudeau *et al.*, 1992).

The multifactorial neuroendocrine regulation of GH secretion, in particular the stimulation of GH secretion, provides a great deal of adaptive opportunity to the fish in terms of which factors may be involved in the regulation of GH secretion at different life cycles stages. From the information available, it seems clear that the stimulatory neuroendocrine factors that predominate in sexually regressed goldfish, with low circulating titers of sex steroids, would be different from the factors that would dominate in sexually mature fish, with high circulating levels of sex steroids. Prepuberal fish are also likely to differ from sexually regressed post-puberal fish in this regard. From the information available in goldfish, the predominate stimulatory factors in sexually regressed fish are DA, CCK and GRF. In sexually mature goldfish GnRH, NPY and TRH would be the predominate factors.

Sexual dimorphism of the galanin-like immunoreactivity in brain and pituitary of goldfish, with males showing greater prominence of galanin immunoreactivity in the preoptic, posterior-ventrobasal hypothalamic and telencephalic regions, and the co-localization of galanin in GH cells, suggests that this peptide may have a hypophysiotropic role in the regulation of GH secretion (Prasada Rao *et al.*, 1996). Possible changes in GH responsiveness to galanin and other neuroendocrine factors have not been explored. While the cellular basis for these changes in responsiveness to neuroendocrine factors remain to be investigated further, such changes must be part of the explanation for seasonal changes in GH secretion.

3.4. Signal Transduction in Multifactorial Control of GH Secretion

The final secretory response to multifactorial control of somatotrope activity is determined by the integrated output of signal transduction events mediating the action of individual regulators. In goldfish and tilapia, GH secretion is stimulated by at least two major signal transduction cascades, the protein kinase C (PKC) and adenylate cyclase-cAMP-PKA systems (Figure 1).

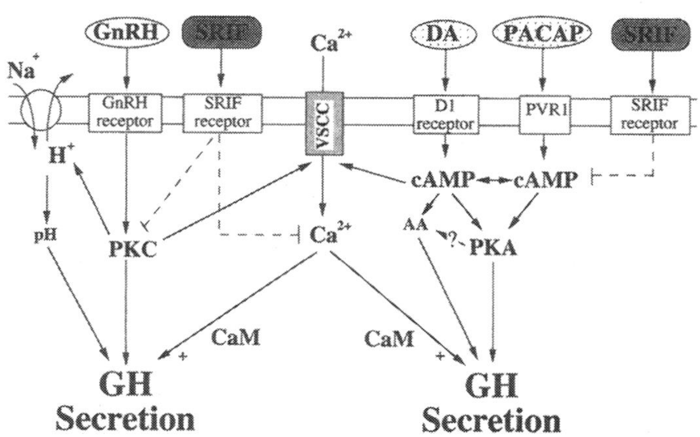

Figure 1. Diagrammatic summary of the signal transduction pathways mediating the neuroendocrine regulation of GH secretion by GnRH, DA, PACAP and SRIF in goldfish. Releasable GH is separated into protein kinase C (PKC)- and cAMP/PKA-sensitive pools. GnRH utilizes PKC-dependent mechanisms; in contrast, DA via D1 receptors and PACAP via type 1 receptors (PVR1) activate cAMP/PKA-sensitive mechanisms. Activation of PKC and cAMP both lead to increases in extracellular Ca^{2+} entry through voltage-sensitive Ca^{2+} channels (VSCC). In addition to the involvement of calmodulin (CAM), stimulation of PKC also activates a Na^+/H^+ antiport system that may affect intracellular pH, whereas cAMP-dependent pathways leads to the mobilization of arachidonic acid (AA). Actions of SRIF are mediated via inhibition of PKC-, cAMP- and Ca^{2+}-dependent GH release.

In the goldfish the GH response to the PKC and cAMP-PKA pathways are additive and minimal "cross-talk" between the two pathways exists (for review: Chang *et al.*, 1994, 1996a; Wong *et al.*, 1994). A Na^+/H^+ exchange system is also involved in GH release and its activation is distal to PKC, but not cAMP (Van Goor *et al.*, 1997, and J. Chang, unpublished). Conversely, activation of the cAMP pathway, but not PKC, leads to the mobilization of arachidonic acid through phospholipase A2 (Chang *et al.*, 1996b). However, both PKC- and cAMP/PKA-mediated GH release show a dependence on extracellular Ca^{2+} entry through voltage-sensitive channels (Wong *et al.*, 1994); however, whether more than one population of channels is involved is not known at present.

Among the stimulatory regulators in the goldfish, GnRH stimulates GH release via the PKC-dependent cascade, whereas DA and PACAP utilize cAMP-dependent mechanisms (Chang *et al.*, 1994; Wong *et al.*, 1994, 1998; Leung *et al.*, 1997; Fig. 1). Whether other stimulatory neuroendocrine factors partition themselves along the two (or other) second messenger systems is not known, but is a likely possibility. SRIF inhibition of GH release can be exerted at steps distal to PKC activation, cAMP generation, and Ca^{2+} mobilization (Kwong and Chang, 1997). Whether other inhibitory factors (e.g., NE and 5HT) target these transduction cascades in a similar or different manner compared to SRIF remains to be studied.

Recently, DA D1 receptors have been localized on a morphologically identified goldfish somatotrope subtype which is known to be responsive to GnRH stimulation (D. Taylor, J. Johnson, F. Van Goor and J. Chang, unpublished). Therefore, at least within a certain subpopulation of somatotropes, the releasable GH pool is separated into distinct cAMP- and PKC-sensitive components within an individual cell. This allows for additivity of the GH release response to different stimulatory regulators and provides a cellular basis for multifactorial control of the secretory response. How gonadal steroids and other seasonal reproductive factors influence receptor coupling and signal transduction events leading to changes in the sensitivity of the GH secretory response to different neuroendocrine regulators remains to be investigated. Never-the-less, the presence of PKC- and cAMP-dependent pathways can be demonstrated in goldfish pituitary cells obtained at all times of the reproductive cycle.

4. NEUROENDOCRINE STIMULATION OF GROWTH RATES OF FISH

It has been demonstrated that at least some biologically active proteins and peptides can be absorbed intact from the gastrointestinal tract into the blood of fish (McLean and Donaldson, 1990). The hypothesis to be tested is whether neuroendocrine factors can be added to fish food to stimulate GH secretion, and perhaps food intake in fish, to stimulate growth rates. Since CCK and BBS stimulate GH secretion but inhibit food intake, these are clearly not neuroendocrine factors to be tested for growth stimulation.

Feeding or intraperitoneal injection treatment with apomorphine, a dopamine receptor agonist, stimulates growth rates of goldfish (Wong *et al.*, 1993a, b). Also, treatment with a superactive agonist analog of GnRH by multiple intraperitoneal injection (Marchant *et al.*, 1989a) or by intramuscular injection of a slow release preparation of a superactive agonist analog of GnRH (C. Peng and R.E. Peter, unpublished results) is also effective in stimulating growth rates of goldfish. However, as receptors for GnRH are not present on somatotrophs in all fish, for example the African catfish (Bosma *et al.*, 1997), this peptide may not have application in all species. Using a food carrier vehicle developed by Syndel Laboratories, Ltd., feeding a combination of two neuroendocrine factors, a dopamine D1 agonist drug and a superactive analog of GnRH, added to the food carrier resulted in a highly significant increase in growth hormone secretion and growth rates of goldfish (R.E. Peter, J.P. Chang, C. Neumann, and S. Humphries, unpublished results). Immersion of juvenile tilapia for 15 minutes each week in low dosages of a synthetic analog of enkepalin (Tyr-D-Trp-Ala-D-Phe-NH_2) stimulated food conversion and growth rates (Chang and Lin, 1991). The mechanisms underlying this stimulatory effect on growth rates has not been explored. The results of theses studies indicate that growth rates of farmed fish may be enhanced by such treatment with neuroendocrine factors.

5. CONCLUSIONS

Figure 2 provides a diagrammatic summary of the relationship between brain regulation of food intake and GH secretion. While a number of brain peptides and neurotransmitters have been demonstrated to have actions on food intake in fish, anatomical studies continue to be necessary to demonstrate whether the peptide or neurotransmitter is located in posterior hypothalamic regions implicated in food intake. The combined physiological and anatomical studies on BBS and CCK confirm that these two peptides play an important role in food intake, specifically satiation. While evidence that NPY stimulates food intake in fish is indirect, the anatomical localization in the posterior hypothalamus and gene expression studies provide strong supportive evidence for involvement of NPY in food intake. However, while several additional neuropeptides and neurotransmitters have been shown to have actions on food intake, anatomical evidence demonstrating presence in posterior hypothalamic regions implicated in food intake is wanting, suggesting that actions on food intake are by some indirect mechanism.

ACKNOWLEDGMENTS

Supported by grants from the Natural Sciences and Engineering Research Council of Canada.

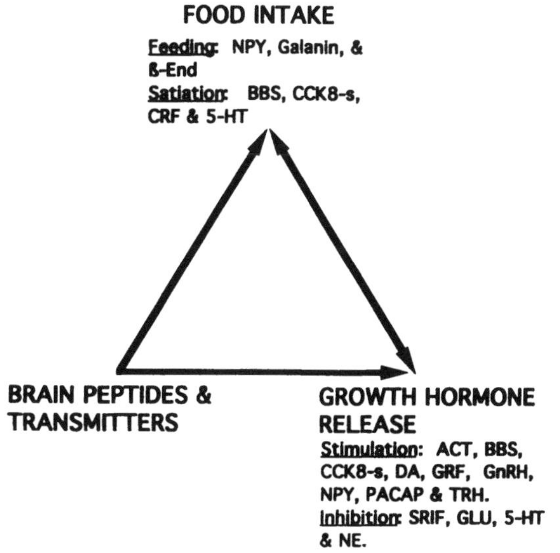

Figure 2. A diagrammatic summary of the relationship between brain regulation of food intake and growth hormone secretion. Abreviations: activin, ACT; bombesin, BSS; cholecystokinin (eight amino acid form - sulfated), CCK8-s; corticotropin-releasing hormone, CRF; dopamine, DA; ß-endorphin, ß-End; glutamate, GLU; gonadotropin-releasing hormone, GnRH; growth hormone-releasing factor, GRF; neuropeptide Y, NPY; norepinephrine, NE; pituitary adenylate cyclase-activating polypeptide, PACAP; thyrotropin-releasing hormone, TRH; serotonin, 5-HT; somatostatin, SRIF.

REFERENCES

Blaise, O., Le Bail, P.-Y., and Weil, C. (1995). Lack of gonadotropin releasing-hormone action on *in vivo* and *in vitro* growth hormone release, in rainbow trout (*Oncorhynchus mykiss*). *Comp. Biochem. Physiol.* **110C**, 133–141.

Blaise, O., Le Bail, P.-Y., and Weil, C. (1997). Permissive effect of insulin-like growth factor I (IGF-I) on gonadotropin releasing-hormone action on the in vitro growth hormone release, in rainbow trout (*Oncorhynchus mykiss*). *Comp. Biochem. Physiol.* **116A**, 75–81.

Bosma, P.T., Kolk, S.M., Rebers, F.E.M., Lescroart, O., Roelants, I., Willems, P.H.G.M., and Schulz, R.W. (1997). Gonadotrophs but not somatotrophs carry gonadotrophin-releasing hormone receptors: receptor localisation, intracellular calcium, and gonadotrophin and GH release. *J. Endocrinol.* **152**, 437–446.

Chang, C.-F. and Lin, S.-J. (1991). An enkephalin analog stimulates growth of tilapia. *Fish Physiol. Biochem.* **9**, 101–106.

Chang, J.P., Abele, J.T., Van Goor, F., Wong, A.O.L., and Neumann, C.M. (1966b). Role of arachidonic acid and calmodulin in mediating dopamine D1- and GnRH-stimulated growth hormone release in goldfish pituitary cells. *Gen. Comp. Endocrinol.* **102**, 88–101.

Chang, J.P., Van Goor, F., Wong, A.O.L., Jobin, R.M., and Neumann, C.M. (1994). Signal transduction pathways in GnRH- and dopamine D1-stimulated growth hormone secretion in the goldfish. *Chin. J. Physiol.* **37**, 111–127.

Chang, J.P., Van Goor, F., Jobin, R.M, and Lo, A. (1996a). GnRH signaling in goldfish pituitary cells. *Biological Signals* **5**, 70–80.

Chang, J.P., Yu, K.L., Wong, A.O.L., and Peter, R.E. (1990). Differential actions of dopamine receptor subtypes on gonadotropin and growth hormone release *in vitro* in goldfish. *Neuroendocrinology* **51**, 664–674.

Cook, A.F. and R.E. Peter. 1984. Effects of somatostatin on serum growth hormone levels in the goldfish, *Carassius auratus*. *Gen. Comp. Endocrinol.* **54**, 109–113.

De Pedro, N., Cèspedes, M.V., Delgado, M.J., and Alonso-Bedate, M. (1995). The galanin-induced feeding stimulation in mediated via α_2-adrenergic receptors in goldfish. *Regul. Peptides* **57**, 77–84.

De Pedro, N., Delgado, M.J., and Alonso-Bedate, M. (1995). Central administration of ß-endorphin increases food intake in goldfish: Pretreatment with the opioid antagonist naloxone. *Regul. Peptides* **55**, 189–195.

De Pedro, N., Gancedo, B., Alonso-Gómez, A.L., Delgado, M.J., and Alonso-Bedate, M. (1993). Role of corticotropin-releasing factor (CRF) as a food intake regulator in goldfish. *Physiol. Behav.* **53**, 517–520.

De Pedro, N., Pinillos, M.L., Valenciano, A.I., Alonso-Bedate, M., and Delgado, M.J. (1998). Inhibitory effect of serotonin on feeding behavior in goldfish: Involvement of CRF. *Peptides* **19**, 505–511.

Donaldson, E.M., Fagerland, U.H.M., Higgs, D.A., McBride, J.R. (1979). Hormonal enhancement of growth. *In* "Fish Physiology, Bioenergetics and Growth" (W.S. Hoar, D.J. Randall and J.R. Brett, Eds.), Vol. VIII, pp. 455–597. Academic Press, New York, NY.

Ge W. and Peter, R.E. (1994). Activin-like peptides in somatotrophs and activin stimulation of growth hormone release in goldfish. *Gen. Comp. Endocrinol.* **95**, 213–221.

Ge W, Peter, R.E. and Nagahama, Y. (1997). Activin and its receptors in the goldfish. *Fish Physiol. Biochem.* **17**, 143–153.

Habibi, H.R., Peter, R.E., Nahorniak, C.S., Milton, R.C., and Millar, R.P. (1992). Activity of vertebrate gonadotropin-releasing hormones and analogs with variant amino acid residues in positions 5, 7 and 8 in the goldfish pituitary. *Regul. Peptides* **37**, 271–284.

Himick, B.A., Golosinski, A.A., Jonsson, A-C, and Peter, R.E. (1993). CCK/gastrin-like immunoreactivity in the goldfish pituitary: regulation of pituitary hormone secretion by CCK-like peptides *in vitro*. *Gen. Comp. Endocrinol.* **92**, 88–103.

Himick, B.A. and Peter, R.E. (1994a). Bombesin acts to suppress feeding behavior and alter serum growth hormone in goldfish. *Physiol. Behav.* **55**, 65–72.

Himick, B.A. and Peter, R.E. (1994b). CCK/gastrin-like immunoreactivity in brain and gut, and CCK suppression of feeding in goldfish. *Am. J. Physiol.* **267**, R841-R851.

Himick, B.A. and Peter, R.E. (1995a). Bombesin-like immunoreactivity in the forebrain and pituitary and regulation of anterior pituitary hormone release by bombesin in goldfish. *Neuroendocrinology* **61**, 365–376.

Himick, B.A. and Peter, R.E. (1995b). Neuropeptide regulation of feeding and growth hormone secretion in fish. *Netherlands J. Zool.* **45**, 3–9.

Himick, BA, Vigna, S.R., and Peter, R.E. (1995). Characterization and distribution of bombesin binding sites in the goldfish hypothalamic feeding center and pituitary. *Regul. Peptides* **60**, 167–176.

Himick, BA, Vigna, S.R., and Peter, R.E. (1996). Characterization of cholecystokinin (CCK) binding sites in goldfish brain and pituitary. *Amer. J. Physiol.* **271**, R137-R143.

Huggard, D., Khakoo, Z., Kassam, G., and Habibi, H.R. (1996). Effect of testosterone on growth hormone gene expression in the goldfish pituitary. *Can. J. Physiol.* **74**, 1039–1046.

Kagabu, Y., Mishiba, T., Okino, T. and Yanagisawa, T. (1998). Effects of thyrotropin-releasing hormone and its metabolites, cyclo(His-Pro)m and TRH-OH, on growth hormone and prolactin synthesis in primary cultured pituitary cells of the common carp, *Cyprinus carpio. Gen. Comp. Endocrinol.* **111**, 395–403.

Kwong, P., and Chang, J.P. (1997). Somatostatin inhibition of growth hormone release in goldfish: Possible targets of intracellular mechanisms of action. *Gen. Comp. Endocrinol.* **108**, 446–456.

Lederis, K., Fryer, J. N., Okawara, Y., Schonrock, C. and Richter, C. (1994). Corticotropin-releasing factors acting on the fish pituitary: Experimental and molecular analysis. *In* "Fish Physiology" (N. M. Sherwood and C. L. Hew, Eds.), Vol. XIII, pp. 67–100. Academic Press, New York.

Leung, M.Y., Chang, J.P., Chow, B.K.C., and Wong, A.O.L. (1997). Pituitary adenylate cyclase-activating polypeptide (PACAP) functions as a novel growth hormone (GH)-releasing factor in the goldfish. In: Abstracts of XIII International Congress of Comparative Endocrinology. (November 16–21, 1997, Yokohama, Japan).

Lin, X.W., Lin, H.R., and Peter, R.E. (1993a). Growth hormone and gonadotropin secretion in the common carp (*Cyprinus carpio* L.): *In vitro* interactions of gonadotropin-releasing hormone, somatostatin, and the dopamine agonist apomorphine. *Gen. Comp. Endocrinol.* **89**, 62–71.

Lin, X.W., Lin, H.R., and Peter, R.E. (1993b). The regulatory effects of thyrotropin-releasing hormone on growth hormone secretion from the pituitary of common carp *in vitro. Fish Physiol. Biochem.* **11**, 71–76.

Mahmoud, S.S., Moloney, M.M., and Hababi, H.R. (1996). Cloning and sequencing of the goldfish growth hormone cDNA. *Gen. Comp. Endocrinol.* **101**, 139–144

Marchant, T.A., Chang, J.P., Nahorniak, C.S., and Peter, R.E. (1989a). Evidence that gonadotropin-releasing hormone also functions as a growth hormone-releasing factor in the goldfish. *Endocrinology* **124**, 2509–2518.

Marchant, T.A., Dulka, J.G. and Peter, R.E. (1989b). Relationship between serum growth hormone levels and the brain and pituitary content of immunoreactive somatostatin in the goldfish, *Carassius auratus* L. *Gen. Comp. Endocrinol.* **73**, 458–468.

Marchant, T.A., Fraser, R.A., Andrews, P.C., and Peter, R.E. (1987). The influence of mammalian and teleost somatostatins on the secretion of growth hormone from goldfish (*Carassius auratus* L.) pituitary fragments *in vitro. Regul. Peptides* **17**, 41–52.

Marchant, T.A. and Peter, R.E. (1989). Hypothalamic peptides influencing growth hormone secretion in the goldfish, *Carassius auratus. Fish Physiol. Biochem.* **7**, 133–139.

Marchant, T.A. and Peter, R.E. (1986). Seasonal variations in body growth rates and circulating levels of growth hormone in the goldfish, *Carassius auratus. J. exp. Zool.* **237**, 231–239.

McLean, E., Donaldson, E. M. (1990). Absorption of bioactive proteins by the gastrointestinal tract of fish: A review. *J. Aquatic Animal Health* **2**, 1–11.

McLean, E., and Donaldson, E.M. (1993). The role of growth hormone in the growth of poikilotherms. *In* "The Endocrinology of Growth, Development, and Metabolism in Vertebrates" (M.P. Schreibman, C.G. Scanes and P.K.T. Pang, Eds). pp. 43–71. Academic Press, New York, NY.

McRory, J.E., Parker, D.B., Ngamvongchon, S ., and Sherwood, N.M. (1995). Sequence and expression of cDNA for pituitary adenylate cyclase activating polypeptide (PACAP) and growth hormone-releasing hormone (GHRH)-like peptide in catfish. *Mol. Cell Endocrinol.* **1085**, 167–177.

Melamed, P., Eliahu, N., Levavi-Sivan, B., Ofir, M., Farchi-Pisanty, O., Rentier-Delrue, F., Smal, J., Yaron, Z., and Naor, Z. (1995). Hypothalamic and thyroidal regulation of growth hormone in tilapia. *Gen. Comp. Endocrinol.* **97**, 13–30.

Melamed, P., Gur, G., Elizur, A., Rosenfeld, H., Sivan, B., Rentier-Delrue, F., and Yaron, Z. (1996). Differential effects of gonadotropin-releasing hormone, dopamine and somatostatin and their second messengers on the mRNA levels of gonadotropin IIß subunit and growth hormone in the teleost fish, tilapia. *Neuroendocrinology* **64**, 320–328.

Murthy, C.K., Nahorniak, C.S., Rivier, J.E., and Peter, R.E. (1993). *In vitro* characterization of gonadotropin-releasing hormone antagonists in goldfish, *Carassius auratus. Endocrinology* **133**:1633–1644.

Murthy, C.K., Nahorniak, C.S., Rivier, J.E., and Peter, R.E. (1994a). Differential actions of a mammalian gonadotropin-releasing hormone antagonist on gonadotropin-II and growth hormone release in goldfish, *Carassius auratus. Neuroendocrinology* **59**, 561–571.

Murthy, C.K., Zheng, W., Trudeau, V.L., Nahorniak, C.S., Rivier, J.E., and Peter, R.E. (1994b). *In vivo* actions of a gonadotropin-releasing hormone (GnRH) antagonist on gonadotropin-II and growth hormone secretion in goldfish, *Carassius auratus. Gen. Comp. Endocrinol.* **96**, 427–437.

Parker, D.B., Coe, I.R., Dixon, G.H., and Sherwood, N.M. (1993). Two salmon neuropeptides encoded by one brain cDNA are structurally related to members of the glucagon superfamily. *Eur. J. Biochem.* **215**: 439–448.

Parker, D.B., Power, M.E., Swanson, P., Rivier, J., and Sherwood, N.M. (1997). Exon skipping in the gene encoding pituitary adenylate cyclase-activating polypeptide in salmon alters the expression of two hormones that stimulate growth hormone release. *Endocrinology* **138**, 414–423.

Peng, C., Chang, J.P., Yu, K.L., Wong, A.O.-L., Van Goor, F., Peter, R.E., and Rivier, J.E. (1993a). Neuropeptide-Y stimulates growth hormone and gonadotropin-II secretion in the goldfish pituitary: Involvement of both presynaptic and pituitary cell actions. *Endocrinology* **132**, 1820–1829.

Peng, C., Gallin, W., Peter, R.E., Blomqvist, A. G., and Larhammar, D. (1994). Neuropeptide-Y gene expression in the goldfish brain: Distribution and regulation by ovarian steroids. *Endocrinology* **134**, 1095–1103.

Peng, C., Huang, Y.P., and Peter, R.E. (1990). Neuropeptide Y stimulates growth hormone and gonadotropin release from the goldfish pituitary *in vitro*. *Neuroendocrinology* **52**, 28–34.

Peng, C., Humphries, S., Peter, R.E., Rivier, J.E., Blomqvist, A.G., and Larhammar, D. (1993b). Actions of goldfish neuropeptide Y on the secretion of growth hormone and gonadotropin-II in female goldfish. *Gen. Comp. Endocrinol.* **90**, 306–317.

Peng, C. and Peter, R.E. (1997). Neuroendocrine regulation of growth hormone secretion and growth in fish. *Zool. Studies* **36**: 79–89.

Peng, C., Trudeau, V.L., and Peter, R.E. (1993c). Seasonal variation of neuropeptide Y actions on growth hormone and gonadotropin-II secretion in the goldfish: effects of sex steroids. *J. Neuroendocrinol.* **5**, 273–280.

Peter, R.E. (1979). The brain and feeding behavior. *In* "Fish Physiology, Bioenergetics and Growth" (W.S. Hoar, D. J. Randall and J.R. Brett, Eds.), Vol VIII. Bioenergetics and Growth. pp. 121–159. Academic Press, New York, NY.

Peter, R.E., and Marchant, T.A. (1995). The endocrinology of growth in carp and related species. *Aquaculture* **129**, 299–321.

Peyon, P, Lin, X.W., Himick, B.A., and Peter, R.E. (1997). Molecular cloning and expression of cDNA encoding brain preprocholecystokinin in goldfish. *Peptides* 19, 199–210.

Peyon, P. and Peter, R.E. (1997). Phylogenetic and functional aspects of the neuropeptide cholecystokinin in goldfish. *In* "Advances in Comparative Endocrinology. Proceedings of the Thirteenth International Congress of Comparative Endocrinology" (S. Kawashima and S. Kikuyama, Eds.). pp. 577–580. Monduzzi Editore, Bologna, Italy.

Pontet, A., Danger, J.M., Dubourg, P., Pelletier, G., Vaudry, H., Calas A., Kah, O. (1989). Distribution and characterization of neuropeptide Y-like immunoreactivity in the brain and pituitary of the goldfish. *Cell Tissue Res.* **255**, 529–538.

Prasada Rao, P.D., Murthy, C.K., Cook, H., and Peter, R.E. (1996). Sexual dimorphism of galanin-like immunoreactivity in the brain and pituitary of goldfish, *Carassius auratus*. *J. Chem Neuroanat.* **10**, 119–135.

Rao, S.D., Prasada Rao, P.D., and Peter, R.E. (1996). Growth hormone-releasing hormone immunoreactivity in the brain, pituitary, and pineal of the goldfish, *Carassius auratus*. *Gen. Comp. Endocrinol.* **102**, 210–220.

Silverstein, J.T., Breininger, J., Baskin, D.G. and Plisetskaya, E.M. (1998). Neuropeptide Y-like gene expression in the salmon brain increases with fasting. *Gen. Comp. Endocrinol.* **110**, 157–165.

Trudeau, V.L., Sloley, B.D., Kah, O., Mons, N., Dulka, J.G., and Peter, R.E. (1996). Regulation of growth hormone secretion by amino acid neurotransmitters in the goldfish (I): Inhibition by N-methyl-D,L-aspartic acid. *Gen. Comp. Endocrinol.* **103**, 129–137.

Trudeau, V.L., Somoza, G.M., Nahorniak, C.S., and Peter, R.E. (1992). Interactions of estradiol with gonadotropin-releasing hormone and thyrotropin-releasing hormone in the control of growth hormone secretion in the goldfish. *Neuroendocrinology* **56**, 483–490.

Van Goor, F., Goldberg, J.I., and Chang, J.P. (1997). Extracellular sodium dependence of GnRH-stimulated growth hormone release in goldfish pituitary cells. *J. Neuroendocrinology* **9**, 207–216.

Vaughan, J.M., Rivier, J., Spiess, J., Peng, C., Chang, J.P., Peter, R.E., and Vale, W. (1992). Isolation and characterization of hypothalamic growth-hormone releasing factor from common carp, *Cyprinus carpio*. *Neuroendocrinology* **56**, 539–549.

Winbeg, S., Carter, C.G., McCarthy, I.D., He, Z.-Y., Nilsson, G.E., and Houlihan, D.F. (1993). Feeding rank and brain serotonergic activity in rainbow trout *Oncorhynchus mykiss*. *J. Exp. Biol.* **179**, 197–211.

Wong, A.O.L. (1993). Dopamine D1 regulation of growth hormone release in the goldfish. Ph.D. thesis, University of Alberta, 277 pp.

Wong, A.O.-L., Chang, J.P., and Peter, R.E. (1993a.) Characterization of D1 receptors mediating dopamine-stimulated growth hormone release from pituitary cells of the goldfish, *Carassius auratus*. *Endocrinology* **133**, 577–584.

Wong, A.O.-L., Chang, J.P., and Peter, R.E. (1993b). Dopamine functions as a growth hormone-releasing factor in the goldfish, *Carassius auratus*. *Fish Physiol. Biochem.* **11**, 77–84.

Wong, A.O.-L., Chang, J.P., and Peter, R.E. (1993c). *In vitro* and *in vivo* evidence that dopamine exerts growth hormone-releasing activity in goldfish. *Amer. J. Physiol.* **264**, E925-E932.

Wong, A.O.-L., Chang, J.P., and Peter, R.E. (1993d). Interactions of somatostatin, gonadotropin-releasing hormone, and the gonads on dopamine-stimulated growth hormone release in the goldfish. *Gen. Comp. Endocrinol.* **92**, 366–378.

Wong, A.O.-L., Chang, J.P., and Peter, R.E. (1992). Dopamine stimulates growth hormone release from the pituitary of goldfish, *Carassius auratus*, through the dopamine D1 receptors. *Endocrinology* **130**, 1201–1210.

Wong, A.O.L., Van Goor, F., Jobin, R.M., Neumann, C.M., and Chang, J.P. (1994). Interactions of cyclic adenosine 3',5'-monophosphate, protein kinase-C, and calcium in dopamine- and gonadotropin-releasing hormone-stimulated growth hormone release in the goldfish. *Endocrinology* **135,** 1593–1604.

Yada, T., and Hirano, T. (1992). Inhibition of growth hormone synthesis by somatostatin in cultured pituitary of rainbow trout. *J. Comp. Physiol.* **162**, 575–580.

Zou, J.J., Trudeau, V.L., Cui, Z., Brechin, J. MacKenzie, K., Zhu, Z., Houlihan, D.F., and Peter, R. E. (1997). Estradiol stimulates growth hormone production in female goldfish. *Gen. Comp. Endocrinol.* **106**, 102–112.

5

GHRH

A Growth Hormone-Releasing Factor in Birds?

Stephen Harvey

Department of Physiology
University of Alberta
Edmonton, Alberta, T6G 2H7, Canada

1. INTRODUCTION

A number of hypothalamic neuropeptides stimulate growth hormone (GH) secretion in mammals, although one, GH-releasing hormone (GHRH), is thought to be primarily responsible for stimulating pituitary somatotrophs (Harvey, 1995). Indeed, the hypersecretion of GHRH results in excessive GH secretion and acromegaly (Kovacs *et al.*, 1997), whereas a GHRH deficiency impairs GH release and can result in dwarfism (Flavell *et al.*, 1996). It is, however, uncertain if GHRH is a physiologically important GH-releasing factor in vertebrates other than mammals, in which other hypophysiotropic peptides, such as thyrotropin-releasing hormone (TRH) in birds and gonadotropin-releasing hormone (GnRH) in fish, have been shown to be potent GH-releasing factors (GRFs) (Harvey, 1990a, 1993; Lovejoy *et al.*, 1992). In this brief review, the possibility that GHRH may be a GRF in birds is considered.

2. GH SECRETION IN BIRDS

2.1. Hypothalamic Dependence

Avian somatotrophs are dependent upon hypothalamic stimulation. They become reduced in size and number and degranulated within days of the ectopic transplantation of pituitary tissue (Harvey *et al.*, 1991). In contrast, pituitary somatotrophs remain viable for at least 30 days when transplanted intracranially (Vasilatos-Younkens & Hibbard, 1987), presumably as a result of continued hypothalamic stimulation. Similarly, in the absence of hypothalamic tissue, pituitary glands release very little GH *in vitro*. Hypothalamic tissue therefore contains factors that stimulate GH release and extracts of chicken, turkey, pigeon

Neural Regulation in the Vertebrate Endocrine System, edited by Prasada Rao and Peter, Kluwer Academic / Plenum Publishers, New York, 1999.

Table 1. GHRH-induced GH release in birds: *in vivo*

Peptide	Species	References
$hGHRH_{(1\text{-}44)}NH_2$	chicken	Leung & Taylor 1983, 1985
		Scanes & Harvey 1984a/b, 1985, 1986, 1989a/b
		Harvey & Scanes 1984, 1985
		Harvey *et al.*, 1984a,b, 1985, 1991
		Scanes *et al.*, 1984, 1986a
		Huybrechts *et al.*, 1985
		Taylor *et al.*, 1986
		Leung *et al.*, 1986
		Hoshino *et al.*, 1986
		Buonomo *et al.*, 1987
		Johnson & McMurty 1989
		Vasillatos-Younkens *et al.*, 1992
		Moellers & Cogburn 1994
		Radecki *et al.*, 1994
	turkey	Proudman 1984
	duck	Foltzer-Jourdainne *et al.*, 1987
$rGHRH_{(1\text{-}44)}NH2$	turkey	Huybrechts *et al.*, 1984
$hGHRH_{(1\text{-}40)}$ OH	chicken	Scanes *et al.*, 1984, 1986
		Buonomo & Baile 1986
		Huybrechts *et al.*, 1985
	turkey	Huybrechts *et al.*, 1984
$hGHRH_{(1\text{-}32)}$ OH	chicken	Scanes *et al.*, 1984
$hGHRH_{(1\text{-}32)}NH_2$	turkey	Huybrechts *et al.*, 1984
$(Nle_{27})rGHRH_{(1\text{-}32)}NH_2$	turkey	Huybrechts *et al.*, 1984
$(Nle_{27})rGHRH_{(1\text{-}32)}NH_2$	turkey	Huybrechts *et al.*, 1986
$rGHRH(1\text{-}29)NH_2$	chicken	Scanes *et al.*, 1986a
	turkey	Huybrechts *et al.*, 1984
$(Nle_{27})rGHRH_{(1\text{-}29)}NH_2$	chicken	Scanes *et al.*, 1986a
	chicken	Scanes *et al.*, 1986a
$pGHRH_{(1\text{-}29)}$	chicken	Harvey 1997*

Abbreviations: GHRH, growth hormone-releasing hormone; h, human; r, rat; p, porcine; *, unpublished

and duck hypothalami have been shown to possess GH-releasing activity when incubated with pituitary tissue (Harvey *et al.*, 1979a,b; Harvey, 1983).

The GH-releasing activity of the chicken hypothalamus has also been demonstrated *in vivo* by the suppression of GH secretion following the electrical lesioning of hypothalamic nuclei (Rabii *et al.*, 1984). The destruction of neurons in the arcuate nucleus (a site of GHRH production in mammals) by monosodium glutamate similarly lowers plasma GH concentrations in female chickens (Scanes and Cammaratto, 1982) and suggests that a GHRH moiety may similarly stimulate GH release in birds.

2.2. Stimulation by GHRH *in Vivo*

Human, pig and rat GHRH peptides increase circulating GH levels when systemically administered to chickens, turkeys and ducks (Table 1), although higher dose-levels are required than in mammalian species (Harvey *et al.*, 1991). The maximal GH response occurs within 5 to 10 minutes of i.v. GHRH injection (Figure 1) and results in the degranulation of somatotrophs (Hull *et al.*, 1997). The rapidity of this GH response suggests a pituitary site

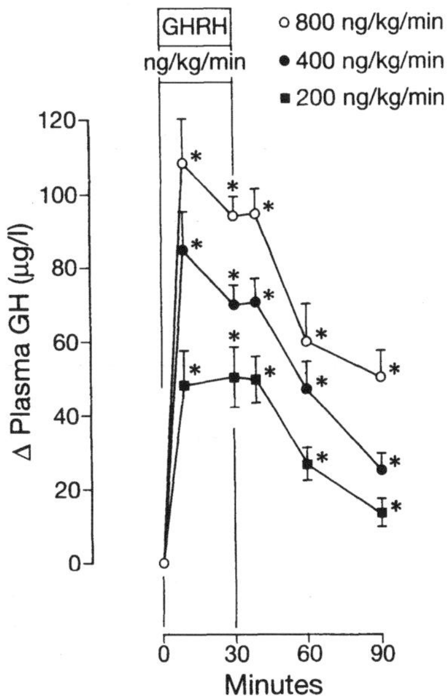

Figure 1. Change in growth hormone (GH) concentrations (μg/ℓ) in the plasma of immature ducks during and after the intravenous infusion of $hGHRH_{1-44}$ at doses of 200, 400 or 800 ng/kg body weight per min. Data are expressed relative to the pre-infusion GH concentration. Means±SEMs (n=5–10). *P<0.05 compared with pre-infusion concentration. Redrawn from Foltzer-Jourdainne *et al.*, 1987.

of action. Indeed, when administered centrally, human (h)GHRH induces inhibitory hypothalamic tone, by increasing somatostatin (SRIF) release (Ahene *et al.*, 1991).

The *in vivo* GH-releasing activity of GHRH is dose-related, with maximal stimulation occurring at dose between 1 and 10 μg kg (Harvey and Scanes, 1984). The sensitivity of the GH response is the same in young and old birds, although the magnitude of the response is greater in young birds than in adults, (Harvey, 1984; Scanes and Harvey, 1984), reflecting an increase in inhibitory hypothalamic control with age (Harvey *et al.*, 1991). Juvenile patterns of GH secretion (basal GH concentrations and pulse amplitude) can, however, be maintained in old birds chronically infused with synthetic GHRH (Vasilatos-Younkens *et al.*, 1992). This treatment increases pituitary size and GH mRNA content (Vasilatos-Younkens *et al.*, 1992) and allows circulating GH levels to be maintained at an elevated level (Moellers and Cogburn, 1994) for at least 60 minutes after the withdrawal of GHRH infusion (Foltzer-Jourdainne *et al.*, 1987).

In addition to age, the GH-releasing activity of mammalian GHRHs is modulated by a variety of other factors (Table 2). The GH responses to mammalian GHRHs are, for instance, suppressed by SRIF (Scanes and Harvey, 1989a), a physiologically important inhibitor of GH secretion (Harvey *et al.*, 1991a). The GH response is similarly suppressed by thyroid hormones (Scanes *et al.*, 1989b), and insulin-like growth factor (IGF-1) (Buonomo *et al.*, 1987), both of which provide inhibitory long-loop feedback in the control of GH release (Harvey, 1990a,b; Harvey *et al.*, 1991). It is, therefore, possible that GHRH is a physiologically important regulator of GH secretion in birds. This possibility is supported by the synergistic interactions between mammalian GHRH and TRH in the stimulation of GH release (Harvey and Scanes, 1985; Taylor *et al.*, 1995; Foltzer-Jourdainne *et al.*, 1987; Buonomo *et al.*, 1987), because TRH is thought to be a physiologically important GRF in birds. The ability of TRH to stimulate GH release *in vivo* is also mediated by central actions that were thought to result in the release of endogenous GHRH from hypothalamic nuclei (Harvey *et al.*, 1990). This central mechanism was thought to account for the potent GH-releasing activity of TRH *in vivo*, which is greater than mammalian GHRH *in vivo*, but not *in vitro* (Harvey and Scanes, 1984).

Table 2. GHRH-induced GH release in birds: *Modulation in vivo*

Factor	Effect	Reference
GHRH	↓	Harvey *et al.*, 1985
TRH	↑	Harvey & Scanes 1984
		Scanes & Harvey 1985
		Buonomo *et al.*, 1987
		Leung *et al.*, 1985
		Taylor *et al.*, 1986
Galanin	↑	Cheung & Hall 1990
SRIF	↓	Taylor *et al.*, 1986
		Harvey & Scanes 1987
		Scanes & Harvey 1989a
IGF-1	↓	Lauterio *et al., 1986*
		Perez *et al.*, 1985
		Buonomo *et al.*, 1987
		Scanes et al., 1986c
T3/T4	↓	Scanes *et al.*, 1986b
		Harvey 1990
		Donoghue *et al.*, 1990
		Scanes *et al.*, 1989
		Harvey *et al.*, 1991
Calcium	0	Harvey *et al.*, 1984
Age	↓	Harvey & Scanes 1984
		Foltzer-Jourdainne *et al.*, 1987
		Scanes & Harvey 1984
		Scanes *et al.*, 1984
		Johnson & McMurtry 1989

Abbreviations: GHRH, growth hormone-releasing hormone; TRH, thyrotropin-releasing hormone; SRIF, somatostatin; IGF-1, insulin-like growth factor-1; T3/T4, triiodothyronine/thyroxine; ↑, increase; ↓, decrease; 0, no effect.

2.3. Stimulation by GHRH *in Vitro*

The stimulation of GH release by mammalian GRFs is due to direct pituitary actions (Table 3), indicating the presence of GHRH receptors on the membranes of pituitary cells. These membranes are depolarized for 3 sec following GHRH stimulation, as a result of an influx of extracellular Ca^{2+} and Na^{+} (Robson *et al.*, 1997a/b). The induction of GH release occurs within 5 min (Figure 2) and appears to be mediated by cyclic adenosine 3,5'-monophosphate, and is potentiated in the presence of a phosphodiesterase inhibitor (Perez *et al.*, 1989a,b). The GH response may also involve protein kinase activation and diacylglycerol (Donoghue and Scanes, 1991b).

The *in vitro* stimulation of GH release by GHRH also appears to be physiological, since it is also antagonized by SRIF, IGF-1 and thyroid hormones (Table 4). It is also potentiated by TRH, through a Ca^{2+}-dependent mechanisms that results in increased protein synthesis.

In addition to inducing GH release in chickens, mammalian GHRH preparations stimulate GH gene expression in somatotrophs in the caudal lobe of the pituitary gland, although somatotrophs in the cephalic lobe are insensitive to GHRH stimulation (Radecki *et al.*, 1994; Kansaku *et al.*, 1995). The synthesis and release of immunoreactive GH is simi-

Table 3. GHRH-induced release in birds: *in vitro*

Peptide	Species	References
$hGHRH_{(1-44)}NH_2$	chicken	Leung & Taylor 1983
		Harvey & Scanes 1984, 1986
		Scanes *et al.*, 1984, 1986a,b
		Huybrechts *et al.*, 1985
		Perez *et al.*, 1985, 1987, 1989a,b
		Aramburo *et al.*, 1990
		Donoghue *et al.*, 1990
		Harvey 1990a,b
		Donoghue & Scanes 1991a,b
		Harvey *et al.*, 1991
		Denver & Harvey 1991
		Porter *et al.*, 1995
		Kansaku *et al.*, 1995
		Robson *et al.*, 1997a,b
$hGHRH_{(1-40)}OH$	chicken	Leung & Taylor 1983
		Scanes *et al.*, 1984
		Huybrechts *et al.*, 1984
		Porter *et al.*, 1995
		Lopez *et al.*, 1995
		Piper & Porter 1997
	ducks	Foltzer-Jourdainne *et al.*, 1988
$hGHRH_{(1-32)}OH_2$	chicken	Scanes *et al.*, 1984
$hGHRH_{(1-29)}NH_2$	chicken	Cheung & Hall 1990
		Darras *et al.*, 1994

Abbreviations: GHRH, growth hormone-releasing hormone; h, human

larly increased by human GHRH *in vitro* (Figure 3), including a phosphorylated GH moiety (Aramburo *et al.*, 1990). 'Silent' somatotrophs may also become GH-secretory cells when exposed to GHRH, which appears to participate in the recruitment or differentiation of somatotrophs in the embryonic pituitary gland (Porter *et al.*, 1995; Piper and Porter, 1997).

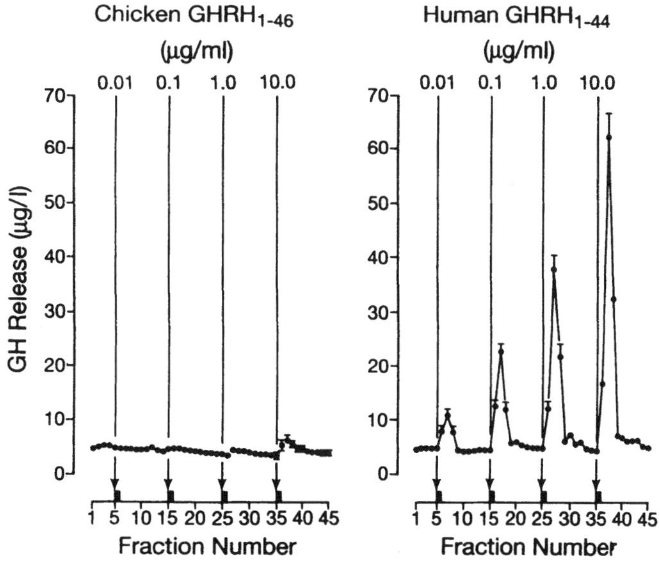

Figure 2. Growth hormone (GH) release from chicken hemipituitary glands perifused with chicken GH-releasing hormone ($GHRH_{1-46}$) in comparison with the contralateral glands perifused with human $GHRH_{1-44}$. Each fraction represents the elution of 1.5 ml of perfusate during intervals of 5 min. Means ± SEMs (n=8) (see Harvey, 1990b, for experimental details).

Table 4. GHRH-induced GH release in birds: *modulation in vitro*

Factor	Effect	Reference
TRH	↑	Leung & Taylor 1983
		Foltzer-Jourdainne *et al.*, 1988
		Perez *et al.*, 1987,1989a,1989b
		Donoghue *et al.*, 1990
SRIF	↓	Leung & Taylor 1983
		Harvey & Scanes 1984
		Perez *et al.*, 1985,1987, 1989a, 1989b
		Donoghue& Scanes 1991b
DA	↓	Donoghue *et al.*, 1990
NE	↓	Donoghue *et al.*, 1990
E	↓	Donoghue *et al.*, 1990
IGF-1	↓	Perez *et al.*, 1985
		Scanes *et al.*, 1986a
T3	↓	Harvey 1990
		Harvey *et al.*, 1991
		Denver & Harvey 1991
		Scanes *et al.*, 1986b
		Donoghue *et al.*, 1990
		Donoghue & Scanes 1991a
		Proudman 1984
PGE1	0	Donoghue *et al.*, 1990
PGE2	0	Donoghue *et al.*, 1990
$PGE_{f2\lambda}$	0	Donoghue *et al.*, 1990
Age	↓	Foltzer-Jourdainne 1988
		Scanes *et al.*, 1984

Abbreviations: GHRH, growth hormone-releasing hormone; TRH, thyrotropin-releasing hormone; SRIF, somatostatin; IGF-1, insulin-like growth factor-1; T3, triiodothyronine; DA, dopamine; NE, norepinephrine; E, epinephrine; PG, prostaglandin; ↑, increase; ↓, decrease; 0, no effect

3. GHRH: A GH-RELEASING FACTOR IN BIRDS?

Although mammalian GHRHs stimulate GH release in birds, the inability of antibodies for human, bovine, porcine and rat GHRH preparations to immunocytochemically label chicken hypothalamic cells (Harvey *et al.*, 1991) indicates that structurally similar peptides are not endogenous GRFs in birds. This possibility is also supported by the recent demonstration that active immunization with porcine (p)$GHRH_{1-32}$ has no significant effect on growth or circulating IGF-1 or IGF-1-binding protein levels in immature chickens (Leili, 1997). In these studies, antibodies were generated against (p)GHRH in the immunized birds, but Leili (1997) concluded that there was no evidence for the presence of a GH-releasing GHRH in the chicken hypothalamus.

The induction of GH release by human GHRH has a bell-shaped dose-response curve, with high doses being less effective than intermediate dose levels (Harvey and Scanes, 1984). The induction of GH release is also followed by a period of refractoriness, during which repeated stimulation with human GHRH does not induce GH secretion (Harvey *et al.*, 1985; Scanes and Harvey, 1985) . These observations suggest the presence of receptors for GHRH-like peptides on avian pituitary membranes that down-regulate after occupancy at high dose levels.

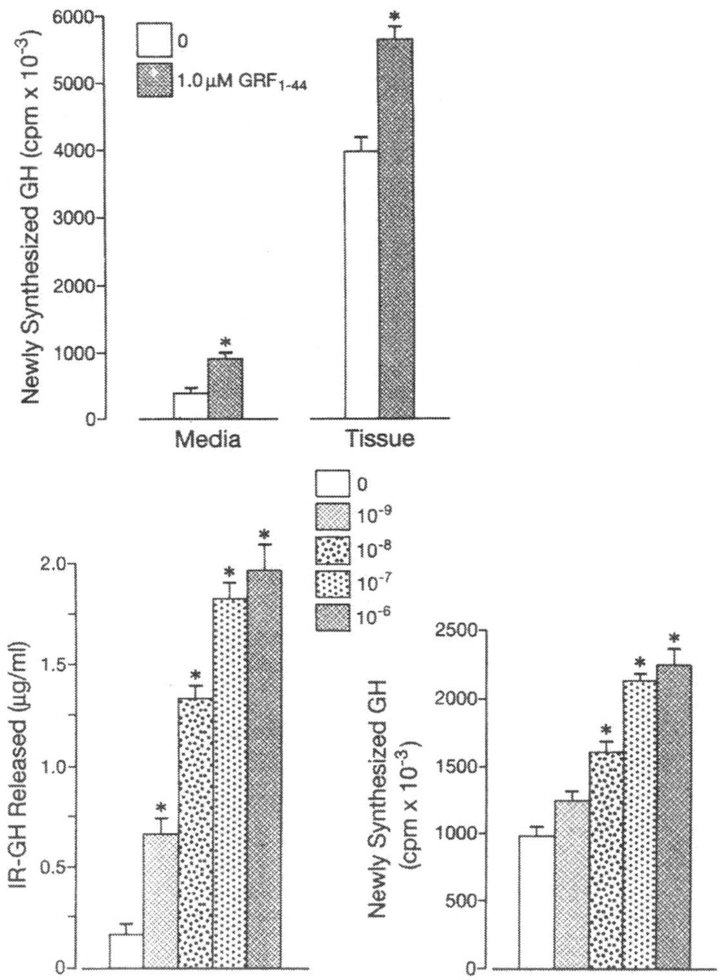

Figure 3. Influence of human growth hormone (GH) releasing hormone ($GHRH_{1-44}$) on the synthesis and release of chicken pituitary GH. *In vitro,* (a) immuno precipitable I^{125} GH (newly synthesized GH) in incubation media and chicken pituitary glands following a 2h incubation of the glands with I^{125}-lysine and 1.0 μm hGHRH (see Denver and Harvey, 1990, for experimental details). (b) Release of Immunoreactive GH and newly synthesized GH from chicken pituitary glands, after a 2h incubation of the tissue with I^{125} lysine and 10^{-9} -10^{-6} M hGHRH, Means ± SEMs (n=6) *P<0.05, compared with controls.

These receptors are, however, likely to differ from mammalian GHRH receptors in their affinities for mammalian GHRHs. The GH-releasing activity of mammalian GHRH is thought to reside in the first 29 amino acids, but $hGHRH_{1-40}$ is more potent than $hGHRH_{1-29}$ in chickens (Scanes *et al.,* 1986a), and $hGHRH_{1-40}$ is equipotent with $hGHRH_{1-32}$ in both chickens (Scanes *et al.,* 1986a) and turkeys (Huybrechts *et al.,* 1984). The GH-releasing activity of $hGHRH_{1-40}$ in chickens is also less than 55% of that induced by $hGHRH_{1-44}$ (Leung and Taylor, 1983), suggesting an importance of C-terminal residues in the stimulation of GH release in birds.

The avian receptors that mediate the stimulatory effects of mammalian GHRHs on GH release are, axiomatically, likely to normally interact with an endogenous GHRH-like ligand. The actions of mammalian GHRH's or GH release are thus unlikely to be mediated through receptors for other GHRH-like peptides. GHRH is a member of the glucagon peptide superfamily, that includes vasoactive intestinal peptide (VIP), and secretin. In contrast with GHRH, secretin has no effect on GH secretion in chickens (Leung, 1990) and glucagon inhibits GH release in both chickens (Kurima *et al.*, 1994) and ducks (Foltzer *et al.*, 1981). Moreover, porcine VIP stimulates GH release from the pituitary glands of immature broiler fowl (Hall *et al.*, 1988) and adult bantam hens (Kansaku *et al.*, 1995), it does not stimulate GH release in chick embryos *in vitro* (Porter *et al.*, 1995), nor the *in vivo* release of GH in immature turkeys (Huybrechts *et al.*, 1984) or adult bantam fowl (MacNamee *et al.*, 1986). Similarly, while mammalian GHRH induces GH gene expression in chicken pituitaries (Foster *et al.*, 1990), the GH-releasing actions of VIP are independent of gene transcription (Kansaku *et al.*, 1995). The GH-releasing activity of mammalian GHRH in birds is, therefore, unlikely to be mediated through glucagon, secretin or VIP receptors.

4. CHICKEN GHRH

Peptides with immunoreactivity to mammalian GHRH antisera have not been detected in the avian hypothalamus (Harvey *et al.*, 1991). However, the presence of a GHRH gene in birds was first indicated by the hybridization of chicken hypothalamic mRNA with a cRNA probe for rat (r)GHRH (Harvey *et al.*, 1991). One mRNA was comparable in size to rGHRH mRNA, whereas other moieties of greater abundance and larger size were also present. It was therefore concluded that two or more GHRH genes maybe present in the chicken hypothalamus or a single GHRH gene may undergo differential splicing. In more recent studies, a novel chicken GHRH gene has been cloned and sequenced (McRory *et al.*, 1997). This gene can be alternatively spliced to produce mRNAs that code for several peptides. The most abundant mRNA encodes for a 46 amino acid GHRH peptide that is one amino acid larger in size than fish GHRHs. This peptide has only 42% sequence identity with human GHRH (42%) and 47% with rat GHRH, but it has 74% sequence identity with salmon and carp $GHRH_{1-45}$ (Figure 4).

Exon sliding also occurs within the chicken (c)GHRH gene, at an intron/exon splice site 9 bases downstream from the usual site, resulting in a transcript that encodes for a 43 amino acid peptide. This transcript appears to be far less abundant than the transcript encoding $cGHRH_{1-46}$. A third transcript, which is present in adults but not in embryos, lacks exon 4, and encodes $cGHRH_{33-46}$, lacking the biologically active core located within $GHRH_{1-32}$. The gene further differs from mammalian GHRH genes in that it also codes for PACAP (pituitary adenylate cyclase activating peptide), another member of the glucagon peptide superfamily. PACAP is coded by a separate gene in mammals, although PACAP and GHRH are similarly coded by a single gene in salmon (Parker *et al.*, 1996). The deduced amino acid sequence of cPACAP is 97% identical to the human sequence and its mRNA shares 80% identity with nucleotides encoding chicken VIP (Talbot *et al.*, 1995). The amino acid identity between cGHRH and cPACAP is, however, only 30%.

4.1. GH-Releasing Activity of Chicken $GHRH_{1-46}$

In preliminary studies with anesthetized chickens, systemic injections of chicken GHRH ($cGHRH_{1-29}$ and $cGHRH_{1-46}$), at doses of up to 100 μg/kg had very little effect on plasma GH concentrations (Figure 5), although, the plasma GH level was significantly elevated 10 and 20 min after the administration of the highest dose of $cGHRH_{1-46}$ (Fig-

	10 ↓ 20 ↓ 30 ↓ 40 ↓
human[1,2,3]	Y A D A I F T N S Y R K V L G Q L S A R K L L Q D I M S R Q Q G E S N Q E R G A R A R L
pig[4]	Y A D A I F T N S Y R K V L G Q L S A R K L L Q D I M S R Q Q G E R N Q E Q G A R V R L
cattle/goat[5,6]	Y A D A I F T N S Y R K V L G Q L S A R K L L Q D I M N R Q Q G E R N Q E Q G A K V R L
sheep[6]	Y A D A I F T N S Y R K I L G Q L S A R K L L Q D I M N R Q Q G E R N Q E Q G A K V R L
hamster[7]	Y A D A I F T S S Y R K V L G Q L S A R K L L Q D I M S R Q Q G E R N Q E Q G P R V R L
rat[8]	H A D A I F T S S Y R R I L G Q L Y A R K L L H E I M N R Q Q G E R N Q E Q R S R F N
mouse[9]	H V D A I F T T N Y R K L L S Q L Y A R K V I Q D I M N K Q G E R I Q E Q R A R L S
chick $_{1\text{-}46}$ [10]	H A D G I F S K A Y R K L L G Q L S A R N Y L H S L M A K R V G G A S S G L G D E A E P L S
chick $_{1\text{-}43}$ [10]	H A D G I F S K A Y R K L L G Q L S A R N Y L H S L M A K R V G S G L G D E A E P L S
salmon[11,12]	H A D G M F N K A Y R K A L G Q L S A R K Y L H S L M A K R V G G G S T M E D D T E P L S
carp[13]	H A D G M F N K A R K A Y L G Q L S A R K Y L H T L M A K R V G G G S M I E D D N E P L S
catfish[14]	H A D G L L D R A L R D I L V Q L S A R K Y L H S L T A V R V G E E E E D E E D S E P L S
tunicate -1[15]	H S D G I F T K D Y R K Y L G Q L R A Q K F L Q W L M
tunicate -2[15]	H S D G I F T S D Y R R Y L G Q L S A Q K F L Q W L M

References: [1]Guillemin *et al* 1982; [2]Rivier *et al* 1982; [3]Ling *et al* 1984; [4]Böhlen *et al* 1983; [5]Esch *et al* 1983; [6]Brazeau *et al* 1984; [7]Ono *et al* 1994; [8]Speiss *et al* 1983; [9]Frohman *et al* 1989; [10]McRory *et al* 1996; [11]Parker *et al* 1993; [12]Parker *et al* 1997; [13]Vaughan *et al* 1992; [14]McRory *et al* 1995; [15]McRory and Sherwood 1996

Figure 4. Amino acid sequence identity between GHRH peptides.

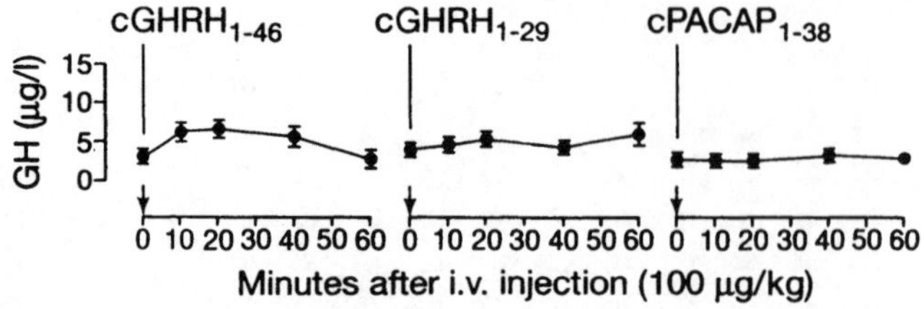

Figure 5. Concentrations of growth hormone (GH) in the plasma of immature chickens at intervals after the i.v. administration of chicken GH-releasing hormone (cGHRH$_{1-46}$ and cGHRH$_{1-29}$) and chicken PACAP (pituitary adenylate cyclase activating peptide) Means± SEMs (n=5).

ure 6). These peptides similarly had very little effect on the release of GH from perfused chicken pituitary glands, compared with the effect of hGHRH (Figure 2). Chicken PACAP$_{1-38}$, at these doses, similarly had little *in vivo* or *in vitro* GH-releasing activity in chickens. In contrast, carp (cp)GHRHs (cpGHRH$_{1-45}$ and cpGHRH$_{1-29}$, at doses of 10–30 µg/kg *in vivo* and 1–3 µg/ml *in vitro*) were potent *in vivo* and *in vitro* GH secretgogues in chickens (data not shown), despite having 70% structural identity with cGHRH.

5. CONCLUSIONS

GHRH gene expression occurs in the chicken brain, but cGHRH$_{1-46}$ appears to have very little GH-releasing activity, unlike mammalian GHRHs. The GH-releasing potency of cGHRH$_{1-43}$ is unknown but its C-terminally truncated synthetic peptide (cGHRH$_{1-29}$) does not appear to be a potent GH secretagogue in chickens. It is therefore possible that cGHRH is not a physiologically important GH-releasing factor in birds, despite its nomenclature. The low sequence identity between chicken GHRH and the sequences of human and rat GHRH that stimulate GH release support this view. The differences between these peptides in GH-releasing activity is, however, more surprising than their low sequence identity, as even human and mouse GHRH have only 57% sequence identity.

The release of GH from the chicken pituitary gland is likely to comprise many variant forms and isomers (Aramburo *et al.*, 1997). It is therefore possible that cGHRH may

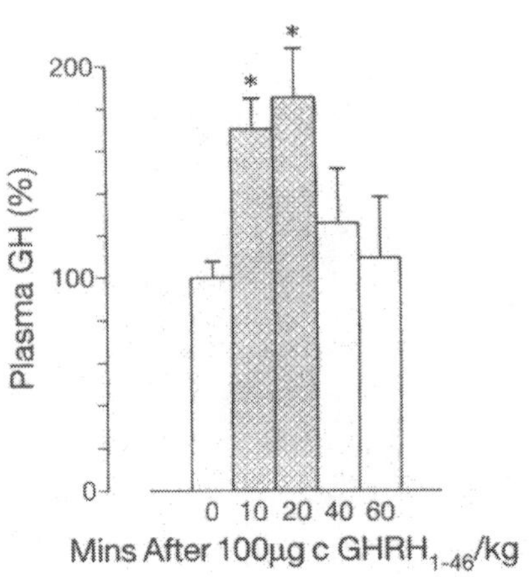

Figure 6. Concentrations of plasma GH, relative to pretreatment levels, in immature chickens at intervals after the i.v. administration of cGHRH$_{1-46}$. *P<0.05, compared with pretreatment levels (paired t test) Means±SEM's (n=5).

stimulate the release of GH moieties that have poor immunoreactivity with the radioimmunoassay we used to measure GH concentrations. This is, however, unlikely because our antibody cross-reacts with all the forms of chicken GH and with the presumptive GH proteins that have so far been separated from pituitary extracts by gel electrophoresis (Harvey, 1977; Aramburo *et al.*, 1990, 1997).

If cGHRH is not a physiologically important GRF in birds, it is, however, possible that it has roles other than in GH regulation. Indeed mammalian GHRHs have been shown to act as neural growth factors in the embryonic growth of the chick brain (Kentroti and Vernadakis, 1989;1990). The expression of the GHRH/PACAP gene in tissues other than the brain (in the ovary, oviduct and testis; McRory *et al.,* 1997) also suggests that these peptides, like mammalian GHRHs, have roles unrelated to the secretion of pituitary GH.

The ability of mammalian GHRH to induce GH-release in birds does, however, suggest the presence of pituitary receptors for an endogenous ligand that more resembles mammalian GHRHs than cGHRH. It is therefore possible that other GHRH genes may be additionally present in the chicken hypothalamus. It is, alternatively, possible that this paradox may reflect receptor-ligand differences in their sensitivity to evolutionary change and that the receptor and its ligand have different mutational rates and evolutionary history.

ACKNOWLEDGMENT

This work was supported, in part, by a grant from the Natural Sciences and Engineering Research Council. cGHRH and cPACAP were kindly provided by Dr. N.M. Sherwood and Dr. J.E. McRory (Department of Biology, University of Victoria, British Columbia) and Dr. J. Rivier (Salk Institute, San Diego).

REFERENCES

Ahene, C.A., Lea, R.W. and Harvey, S. (1991). Thyrotropin-releasing hormone-induced growth hormone (GH) secretion in anaesthetized chickens: inhibition by GH-releasing factor at central sites. *J. Endocrinol.* 128, 13–19.

Aramburo, C., Carranza, M., Martinez, H. and Luna, M. (1979). A comparative overview of growth hormone: proteins and gene. In *Adv. Comp. Endocrinol.* (Ed. Kawashima S and Kikuyama S) Monduzzi Editore, Bologna. pp. 893–898., 1997.

Aramburo, D., Donoghue, D., Montiel, J.L., Berghman, L.R. and Scanes, C.G. (1990). Phosphorylation of chicken growth hormone. *Life Sci.* 47, 945–952.

Böhlen, P., Esch, F., Brazeau, P., Ling, N. and Guillemin, R. (1983). Isolation and characterization of the porcine hypothalamic growth hormone releasing-factor. *Biochem. Biophys. Res. Commun.* 116, 726–734.

Brazeau, P., Böhlen, P., Esch, F., Ling, N., Wehrenberg, W.B. and Guillemin, R. (1984). Growth hormone-releasing factor from ovine and caprine hypothalamus: Isolation, sequence analysis and total synthesis. *Biochem. Biophys. Res. Commum.* 125. 606–614.

Buonomo, F.C. and Baile, C.A. (1986). Effect of daily injections of growth hormone-releasing factor and thyrotropin-releasing hormone on growth and endocrine parameters in chickens. *Domest. Anim. Endocrinol.* 3, 269–276.

Buonomo, F.C., Lauterio, T.J., Baile, C.A. and Daughaday, W.H. (1987). Effects of insulin-like growth factor I (IGF-I) on growth hormone releasing factor (GRF) and thyrotropin-releasing hormone (TRH) stimulation of growth hormone (GH) secretion in the domestic fowl *(Gallus domesticus). Gen. Comp. Endocrinol.* 66, 274–279.

Cheung, A. and Hall, T.R. (1990). Direct stimulation of growth hormone secretion by galanin in the domestic fowl. *J. Neuroendocrinol.* 2, 285–289.

Darras, V.M., Finne, M.F., Berghman, L.R. and Kühn, E.R. (1994). Ontogeny of the sensitivity of the somatotrophs to thyrotropin releasing hormone (TRH) and growth hormone releasing factor (GRF) in the embryonic and post-hatch chick. *Ann. d'Endocrinol.* 55, 255–260.

Denver, R.J. and Harvey, S. (1991). Thyroidal inhibition of chicken pituitary growth hormone: alterations in secretion and accumulation of newly synthesized hormone. *J. Endocrinol.* 131, 39–48.

Donoghue, D.J. and Scanes, C.G. (1991a). Triiodothyronine (T_3) inhibition of growth hormone secretion by chicken pituitary cells *in vitro. Gen. Comp. Endocrinol.* 84, 344–354.

Donoghue, D.J. and Scanes, C.G. (1991b). Possible involvement of adenylyl cyclase-cAMP-protein kinase A pathway in somatostatin inhibition of growth hormone release from chicken pituitary cells. *Gen. Comp. Endocrinol.* 81, 113–119.

Donoghue, D.J., Perez, F.M., Diamante, B.S.A., Malamed, S. and Scanes, C.G. (1990). Influence of catecholamines, prostaglandins and thyroid hormone on growth hormone secretion by chicken pituitary cells *in vitro. Domest. Anim. Endocrinol.* 7, 35–42.

Esch, F., Böhlen, P., Ling, N., Brazeau, P. and Guillemin, R. (1983). Isolation and characterization of the bovine hypothalamic growth hormone releasing-factor. *Biochem. Biophys. Res. Commun.* 117, 772–779.

Flavell, D.M., Wells, T., Wells, S.E., Carmignac, D.F., Thomas, G.B. and Robinson K. (1996). Dominant dwarfism in transgenic mice by targeting human growth hormone (GH) expression to hypothalamic GH-releasing factor neurons. *EMBO Journal* 15, 3871–3879.

Foltzer, C., Harvey, S., Strosser, M.T. and Mialhe, P. (1981). Influence of insulin and glucagon on growth hormone secretion in growing ducks (*Anas platyrhynchos). J. Endocrinol.* 91, 189–196.

Foltzer-Jourdainne, C., Harvey, S., Karmann, H. and Mialhe, P. (1987). Plasma growth hormone, somatostatin, insulin and glucagon interrelationships during infusion of human pancreatic growth hormone-releasing factor in young and adult ducts (*Anas platyrhynchos). J. Endocrinol.* 114, 25–32.

Foltzer-Jourdainne, C., Harvey, S. and Mailhe, P. (1988) *In vitro* control of growth hormone secretion by synthetic releasing factors in young and adult ducks. (*Anas platyrhynchos). J. Endocrinol.* 109, 421–429.

Foster, D.N., Kim, S.U., Enyeart, J.J. and Foster, L.K. (1990). Nucleotide sequence of the complimentary DNA for the turkey growth hormone. *Biochem. Biophys. Res. Commun.* 173, 967–975.

Frohman, M.A., Downs, T.R., Chomczynski, P. and Frohman, L.A. (1989). Cloning and characterization of mouse growth hormone-releasing hormone (GRF) complementary DNA: Increase GRF messenger RNA levels in the growth hormone deficient lit/lit mouse. *Mol. Endocrinol.* 3, 1529–1536.

Guillemin, R., Brazeau, P., Böhlen, P., Esch, F., Ling, N. and Wehrenberg, W. (1982). Growth hormone releasing-factor from a human pancreatic tumor that caused acromegaly. *Science* 218, 585–587.

Hall, T.R., Cheung, A. and Griffiths, E.C. (1988). Peptide effects on GH release in domestic fowl. *Prog. Abst. 8th Int. Cong. Endocrinol.* Tokyo, Japan.

Harvey, S. (1977) *Purification and Radioimmunoassay of Chicken Growth Hormone.* Ph.D. Thesis, University of Leeds.

Harvey, S. (1983). Neuroendocrine control of growth hormone secretion in birds. *In* "Progress in Non-Mammalian Brain Research" (G. Nistico and L. Bolis, Eds.) Vol. 3, p. 307–327. CRC Press, Boca Raton.

Harvey, S. (1990a). Thyrotropin-releasing hormone: a growth hormone-releasing factor. *Endocrinol.* 125, 345–358.

Harvey, S. (1990b). Triiodothyronine inhibition of TRH-induced GH secretion from fowl pituitary glands *in vitro. J. Endocrinol.* 126, 75–81.

Harvey, S. (1990c). Thyroidal inhibition of growth hormone secretion: negative feedback? *In* "The Endocrinology of Growth, Development and Metabolism in Vertebrates" (M.P. Schreibman, C.G. Scanes and P.K.T. Pang, Eds.), pp. 151–182. Academic Press, New York.

Harvey, S. (1993). Growth hormone secretion in poikilotherms and homeotherms. *In* "The Endocrinology of Growth, Development and Metabolism in Vertebrates" (M. Wada, S. Ishii and C.G. Scanes, Eds.), pp. 111–128. Jap. Sci. Soc. Press Tokyo.

Harvey, S. (1995). Growth hormone release: stimulatory hypothalamic control . *In* "Growth Hormone" (S. Harvey, C.G. Scanes and W.H. Daughaday, Eds.), pp. 97–114. CRC Press, Boca Raton.

Harvey, S. and Scanes, C.G. (1985). Interaction between human pancreatic growth hormone-releasing factor (hpGRF) and thyrotropin-releasing hormone (TRH) on *in vivo* growth hormone secretion in the domestic fowl. *Horm. Metab. Res.* 17, 113–114.

Harvey, S. and Scanes, C.G. (1984). Comparative stimulation of growth hormone secretion in anaesthetized chickens by human pancreatic growth hormone-releasing factor (hpGRF) and thyrotropin-releasing hormone (TRH). *Neuroendocrinology* 39, 314–320.

Harvey, S. and Scanes, C.G. (1987). Somatostatin inhibition of growth hormone secretion in an adult bird: the domestic fowl. *Comp. Biochem. Physiol.* 87A, 315–318.

Harvey, S., Decuypere, E., Darras, V.M. and Berghman, L. (1991). Differential effects of T_4 and T_3 on TRH and GRF-induced GH secretion in domestic fowl. *Reprod. Nutr. Dev.* 31, 451–460.

Harvey, S., Fraser, R.A. and Lea, R.W. (1991). Growth hormone secretion in poultry. *Crit. Rev. Poult.Biol.* 3, 239–282.

Harvey, S., Huybrechts, L.M. and Scanes, C.G. (1984). Growth hormone secretion in chicks during dietary calcium deficiency. *Domest. Anim. Endocrinol.* 1, 269–277.

Harvey, S., Lea, R.W. and Ahene, C. (1990). Central action of thyrotropin-releasing hormone on growth hormone secretion in domestic fowl. *J. Endocrinol.* 126, 83–88.

Harvey, S., Scanes, C.G., Chadwick, A. and Bolton, N.J. (1979a). Effect of chicken hypothalamus on prolactin and growth hormone secretion in male chickens. *J. Endocrinol.* 82, 193–197.

Harvey, S., Scanes, C.G., Chadwick, A. and Bolton, N.J. (1979b). *In vitro* stimulation of chicken pituitary growth hormone and prolactin secretion by chicken hypothalamic extract. *Experientia,* 35, 694–695.

Harvey, S., Scanes, C.G. and Marsh, J.A. (1984). Stimulation of growth hormone secretion in dwarf chickens by thyrotropin-releasing hormone (TRH) and human pancreatic growth hormone-releasing factor (hpGRF). *Gen. Comp. Endocrinol.* 55, 493–497.

Harvey, S., Scanes, C.G. and Phillips, J.G. (1985). Growth hormone secretion in anaesthetized fowl. 1. Refractoriness to repeated stimulation by human growth hormone-releasing factor (hpGRF) or thyrotropin releasing hormone (TRH). *Gen. Comp. Endocrinol.* 59, 1–9.

Hoshino, S., Suzuki, M., Nakegawa, J., Wakita, M. and Kobayashki, Y. (1986). Stimulation of growth hormone release in dwarf and normal chickens by thyrotropin-releasing hormone (TRH) or human pancreatic growth hormone-releasing factor (hpGRF). *Endocrinol. Jpn.* 33, 803–811.

Hull, K.L., Janssens, W.C.J. and Harvey, S. (1997). Growth hormone: local actions in sites of synthesis. *In* "Perspective Avian Endocrinology" (S. Harvey and R.J. Etches, Eds.), pp.375–386. Journal of Endocrinology Ltd., Bristol.

Huybrechts, L.M., Decuypere, E., Scanes, C.G., Callewaert, P., Peys, E. and Kühn, E.R. (1985). Human pancreatic growth hormone releasing factor stimulates growth hormone secretion in perinatal dwarf and control chickens. *Horm. Metab. Res.* 17, 690–691.

Huybrechts, L.M., Scanes, C.G., Rivier, J. and Vale, W. (1984). Synthetic human pancreatic growth hormone releasing factor and rat hypothalamic growth hormone releasing factor stimulate *in vivo* secretion of growth hormone in domestic turkey. *IRCS Med. Sci.* 12, 908–909.

Johnson, R.J. and McMurtry, J.P. (1989). The effects of age and sex on the growth hormone response of broiler chickens to exogenous growth hormone-releasing factor ($hpGRF_{1-44}$). *Poult. Sci.* 68, 75.

Kansaku, N., Shimada, K. and Saito, N. (1995) Regionalized gene expression of prolactin and growth hormone in the chicken anterior pituitary gland. *Gen. Comp. Endocrinol.* 99, 60–68.

Kentroti, S. and Vernadakis, A. (1989). Growth hormone-releasing hormone influences neuronal expression in the developing chick brain. I. Catecholaminergic neurons. *Dev. Brain Res.* 49, 275–280.

Kentroti, S. and Vernadakis, A. (1990). Growth hormone-releasing hormone and somatostatin influence neuronal expression in developing chick brain. II. Cholinergic neurons. *Brain Res.* 512, 297–303.

Klandorf, H., Harvey, S. and Fraser, H.M. (1985). Physiological control of growth hormone secretion by thyrotropin-releasing hormone in the domestic fowl. *J. Endocrinol.* 105, 351–355.

Kovacs, M., Kineman, R.D., Schally, A.V., Zarandi, M., Groot, K. and Frohman, L.A. (1997). Effects of antagonists of growth hormone-releasing hormone (GHRH) on GH and insulin-like growth factor-1 levels in transgenic mice over-expressing the human GHRH gene, an animal model of acromegaly. *Endocrinology* 138, 4536–4542.

Kühn, E.R., Vanderpooten, A., Huybrechts, L.M., Decuypere, E., Darras, V. and Sharp, P.J. (1988). Hypothalamic hormones that release growth hormone stimulate hepatic 5'-monodeiodination activity in the chick embryo. *J. Endocrinol.* 118, 233–236.

Kurima, R., Bacon, W.L. and Vasilatos-Younkens, R. (1994). Effects of glucagon infusion, alone or in combination with somatostatin, on plasma growth hormone and metabolite levels in young female turkeys under different feeding regimens. *Poult. Sci.* 73, 704–713.

Lauterio, T.J., Daughaday, W.H., Buonomo, F.C. and Baile, C.A. (1986). The possible roles of triiodothyronine and insulin-like growth factor I on GRF or TRH stimulated growth hormone release in the domestic fowl *(Gallus domesticus). Poult. Sci.* 65 (Suppl. 1), 75.

Leili, S. (1997). Effect of active immunization against growth hormone-releasing factor on growth, insulin-like growth factor-1, IGF-1 and IGF-1 binding proteins in chickens. *PhD Thesis*, Rutgers University.

Leung, F.C. (1986). Hormonal regulation of growth in chickens. In *Control and Manipulation of Animal Growth.* ed. PJ Buttery, DB Linday, and NB Haynes Butterworths, London. p. 223–235.

Leung, F.C. and Taylor, J.E. (1983). *In vivo* and *in vitro* stimulation of growth hormone release in chickens by synthetic human pancreatic growth hormone-releasing factors (hpGRFs). *Endocrinology* 113, 1913–1915.

Leung, F.C, Taylor, J.E. and Ball, C.A. (1985). Potent interaction between thyrotropin-releasing hormone (TRH) and human pancreatic growth hormone-releasing factor (hpGRF) in stimulating chicken growth hormone (cGH) *in vivo:* hypothalamic noradrenergic mediation in TRH stimulation of cGH release. *Domest. Anim. Endocrinol.* 2, 183–190.

Leung, F.C., Taylor, J.E., Wien, S. and Van Iderstine, A. (1986). Purified chicken growth hormone (GH) and human pancreatic GH-releasing hormone increase body weight gain in chickens. *Endocrinology* 118, 1961–1965.

Ling, N., Esch, F., Böhlen, P., Brazeau, P., Wehrenberg, W. and Guillemin, R. (1984). Isolation, primary structure and synthesis of human hypothalamic somatocrinin: growth hormone releasing-factor. *Proc. Natl. Acad. Sci.* 81, 4302–4306.

Lopez, M.E., Hargis, B.M., Dean, C.E. and Porter, T.E. (1995). Uneven regional distributions of prolactin- and growth hormone- secreting cells and sexually dimorphic proportions of prolactin secretions in the adenohypophysis of adult chickens. *Gen. Comp. Endocrinol.* 100, 246–254.

Lovejoy, D.A., Fischer, W.H., Ngamvongchon, S., Craig, A.G., Nahorniak, C.S., Peter, R.E., Rivier, J.E. and Sherwood, N.M. (1992). Distinct sequence of gonadotropin-releasing hormone (GHRH) in dogfish brain provides insight into GnRH evolution. *Proc. Nat. Acad. Sci.* 89, 6373–6377.

MacNamee, M.E., Sharp, P.J., Lea, R.W., Sterling, R.J. and Harvey, S. (1986). Evidence that vasoactive intestinal polypeptide is a physiological prolactin-releasing factors in the bantam hen. *Gen. Comp. Endocrinol.* 62, 470–478.

McRory, J.E., Parker, R.L. and Sherwood, N.M. (1997). Expression and alternative processing of a chicken gene encoding both growth hormone-releasing hormone and pituitary adenylate cyclase-activating polypeptide. *DNA and Cell Biol.* 16, 95–102.

McRory, J.E., Parker, D.B., Ngamvongchon, S. and Sherwood, N.M. (1995). Sequence and tissue expression of growth hormone-releasing hormone (GHRH) and pituitary adenylate-activating polypeptide (PACAP) from brains of catfish *(Clarias macrocephalus). Mol. Cell. Endocrinol.* 108, 169–177.

McRory, J.E. and Sherwood, N.M. (1997). Evolution of the glucagon superfamily: Sequence and expression of tunicate prepro$PACAP_{1-27}$ and VIP_{1-27}. *Endocrinology* 138, 2380–2390.

Moellers, R.F. and Cogburn, L.A. (1994). Pulsatile infusion of growth hormone-releasing factor depresses growth of young broiler chickens. *Comp. Biochem. Physiol.* 107A, 665–672.

Ono, M., Miki, N., Demura, H., Tadokoro, K., Nagafuchi, S. and Yamada, M. (1994). Molecular cloning of cDNA encoding the precursor for hamster hypothalamic growth hormone-releasing factor. *DNA Seq.* 5, 93–102.

Parker, D.B., Coe, I.R., Dixon, G.H. and Sherwood, N.M. (1993). Two salmon neuropeptides encoded by one brain cDNA are structurally related to members of the glucagon superfamily. *Eur. J. Biochem.* 215, 439–448.

Parker, D.B., Powers, M.E., Swanson, P., Rivier, J. and Sherwood, N.M. (1997). Exon skipping in the gene encoding pituitary adenylate cyclase-activating polypeptide (PACAP) in salmon alters the expression of two hormones that stimulate growth hormone release. *Endocrinology* 138, 414–423.

Perez, F.M., Malamed, S. and Scanes, C.G. (1985). Biosynthetic human somatomedin C inhibits $hpGRF_{(1-44)}$ NH_2 induced and TRH-induced GH release in a primary culture of chicken pituitary cells. *IRCS Med. Sci.* 13, 871–872.

Perez, F.,M., Malamed, S. and Scanes, C.G. (1987). Growth hormone secretion from chicken adenohypophyseal cells in primary culture: effects of human pancreatic growth hormone-releasing factor, thyrotropin-releasing hormone and somatostatin on growth hormone release. *Gen. Comp. Endocrinol.* 65, 408–414.

Perez, F.M., Malamed, S. and Scanes, C.G. (1989a). Possible participation of calcium in growth hormone release and in thyrotropin-releasing hormone and human pancreatic growth hormone-releasing factor synergy in a primary culture of chicken pituitary cells. *Gen. Comp. Endocrinol.* 75, 481–491.

Perez, F.M., Malamed, S. and Scanes, C.G. (1989b). Growth hormone release from chicken anterior pituitary cells in primary culture: TRH and hpGRF synergy, protein-synthesis and cyclic adenosine 3' 5' monophosphate. *Gen. Comp. Endocrinol.* 73, 12–20.

Piper, M.M. and Porter, T.E. (1997). Responsiveness of chicken embryonic somatotropin to somatostatin (SRIF) and IGF-1. *J. Endocrinol.* 154, 303–310.

Porter, T.E., Couger, G.S. and Morpurgo, B. (1995). Evidence that somatotroph differentiation during chicken embryonic development is stimulated by a blood-borne signal. *Endocrinology* 136, 3721–3728.

Porter, T.E., Cougar, G.S., Dean, C.E. and Hargis, B.M. (1995). Ontogeny of growth hormone (GH)-secreting cells during chicken embryonic development: initial somatotrophs are responsive to GH-releasing hormones. *Endocrinology* 136, 1850–1856.

Proudman, J.A. (1984). Growth hormone and prolactin response to thyrotropin releasing hormone and growth hormone releasing factor in the immature turkey. *Proc. Soc. Exp. Biol.* 175, 79–83.

Rabii, J., Knapp, L., De La Guardia, A., Zafian, P., Lauterio, T.J. and Scanes, C.G. (1984). Brain sites involved in the regulation of growth hormone secretion in the young male domestic fowl. *J. Endocrinol.* 103, 327–332.

Radecki, S.V., Deaver, D.R. and Scanes, C.G. (1994). Triiodothyronine reduces growth hormone secretion and pituitary growth hormone mRNA in the chicken, *in vivo* and *in vitro Proc. Soc. Exp. Biol. Med.* 205, 340–346.

Rivier, J., Speiss, J., Thorner, M. and Vale, W. (1982). Characterization of growth hormone releasing factor from a human pancreatic islet tumor. *Nature* 300, 276–278.

Robson, H., Harvey, S., Lea, R.W. and Brooks, D.D. (1997a). GRF-and induced depolarization of chicken somatotrophs: a Na^{+} and Ca^{2+} dependent response. *Bio. Amines* 13, 401–423.

Robson, H., Lea, R.W., Harvey, S. and Brooks, D.D. (1997b). Extracellular calcium mediates both GRF- and TRH- induced GH-release from chicken pituitary cells. *Bio. Amines* in press.

Scanes, C.G. and Cammaratto, A. (1982). Effect of monosodium glutamate on circulating concentrations of luteinizing hormone and growth hormone in young growing domestic fowl. *Experientia*, 38, 1246–1247.

Scanes, C.G. and Harvey, S. (1984). Stimulation of growth hormone secretion by human pancreatic growth hormone-releasing factor and thyrotropin-releasing hormone in anaesthetized chickens. *Gen. Comp. Endocrinol.* 56, 198–203.

Scanes, C.G. and Harvey, S. (1985). Growth hormone secretion in anaesthetized fowl. II. Influence of heterologous stimuli in birds refractory to human pancreatic growth hormone-releasing factor (hpGRF) or thyrotropin releasing hormone. *Gen. Comp. Endocrinol.* 59, 10–14.

Scanes, C.G. and Harvey, S. (1989a). Somatostatin (SRIF) inhibition of thyrotropin-releasing hormone (TRH) and growth hormone (GH)-releasing factor (GRF)-induced GH secretion in young and adult anaesthetized chickens. *Gen. Comp. Endocrinol.* 75, 256–264.

Scanes, C.G. and Harvey, S. (1989b). Triiodothyronine (T_3) inhibition of thyrotropin-releasing hormone (TRH) and growth hormone-releasing factor (GRF)-induced secretion of chicken hormone (GH) *in vivo*. *Gen. Comp. Endocrinol.* 73, 477–484.

Scanes, C.G., Carsia, R.V., Lauterio, T.J., Huybrechts, L., Rivier, J. and Vale, W. (1984). Synthetic human pancreatic growth hormone releasing factor (GRF) stimulates growth hormone secretion in the domestic fowl (*Gallus domesticus*). *Life Sci.* 34, 1127–1134.

Scanes, C.G., Harvey, S., Rivier, J. and Vale, W. (1986a). Stimulation of *in vivo* growth hormone secretion in young chickens by rat hypothalamic growth hormone-releasing factor and synthetic analogues. *J. Endocrinol.* 108, 413–416.

Scanes, C.G., Klandorf, H., Carsia, R.V. and Perez, F.M. (1986b). Growth-hormone releasing factor and thyrotropin induced GH release from perifused chicken pituitary cells *in vitro*. *IRCS Med. Sci.* 14, 920–921.

Scanes, C.G., Lauterio, T.J. and Perez, F.M. (1986c). Somatomedins in domestic animals. *IRCS Med. Sci.* 14, 515–518.

Speiss, J., Rivier, J. and Vale, W. (1983). Characterization of rat hypothalamic growth hormone releasing-factor. *Nature* 303, 532–534.

Talbot, R.T., Dunn, K., Wilson, P.W., Sand, H.M. and Sharp, P.J. (1995). Evidence for alternative splicing of the chicken vasoactive intestinal peptide gene transcript. *J.Mol. Endocrinol.* 15, 81–91.

Taylor, J.E., Ball, C.A. and Leung, F.C. (1986). Interaction of human pancreatic growth hormone-releasing factor, thyrotropin-releasing hormone and somatostatin on growth hormone release in chickens. *Proc. Soc. Exp. Biol. Med.* 183, 363–367.

Vasilatos-Younkens, R., Tsao, P.H., Foster, D.N., Smiley, D.L. and Heiman, M.L. (1992). Restoration of juvenile baseline growth hormone secretion with preservation of the ultradian growth hormone rhythm by continuous delivery of growth hormone-releasing factor. *J. Endocrinol.* 135, 371–382.

Vasilatos-Younkens, R. and Hibbard, E. (1987). Hollow fibre-encapsulated pituitary cells for the study of adenohypophyseal regulation of growth in poultry. 1. Preparation and use. *Poultry Sci.* 66, 881–888.

Vaughan, J.M., Rivier, J., Speiss, J., Peng, C., Chang, J.P., Peter, R.E. and Vale, W. (1992). Isolation and characterization of hypothalamic growth hormone-releasing factor from common carp, *Cyprinus carpio*. *Neuroendocrinology* 56, 539–549.

6

GONADOTROPIN-RELEASING HORMONE NEURONAL SYSTEMS IN THE TELEOSTEAN BRAIN AND FUNCTIONAL SIGNIFICANCE

Koichi Okuzawa[1] and Makito Kobayashi[2]

[1]National Research Institute of Aquaculture
Tamaki, Mie 519-0423, Japan
[2]Department of Aquatic Bioscience
Graduate School of Agricultural and Life Sciences
The University of Tokyo
Bunkyo-ku, Tokyo 113-8657, Japan

1. INTRODUCTION

Gonadotropin-releasing hormone (GnRH) is well known as the most important regulator of gonadotropin (GtH) secretion. An understanding of morphological and molecular characteristics of the GnRH neuronal system, its regulatory mechanisms and of its role in teleost reproduction are very important issues not only for comparative biology but also for aquaculture. However, the information generated on GnRH in teleosts is not similar to that in mammalian species because of some major differences which are unique to teleosts. A few major differences are listed below:

1. In teleosts there occur multiple types of GnRH neuronal system, whereas it is not so in most mammalian species. This characteristic makes the research on teleostean GnRH system challenging since the antibody or molecular probe may cross-react with multiple molecular forms of GnRH produced in fish brain. One should therefore be very careful for the molecular form of GnRH which is under investigation. We discuss this issue in the following section.
2. Unlike the condition in tetrapods, there is no hypothalamo-hypophysial portal vascular system in teleosts (Figure 1). The GnRH neurons directly innervate the pituitary and GnRH is considered to be released from nerve terminals located near or in direct synaptoid contact with the pituitary cells, or from terminals separated from the pituitary cells by a basement membrane (Peter *et al.*, 1990). Therefore, it is not possible either to measure the levels of GnRH in hypothalamo-hypophysial portal blood, or to study the release of GnRH *in vivo*.

Neural Regulation in the Vertebrate Endocrine System, edited by Prasada Rao and Peter, Kluwer Academic / Plenum Publishers, New York, 1999.

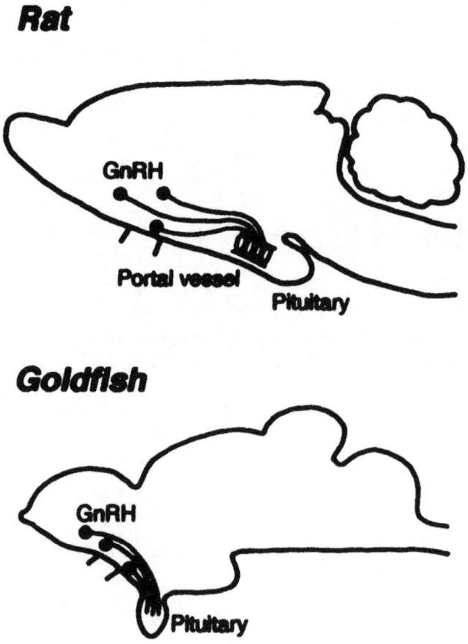

Figure 1. Hypothalamo-hypophysial system of rat and goldfish. In rat, GnRH is released from the terminals of GnRH neurons into capillaries in the median eminence and reaches the pars distalis of the pituitary through the pituitary portal vessels, while in goldfish, the portal vessels do not exist and fibers of GnRH neurons directly innervate the pituitary. GnRH is released from the nerve terminals in the pituitary of goldfish.

3. Although two GtH molecules, *viz*., GtH-I and GtH-II (Kawauchi *et al*., 1989) comparable to mammalian follicle stimulating hormone (FSH) and luteinizing hormone (LH) respectively, are produced in the pituitary gland of many teleosts, information available about the regulation and function of GtH-I is comparatively meager; this is partly because the quantification of GtH-I is possible only in salmonids (Suzuki *et al*., 1988; Swanson *et al*., 1989; Prat *et al*., 1996). In salmonids, GtH-I and GtH-II are considered respectively as FSH- and LH-like, because GtH-I is secreted during the period of gonadal growth, and play a role in stimulation of ovarian growth and steroidogenesis, while during the period of final maturation plasma levels of GtH-II show an increase to promote gamete maturation and release (Swanson, 1991). In other teleosts, however, the difference in biological function between GtH-I and II is unclear (Quérat, 1995; Kobayashi *et al*., 1997). Administration of GnRH or its agonist increased blood GtH-II levels in several teleosts (Peter *et al*., 1991), however, studies on GtH-I secretion by GnRH are restricted to salmonids (Kawauchi *et al*., 1989). Therefore, development of an assay for GtH-I in other teleosts will prove valuable.

There is a great need to overcome these shortcomings and achieve progress in the knowledge about GnRH in teleosts. In this article, we describe the present status of GnRH research in teleosts, the multiplicity of GnRH peptides and GnRH neuronal systems, and possible functions of each molecular form of GnRH neuronal system in the species investigated. Special attention will be paid to the GnRH form and GnRH neuronal system that is most important for regulation of GtH secretion and consequent gonadal development. In the end, we propose future directions for GnRH research in teleosts.

2. MOLECULAR FORMS OF GnRH PRODUCED IN TELEOST BRAIN

Twelve distinct molecular forms of GnRH have been identified to date from vertebrates and an invertebrate (Figure 2). All GnRH forms are decapeptides, showing con-

	1	2	3	4	5	6	7	8	9	10
Guinea pig	pGlu	Tyr	Trp	Ser	Tyr	Gly	Val	Arg	Pro	Gly-NH2
Mammalian	pGlu	His	Trp	Ser	Tyr	Gly	Leu	Arg	Pro	Gly-NH2
Chicken-I	pGlu	His	Trp	Ser	Tyr	Gly	Leu	Gln	Pro	Gly-NH2
Seabream	pGlu	His	Trp	Ser	Tyr	Gly	Leu	Ser	Pro	Gly-NH2
Catfish	pGlu	His	Trp	Ser	His	Gly	Leu	Asn	Pro	Gly-NH2
Salmon	pGlu	His	Trp	Ser	Tyr	Gly	Trp	Leu	Pro	Gly-NH2
Dogfish	pGlu	His	Trp	Ser	His	Gly	Trp	Leu	Pro	Gly-NH2
Chicken-II	pGlu	His	Trp	Ser	His	Gly	Trp	Tyr	Pro	Gly-NH2
Lamprey-III	pGlu	His	Trp	Ser	His	Asp	Trp	Lys	Pro	Gly-NH2
Lamprey-I	pGlu	His	Tyr	Ser	Leu	Glu	Trp	Lys	Pro	Gly-NH2
Tunicate-I	pGlu	His	Trp	Ser	Asp	Tyr	Phe	Lys	Pro	Gly-NH2
Tunicate-II	pGlu	His	Trp	Ser	Leu	Cys	His	Ala	Pro	Gly-NH2
	pGlu	His	Trp	Ser	Leu	Cys	His	Ala	Pro	Gly-NH2

Figure 2. Comparison of twelve GnRH peptides. Boxes show conserved regions: all forms have a blocked NH_2 and COOH terminus and invariant amino acids in positions 1, 4, 9, and 10. Solid line in tunicate GnRH-I shows the site of a potential salt bridge (Asp^5-Lys^8). Tunicate GnRH-II is a dimer as shown.

served amino acid residues in positions 1, 4, 9, and 10. Molecular forms of GnRH were named after the animal species in which they were first identified. Ten GnRH forms, *viz.*, mammalian GnRH (mGnRH or LHRH), chicken GnRH-I (cGnRH-I), chicken GnRH-II (cGnRH-II), salmon GnRH (sGnRH), lamprey GnRH-I (lGnRH-I), lamprey GnRH-III (lamprey GnRH-II has not yet been sequenced), catfish GnRH (cfGnRH), dogfish GnRH (dfGnRH), seabream GnRH (sbGnRH) and guinea pig GnRH were isolated from vertebrates, and two forms, *viz.*, tunicate GnRH-I and II were isolated from the sea squirt (Protochordata; Tunicata; *Chelyosoma productum*). The fact that the tunicate has two forms of GnRH suggests that the evolution of GnRH predates the evolution of chordata, and also suggests that GnRH originally acts not via pituitary gland but directly on the gonads (Powell *et al.*, 1996b).

Among twelve known forms, five forms of GnRH, *viz.*, sGnRH, cGnRH-II, cfGnRH, mGnRH and sbGnRH have been identified and sequenced from teleost brain. The primary structure of sGnRH was first determined from chum salmon, *Oncorhynchus keta* (Sherwood *et al.*, 1983). Subsequently, sGnRH peptide was isolated and sequenced from gilthead seabream, *Sparus aurata* (Powell et al., 1994), pacu, *Piaractus mesopotamicus* (Powell *et al.*, 1997) and tilapia, *Oreochromis niloticus* (Weber *et al.*, 1997). The existence of sGnRH was also revealed by molecular cloning of cDNA encoding the sGnRH precursor in several teleosts, *viz.*, three salmonids: Atlantic salmon, *Salmo salar* (Klungland *et al.*, 1992), masu salmon, *Onchorhynchus masou* (Suzuki *et al.*, 1992) and sockeye salmon, *O. nerka* (Ashihara *et al.*, 1995), two cyprinids: goldfish, *Carassius auratus* (Lin and Peter, 1996) and roach *Rutilus rutilus* (Penlington *et al.*, 1997), three perciforms: the African cichlid, *Haplochromis burtoni* (Bond *et al.*, 1991), red seabream, *Pa-*

grus major (Okuzawa *et al.*, 1994a) and gilthead seabream (Gothilf *et al.*, 1996), and one batrachoidiform species: midshipman, *Porichthys notatus* (Grober *et al.*, 1995). Two distinct cDNAs encoding two distinct precursors of sGnRH have been isolated from the brain of sockeye salmon (Ashihara *et al.*, 1995) and goldfish (Suetake *et al.*, 1997). The recent tetraploidization in salmonid and cyprinid fishes is considered as the reason for the presence of duplicate cDNAs for sGnRH peptide.

Chicken GnRH-II has been first isolated and sequenced from chicken hypothalamus (Miyamoto *et al.*, 1984), and later also identified from the brain of the Thai catfish, *Clarias macrocephalus* (Ngamvongchon *et al.*, 1992), the African catfish, *C. gariepinus* (Bogerd *et al.*, 1992), gilthead seabream (Powell *et al.*, 1994), pacu (Powell *et al.*, 1997) and tilapia (Weber *et al.*, 1997). cDNAs for cGnRH-II have been isolated from the African catfish (Bogerd *et al.*, 1994), the African cichlid (White *et al.*, 1995), gilthead seabream (Gothilf *et al.*, 1996), goldfish (Lin and Peter, 1996) and roach *Rutilus rutilus* (Penlington *et al.*, 1997). Two distinct cDNAs for cGnRH-II have been isolated from the brain of goldfish (Lin and Peter, 1997).

Catfish GnRH has been isolated and sequenced only from two species of catfish, the Thai catfish (Ngamvongchon *et al.*, 1992) and the African catfish (Bogerd *et al.*, 1992). Two distinct cDNAs for cfGnRH have been isolated from the African catfish (Bogerd *et al.*, 1994).

A combination of high performance liquid chromatography (HPLC) and radioimmunoassay (RIA) showed the existence of mGnRH in the brain of the European eel, *Anguilla anguilla* (King *et al.*, 1990). Recently, the cDNA encoding mGnRH precursor has been cloned from the brain of the Japanese eel, *Anguilla japonica* (K. Aida, personal communication). Thus, mGnRH seems to be restricted to the brain of eels among teleosts.

The primary structure of sbGnRH was first determined from gilthead seabream (Powell *et al.*, 1994), then this peptide was isolated and sequenced from pacu, (Powell *et al.*, 1997) and tilapia (Weber *et al.*, 1997). cDNA for sbGnRH precursor was first cloned from gilthead seabream (Gothilf *et al.*, 1995), and later from the African cichlid (White *et al.*, 1995) and red seabream (Okuzawa *et al.*, 1997).

It is very interesting to note that different GnRH forms were produced in different teleost species. This indicates that GnRH molecules have evolved together with teleost evolution. Sherwood *et al.* (1997) proposed that sGnRH, cfGnRH and sbGnRH have evolved from mGnRH, cGnRH-II and sGnRH, respectively, while Gothilf *et al.* (1995) suggested that these GnRH forms have evolved from dfGnRH, mGnRH and mGnRH, respectively. Further analyses of cDNA for GnRH precursors will prove useful to clarify the evolution of any status of molecular form of GnRH. It is therefore important that one should be aware of the divergence of GnRH forms in the teleostean brain before initiation of investigation on the GnRH of a new species of teleost fish.

3. MULTIPLE MOLECULAR FORMS OF GnRH IN A SINGLE TELEOST BRAIN

In the brain of all teleostean species examined to date, there occur two or three GnRH forms. Multiple forms of GnRH, *viz.*, sGnRH and cGnRH-II was first demonstrated in the goldfish (Yu *et al.*, 1988) by employing HPLC in combination with RIA (HPLC/RIA). Subsequent investigations revealed the occurrence of these two forms of GnRH in rainbow trout, *O. mykiss* (Okuzawa *et al.*, 1990), masu salmon (Amano *et al.*, 1992a), common carp, *Cyprinus carpio* (Amano *et al.*, 1992b), honmoroko, *Gnathopogon caerulescens* (Okuzawa *et al.*, 1994b), molly, *Poecilia latipinna* (Coe *et al.*, 1992), medaka, *Oryzias latipes*, zebrafish, *Brachydanio rerio* (Powell *et al.*, 1996a) and roach *Rutilus rutilus* (Penlington *et al.*, 1997). In two catfish species, cfGnRH and cGnRH-II were isolated and sequenced (Bogerd *et al.*, 1992; Ngamvongchon *et al.*, 1992). In the European eel, mGnRH and cGnRH-II were detected by an HPLC/RIA study (King *et al.*, 1990).

HPLC/RIA studies have indicated that there are three distinct forms of GnRH, *viz.*, sGnRH, cGnRH-II and an unknown form whose retention time is similar to that of cGnRH-I and lGnRH-I, in the brain of several teleosts which belong to Perciformes (Okuzawa *et al.*, 1993; Sherwood *et al.*, 1993), Pleuronectiformes (Okuzawa *et al.*, 1993) and Characiformes (Somoza *et al.*, 1994). Recently, three forms of GnRH, i.e. sGnRH, cGnRH-II and sbGnRH were isolated and sequenced in the brain of gilthead seabream (Perciformes; Powell *et al.*, 1994), the pacu (Characiformes; Powell *et al.*, 1997) and tilapia (Perciformes; Weber *et al.*, 1997), and now the unknown form of GnRH reported previously is speculated as sbGnRH. cDNAs for sGnRH, cGnRH-II and sbGnRH were found in the brain of the African cichlid (White *et al.*, 1995) and gilthead seabream (Gothilf *et al.*, 1996). HPLC/RIA studies revealed the presence of these three forms in the brains of pumpkinseed fish, *Lepomis gibbosus* (Powell *et al.*, 1995) and rockfish, *Sebastes rastrelliger* (Powell *et al.*, 1996a). Our recent studies using specific RIAs have also demonstrated the presence of these three GnRH forms in red seabream, *Pagrus major*, black seabream, *Acanthopagrus schlegeli* and striped knifejaw, *Oplegnathus fasciatus* (Senthilkumaran *et al.*, 1998).

Stefano *et al.* (1997) revealed that there are three forms of GnRH in the brain of a atherinid fish, pejerrey (*Odontesthes bonariensis*), *viz.*, sGnRH, cGnRH-II and a novel form which is chromatographically different from the other known forms of the vertebrate, and that only this novel form was detected in the pituitary of pejerrey. This study suggests the existence of a novel hypophysiotropic form of GnRH in athernid fish.

Chicken GnRH-II is present in the brains of all teleost species studied to date. Moreover, this GnRH form is also produced in the brains of cartilaginous fish, amphibians, reptiles, birds and some mammalian species including primates (Sherwood *et al.*, 1997). Therefore, cGnRH-II must play some important role which is common to all vertebrates. On the contrary, second or third GnRH forms have evolved among certain vertebrates. One may speculate that emergence of the third GnRH form in the Perciformes, *viz.*, sbGnRH, has perhaps provided the required physiological mechanism to perciforms to spawn large amounts of eggs at a high frequency (even daily) and brought a remarkable success in reproduction. Probably, this reproductive ability made Perciformes the biggest group among teleosts.

4. GnRH NEURONAL SYSTEMS AND THEIR POSSIBLE FUNCTIONS IN DIFFERENT TELEOSTS

4.1. Salmonid

As mentioned above, two GnRH forms, sGnRH and cGnRH-II are produced in the brain of salmonid species. The distribution pattern of sGnRH and cGnRH-II in discrete brain areas was first examined by specific RIAs in the rainbow trout (Okuzawa *et al.*, 1990) and masu salmon (Amano *et al.*, 1992a). It was shown that the distribution pattern of sGnRH and cGnRH-II in the brain was different; concentrations of sGnRH were high in the anterior brain areas, such as the olfactory bulb, telencephalon and hypothalamus, while those of cGnRH-II were high in the posterior brain areas, such as medulla oblongata. In the pituitary gland, only sGnRH was detected in rainbow trout and masu salmon, and the concentration of sGnRH in the pituitary was high in mature fish as compared to that of immature fish (Okuzawa *et al.*, 1990; Amano *et al.*, 1992a). These data clearly show that sGnRH is the pivotal hypophysiotropic form of GnRH and that cGnRH-II may participate in biological functions other than the regulation of pituitary hormones in salmonids.

Immunocytochemistry (ICC) using sGnRH and cGnRH-II antisera (Amano *et al.*, 1991) has revealed the presence of sGnRH-immunoreactive (ir) cell bodies in the ventral forebrain, *viz.*, the olfactory bulb (OB), the terminal nerve (TN), the ventral telencephalon (VT), and the preoptic area (POA), whereas the cGnRH-II-ir cell bodies were found only in the midbrain in masu salmon (Figure 3a). In the pituitary gland, only sGnRH-ir fibers

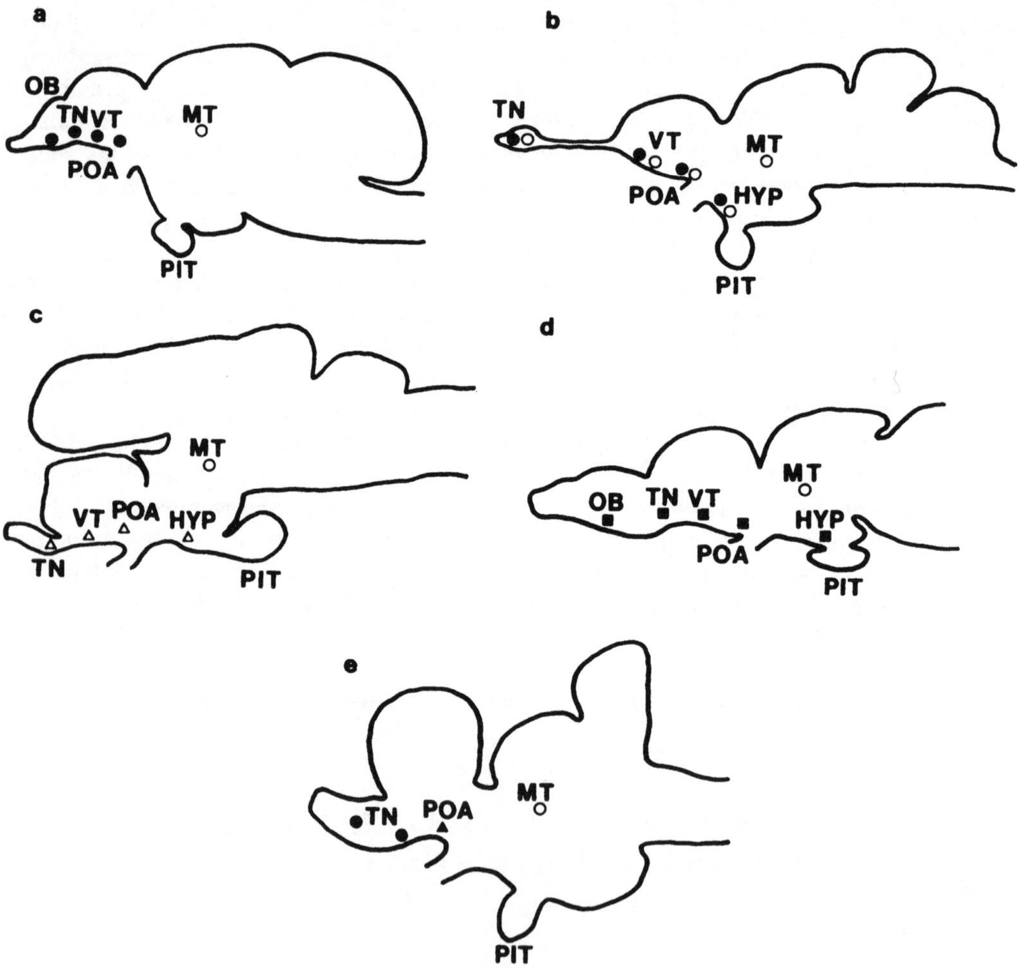

Figure 3. Schematic illustrations of midsagittal brain sections showing GnRH system in teleosts. a, masu salmon (Amano *et al.*, 1991); b, goldfish (Kim *et al.*, 1995b); c, African catfish (Zandbergen *et al.*, 1995); d, European eel (Montero *et al.*, 1994); e, red seabream (Okuzawa *et al.*, 1997). HYP, hypothalamus; MT, midbrain tegmentum; OB, olfactory bulb; PIT, pituitary; POA, preoptic area; TN, terminal nerve; VT, ventral telencephalon. Closed circle, sGnRH; open circle, cGnRH-II; open triangle, cfGnRH; closed square, mGnRH; closed triangle, sbGnRH.

were observed. *In situ* hybridization (ISH) studies showed that the distribution of sGnRH mRNA expressing cells was consistent with that of sGnRH-ir cell bodies in masu salmon (Suzuki *et al.*, 1992); however, ISH for cGnRH-II has not been performed, yet. These ICC and ISH data are in good agreement with the quantitative data generated using RIA.

These results suggest that sGnRH and cGnRH-II belong to distinct GnRH neuronal systems; sGnRH is the component of the ventral forebrain GnRH neuronal system, while cGnRH-II is the component of midbrain GnRH neuronal system. Furthermore, these morphological data suggest that only the ventral forebrain GnRH system (sGnRH-system) functions as a hypophysiotropic regulator, probably of GtH secretion in salmonid species (Amano *et al.*, 1997; Kim *et al.*, 1997).

4.2. Goldfish

Differential distribution of sGnRH and cGnRH-II was also demonstrated in goldfish brain using the same specific RIA used for rainbow trout and masu salmon (Kobayashi *et al.*, 1992). sGnRH concentration was high in anterior brain areas, *viz.*, the olfactory bulb, telencephalon and hypothalamus, whereas levels of cGnRH-II were high in the hypothalamus, optic tectum, thalamus and medulla oblongata. In the pituitary gland, both sGnRH and cGnRH-II were measured and interestingly, sGnRH concentration was much higher than that of cGnRH-II.

Immunocytochemical studies revealed the presence of sGnRH- and cGnRH-II-containing cell bodies in the TN, VT, POA and hypothalamus (Kim *et al.*, 1995). However, the numbers of cGnRH-II-ir cell bodies were fewer than those of sGnRH. In addition, cGnRH-II-ir cell bodies were observed also in the midbrain (Figure 3b). In the pituitary gland of goldfish, fibers of both GnRHs were observed. Recent studies by ISH showed that the distribution of cells expressing sGnRH mRNA was consistent with that of sGnRH-ir cell bodies in goldfish (Suetake *et al.*, 1995). However, no information is available about expression of cGnRH-II mRNA by ISH.

Taken together, in goldfish, both sGnRH and cGnRH-II are the components of the ventral forebrain GnRH neuronal system, and cGnRH-II is the only component of midbrain GnRH neuronal system. Localization of both sGnRH and cGnRH-II in the pituitary suggests that these two GnRHs function as hypophysiotropic factors in goldfish. cGnRH-II production in extra-midbrain areas is known only in goldfish among teleosts examined.

Olfactory tract sectioning, which blocks axonal transport of GnRH from the TN to other brain regions, revealed that most of sGnRH and part of cGnRH-II in the goldfish brain originate from the TN system (Kobayashi *et al.*, 1992, 1994). Moreover, it was also shown that the GnRH of the TN origin is not essential for gonadal maturation, ovulation (Kobayashi *et al.*, 1992, 1994; Kim *et al.*, 1993), female sexual behavior or oviposition (Kobayashi and Kim, unpublished). It is also revealed that olfactory tract sectioning did not affect mRNA levels of GtH-Iβ and GtH-IIβ subunits in the pituitary of goldfish, suggesting that the TN GnRHs are not involved in GtH production (Sohn and Kobayashi, unpublished). Therefore, it seems that the ventral forebrain GnRH neuronal system, excluding the TN GnRH system, regulates GtH secretion in goldfish (Kim *et al.*, 1997; Kobayashi *et al.*, 1997).

4.3. African Catfish

In the African catfish, an HPLC/RIA study revealed that cGnRH-II was primarily located in the brain, and very little was present in the pituitary, whereas cfGnRH was mainly limited to the pituitary and a relatively small amount was present in the brain. In the pituitary ir-cfGnRH was present in 700 to 750 times more than cGnRH-II. Interestingly, more than 90% of the total brain plus pituitary cfGnRH was present in the pituitary itself (Goos *et al.*, 1997).

Zandbergen *et al.* (1995) performed ISH using specific probes for cfGnRH and cGnRH-II mRNAs and ICC with antisera against recombinant GnRH-associated peptides (GAPs) of cfGnRH and cGnRH-II precursors (Fig. 3c). Though the physiological significance of GAP in teleosts is not known, GAP is processed from GnRH precursor and released together with GnRH. GAP sequence is not so conserved as GnRH peptide, thus one can overcome the problem of cross-reaction of GnRH antibodies if GAP antibodies are used for the detection of GnRH neurons. Cell bodies expressing cfGnRH mRNA occurred in the ventral forebrain, such as the medial olfactory tract, VT, POA and medial basal hypothalamus, whereas neurons which express the mRNA for a precursor of cGnRH-II were restricted to the midbrain tegmentum. Numerous processes arising from the cfGnRH neurons project into the rostral pars distalis of the pituitary. Although, cGnRH-II neurons of the midbrain tegmentum revealed a few descending fibers, none were found in the pituitary. The GtH-II releasing potency (GtH-I has not been identified in catfish) of cfGnRH is 150-fold lower than that of cGnRH-II, although its concentration in the pituitary is 700-fold higher than that of cGnRH-II. Thus, despite its low potency cfGnRH appears to be the principal regulator of GtH-II secretion in the catfish (Schulz *et al.*, 1995).

4.4. European Eel

As mentioned above, two GnRH forms, mGnRH and cGnRH-II are produced in the brain of the European eel and levels of mGnRH were higher than those of cGnRH-II in the pituitary, olfactory lobes, tel-, di- and mes-encephalon, while the opposite was found in the posterior part of the brain, *viz.*, met- and myel-encephalon (Dufour *et al.*, 1993).

Using ICC, Montero *et al.* (1994) reported the presence of mGnRH-ir perikarya in the OB, TN, VT, POA and the mediobasal hypothalamus of the European eel brain. These cell bodies were located along a continuum of mGnRH-ir fibers that could be traced from the olfactory nerve to the pituitary; such fibers were also detected in many parts of the brain and in the pituitary. cGnRH-II-ir cell bodies were found in the midbrain tegmentum, but only scattered fibers could be detected in different brain areas (Fig. 3d). The pituitary exhibited very few cGnRH-II-ir fibers, contrasting with an extensive mGnRH innervation.

Administration of carp pituitary extract to female eels resulted in a rapid ovarian maturation, and increased levels of mGnRH in the pituitary and rostral brain areas, while the levels of cGnRH-II decreased (Dufour *et al.*, 1993). Estradiol-17β administration also increased the brain and pituitary levels of mGnRH (Montero *et al.*, 1995). These data indicate that gonadal hormones have a stimulatory effect on mGnRH, and the mGnRH neurons are implicated in the control of reproduction in the European eel.

4.5. Red Seabream

We have developed specific RIAs for three native GnRH forms (sGnRH, cGnRH-II and sbGnRH) produced in the brain of red seabream. Using these RIAs, we have studied quantitative distribution of these three GnRHs in the brain of mature red seabream. Distribution patterns of sGnRH and cGnRH-II were similar to those of masu salmon and goldfish; sGnRH level was high in anterior brain areas and cGnRH-II level was high in posterior areas. In contrast to the wide distribution of sGnRH and cGnRH-II in the brain, localization of sbGnRH was restricted to the telencephalon and hypothalamus. In the pituitary, we have detected sGnRH and sbGnRH, but cGnRH-II levels were below the limit of detection. Moreover, levels of sbGnRH were 2,500-fold higher than levels of sGnRH in the pituitary (Senthilkumaran *et al.*, 1998).

We have done ICC using the same antisera which we used in our RIAs. sbGnRH antiserum does not cross-react with other two forms in ICC, while antisera against sGnRH and cGnRH-II cross-react with the other two forms when high concentrations of the antisera were employed in ICC. Generally speaking, specificity is higher in RIA, while it is lower in ICC in which antigen-antibody reactions take place under non-competitive condition. However, we overcome the problem of cross-reaction by doing ISH and by comparing the results (Fig. 3e).

sbGnRH-ir cell bodies (diameter, 20–30 μm) were found only in the POA. A distinct fiber bundle arising from these cells ran through the hypothalamus and several fibers innervated the pituitary. Only a few sbGnRH-ir fibers were detected in the olfactory bulb and in the telencephalon, but none in other brain areas. ISH studies using a specific probe for sbGnRH precursors of red seabream revealed that the mRNA for sbGnRH was present only in the POA. These data suggest that sbGnRH is produced only in the POA and projected mainly into the pituitary (Okuzawa *et al.*, 1997).

Messenger RNA for sGnRH precursor was detected only in two ganglia of the TN (TN-ggl). ICC using antiserum against sGnRH revealed that the anterior TN-ggl were localized in the medial olfactory bulbs and the posterior TN-ggl were localized in the transitional area between the olfactory bulbs and the telencephalic hemispheres. sGnRH-ir cells in these two TN-ggl were larger (diameter, 40–50 μm) than sbGnRH-ir cells in the POA and form compact cell clusters, especially in the anterior TN-ggl. A distinct bundle of sGnRH-ir fibers was observed between the two TN-ggl, and projecting caudodorsally.

Salmon GnRH and cGnRH-II-ir cells were observed in the midbrain tegmentum. We considered that only cGnRH-II is produced in this midbrain GnRH-ir cells, because there was no expression of sGnRH mRNA in these midbrain cells. An ISH study using a specific probe for cGnRH-II precursor is required to confirm this idea. Processes from these midbrain cGnRH-II-ir cells could, however, be detected for only a short distance.

These ICC and ISH analyses are in good agreement with the results obtained by the RIA and these data suggest that there are three distinct GnRH neuronal systems in the red seabream brain, *viz.*, TN-GnRH system, POA-GnRH system and midbrain-GnRH system in which sGnRH, sbGnRH and cGnRH-II were exclusively produced, respectively.

In the other two perciform species, the African cichlid (White *et al.*, 1995) and gilthead seabream (Gothilf *et al.*, 1996) three GnRH forms, *viz.*, sGnRH, cGnRH-II and sbGnRH having differential localization in brain, were demonstrated by ISH studies; hybridization signals for sGnRH were detected in the TN, and signals for cGnRH-II and sbGnRH were observed in midbrain and the POA, respectively. Therefore, localization of three distinct GnRH forms in three GnRH systems, *viz.*, sGnRH in the TN system, sbGnRH in the POA system and cGnRH-II in midbrain system, seems to be common among perciform species.

Among the three distinct GnRH neuronal systems, only the POA-GnRH system is considered to be involved in the regulation of pituitary hormone(s) in perciforms, because (i) only the POA-GnRH system innervates the pituitary gland (ii) sbGnRH is the only component of the POA-GnRH system and (iii) sbGnRH is the dominant GnRH form detected in the pituitary.

A consideration of the distribution of GnRH neurons (Figure 3) reveals that the nature of midbrain GnRH system (cGnRH-II system) is common among teleost species examined to date, while other GnRH systems have variations. In perciform species examined to date, the TN-GnRH system and the POA-GnRH system are clearly distinguishable and the GnRH forms produced in these two systems are different, while TN- and POA-GnRH systems are undistinguishable in other teleost species examined because GnRH neurons located in the ventral forebrain are consecutive (Amano *et al.*, 1991; Kim *et al.*, 1995) and GnRH form(s) produced in these neurons are the same. From these data, we can speculate that originally only one GnRH system existed in the ventral forebrain of ancient fish, then during evolution, it has diverged into two systems which are morphologically and functionally different. According to this hypothesis, the TN- (sGnRH) and the POA- (sbGnRH) GnRH neuronal systems should share the same ontogenic origin.

It is suggested that sGnRH neurons originate in the olfactory placode and then migrate into the brain in chum salmon (Chiba *et al.*, 1994) and sockeye salmon (Parhar *et al.*, 1995). Recently, Parhar (1997) investigated the ontogeny of GnRH systems in tilapia, *O. mossambicus* and proposed the hypothesis that GnRH cells in the TN-system originate in the olfactory placode while GnRH cells in the POA-system originate from the ventricular zones of the preoptic recess and the floor of the basal telencephalon-anterior preoptic area. However, in teleosts, no study is available on the effects of sectioning the olfactory nerve in the embryo interrupting the migration of GnRH cells from olfactory placode to brain. This experiment is necessary to confirm the ontogenic origin of GnRH neurons in the brain.

5. CHANGES IN GnRH NEURONAL ACTIVITY IN RELATION TO REPRODUCTION IN TELEOSTS

5.1. Masu Salmon

Changes in the activity of GnRH neurons in relation to gonadal maturation is well studied in masu salmon (Amano *et al.*, 1997; Kobayashi *et al.*, 1997). When the levels of sGnRH and cGnRH-II in discrete areas of brain and pituitary of masu salmon were examined from hatching until gonadal maturation, sGnRH in the pituitary increased in parallel with gonadal maturation, while the relationship between levels of sGnRH in brain and gonadal maturation was not clear. On the other hand, cGnRH-II content was undetectable in the pituitary and it showed no changes in the brain areas (Amano *et al.*, 1992a, 1993). The reason for the discrepancy between the levels of brain sGnRH and gonadal maturity may be the heterogeneity of the sGnRH system; both the fibers from the TN-GnRH system which is thought to be unnecessary for GtH secretion and the cell bodies and fibers of the VT-POA-GnRH system intermingle in discrete areas of brain, such as the telencephalon and hypothalamus. However, levels of GnRH gene expression in the VT-POA-GnRH neuronal system seem to be a precise index for an assessment of the relationship between the GnRH and reproductive activities in masu salmon. Changes in expression levels of mRNA for sGnRH precursor were estimated using ISH followed by counting the number of silver grains per neuron and the numbers of neurons expressing sGnRH mRNA during ovarian development in female masu salmon (Amano *et al.*, 1995a). sGnRH mRNA expression in the VT and POA significantly increased during gonadal maturation, whereas no significant changes were observed in the OB and TN.

Synthetic activity of sGnRH in the VT and POA also increased in correlation with gonadal maturation triggered by photoperiod manipulation, while no significant changes were observed in sGnRH synthetic activity of the OB and TN (Amano *et al.*, 1995b).

Amano *et al.* (1994) also examined the effects of sex steroids on the synthetic activity of sGnRH. Male masu salmon treated with 17α–methyltestosterone showed an increase in the number of sGnRH neurons only in the POA, but not in the OB and VT.

These results on the synthetic activity of sGnRH suggest that sGnRH neurons in the VT-POA system are involved in GtH secretion and in turn, in gonadal maturation of masu salmon. Although the same peptide is contained in both the OB-TN- and the VT-POA-GnRH systems of masu salmon, their functional significance appears to be different.

5.2. Red Seabream

We have studied seasonal changes in the three native GnRHs (sGnRH, cGnRH-II and sbGnRH) in the brain and pituitary of female red seabream using a specific RIA for each form of GnRH (Figure 4). In female red seabream, vitellogenesis commences in February and the ovaries attain full maturity in April. Their spawning season begins in April and continues until May. Both brain and pituitary sbGnRH levels increased from October (immature phase) and reached the peak in April (spawning phase) in correlation with an increase in gonadosomatic index (GSI) and vitellogenesis. Brain levels of sbGnRH decreased in June (regressed phase) but the levels were still rather high and similar to levels in March (vitellogenic phase). However, the pituitary levels of sbGnRH did not change and remained high in completely regressed fish in June. On the other hand, levels of sGnRH and cGnRH-II in the brain were high in immature (October) and regressed (June) phases, but remained low during the spawning phase. Pituitary levels of sGnRH were much lower than those of sbGnRH and showed no seasonal change, and pituitary cGnRH-

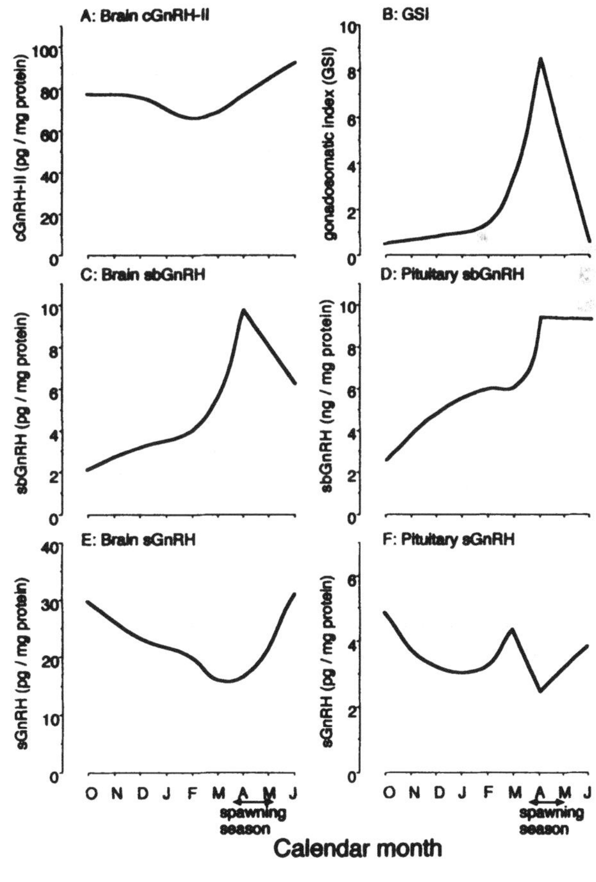

Figure 4. Seasonal changes in gonadosomatic index (GSI) (B), and concentrations of chicken GnRH-II (A), seabream GnRH (C) and salmon GnRH (E) of whole brain and concentrations of seabream GnRH (D) and salmon GnRH (F) of pituitary of female red seabream, *Pagrus major* from October, 1996 to June 1997 (Modified from Senthilkumaran *et al.*, 1998).

II was always below the limit of detection. From these data and from the ICC and ISH analyses mentioned above, it is suggested that the sbGnRH is the pivotal hypophysiotropic form of GnRH in red seabream and probably, in other perciforms also.

Red seabream can be a good model fish for GnRH study, because only sbGnRH seems to be involved in reproduction. We, therefore, propose to conduct studies on the regulation of sbGnRH synthesis and release by several neurotransmitters and neuropeptides using this species.

6. CONCLUSIONS

In all the species examined, the ventral forebrain-GnRH system excluding the TN-GnRH system, seems to be involved in GtH regulation, because 1) this GnRH system directly innervates the pituitary, 2) dominant GnRH form(s) contained in the pituitary is/are produced in this GnRH system, 3) in masu salmon and red seabream, activities in this GnRH system show a marked correlation with reproductive events. However, GnRH molecular forms produced in this system exhibit variations among different species: sGnRH in masu salmon, both sGnRH and cGnRH-II in goldfish, cfGnRH in the African catfish, mGnRH in the European eel and sbGnRH in red seabream. Among them, sbGnRH in red seabream and other perciform species seems to be the specific regulator of GtH, since it is expressed solely in the POA-GnRH system. We could overcome the challenging complexities of the teleost GnRH system (see Introduction) after extensive studies involving seasonal changes in the different GnRHs in brain and pituitary through usage of RIA, ICC and ISH. Further understanding could be achieved through a study of the regulation of GnRH synthesis and release, or control mechanisms of GtH secretion by GnRH through GnRH receptors using red seabream as a model fish. Recently, a cDNA encoding a GnRH receptor has been cloned from the pituitary of the African catfish (Tensen *et al.*, 1997).

However, the significance of the other two GnRH systems, *viz.*, the TN-GnRH system and midbrain-GnRH system, remains unclear. In a perciform, dwarf gourami (*Colisa lalia*), the TN-GnRH cells project widely in the brain and have spontaneous oscillatory activities (Oka and Matsushima, 1993), and Oka (1997) presented the hypothesis that the TN-GnRH system functions as neuromodulator system. A recent study by Kudo *et al.* (1996) suggests the involvement of the TN-GnRH system in the olfactory system during migration in chum salmon.

In all the fish species examined thus far the GnRH observed in the midbrain is cGnRH-II. Although the biological function of this midbrain GnRH system is not known in fish, a recent study in the musk shrew (*Suncus murinus*), a mammal which has midbrain cGnRH-II, suggests its involvement in sexual behavior (Dellovade *et al.*, 1995).

Future studies of teleostean GnRH neuronal systems should be focused on 1) evolution of GnRH molecules in teleost, 2) ontogeny of multiple GnRH neuronal systems, 3) regulation of GnRH synthesis and release, 3) differential regulation of GtH-I and GtH-II secretion by GnRH, 4) biological function of the TN- and midbrain-GnRH systems, 5) GnRH and sexual behavior.

ACKNOWLEDGMENTS

We are grateful to all our colleagues for their contribution to this work. We also would like to express our sincere gratitude to Dr. H. Kagawa and Dr. B. Senthilkumaran of

National Research Institute of Aquaculture for their valuable comments on the manuscript. The research from our laboratories described in this chapter was supported in part by a Grant-in-Aid from the Ministry of Agriculture, Forestry and Fisheries of Japan and by Grants-in-Aid for Scientific Research from the Ministry of Education, Science, Sports, and Culture of Japan.

REFERENCES

Amano, M., Aida, K., Okumoto, N. and Hasegawa, Y. (1992a). Changes in salmon GnRH and chicken GnRH-II contents in the brain and pituitary, and GTH contents in the pituitary in female masu salmon, *Oncorhynchus masou*, from hatching through ovulation. *Zool. Sci.* **9,** 375–386.

Amano, M., Aida, K., Okumoto, N. and Hasegawa, Y. (1993). Changes in levels of GnRH in the brain and pituitary and GTH in the pituitary in male masu salmon, *Oncorhynchus masou*, from hatching to maturation. *Fish Physiol. Biochem.* **11,** 233–240.

Amano, M., Hyodo, S., Kitamura, S., Ikuta, K., Suzuki, Y., Urano, A. and Aida, K. (1995a). Salmon GnRH synthesis in the preoptic area and the ventral telencephalon is activated during gonadal maturation in female masu salmon. *Gen. Comp. Endocrinol.* **99,** 13–21.

Amano, M., Hyodo, S., Kitamura, S., Ikuta, K., Suzuki, Y., Urano, A. and Aida, K. (1995b). Short photoperiod accelerates preoptic and ventral telencephalic salmon GnRH synthesis and precocious maturation in underyearling male masu salmon. *Gen. Comp. Endocrinol.* **99,** 22–27.

Amano, M., Hyodo, S., Urano, A., Okumoto, N., Kitamura, S., Ikuta, K., Suzuki, Y. and Aida, K. (1994). Activation of salmon gonadotropin-releasing hormone synthesis by 17a-methyltestosterone administration in yearling masu salmon, *Oncorhynchus masou*. *Gen. Comp. Endocrinol.* **95,** 374–380.

Amano, M., Oka, Y., Aida, K., Okumoto, N., Kawashima, S. and Hasegawa, Y. (1991). Immunocytochemical demonstration of salmon GnRH and chicken GnRH-II in the brain of masu salmon, *Oncorhynchus masou*. *J. Comp. Neurol.* **314,** 587–597.

Amano, M., Okuzawa, K., Yanagizawa, T., Hasegawa, Y. and Aida, K. (1992b). Immunoreactive gonadotropin-releasing hormone (ir-GnRH) in the carp brain. *Nippon Suisan Gakkaishi* **58,** 593.

Amano, M., Urano, A. and Aida, K. (1997). Distribution and function of gonadotropin-releasing hormone (GnRH) in the teleost brain. *Zool Sci.* **14,** 1–12.

Ashihara, M., Suzuki, M., Kubokawa, K., Yoshiura, Y., Kobayashi, M., Urano, A. and Aida, K. (1995). Two differing precursor genes for the salmon-type gonadotropin-releasing hormone exist in salmonids. *J. Mol. Endocrinol.* **15,** 1–9.

Bogerd, J., Li, K. W., Janssen-Dommerholt, C. and Goos, H. (1992). Two Gonadotropin-releasing hormones from African catfish (*Clarias gariepinus*). *Biochem. Biophys. Res. Comm.* **187,** 127–134.

Bogerd, J., Zandbergen, T., Andersson, E. and Goos, H. (1994). Isolation, characterization and expression of cDNAs encoding the catfish-type and chicken-II-type gonadotropin-releasing hormone precursors in the African catfish. *Eur. J. Biochem.* **222,** 541–549.

Bond, C. T., Francis, R. C., Fernald, R. D. and Adelman, J. P. (1991). Characterization of complementary DNA encoding the precursor for gonadotropin-releasing hormone and its associated peptide from a teleost fish. *Mol. Endocrinol.* **5,** 931–937.

Chiba, A., Oka, S. and Honma, Y. (1994). Ontogenic development of Gonadotropin-releasing hormone-like immunoreactive neurons in the brain of the chum salmon, *Oncorhynchus keta*. *Neurosci. Lett.* **178,** 51–54.

Coe, I. R., Grier, H. J. and Sherwood, N. M. (1992). Gonadotropin-releasing hormone in the molly *Poecilia latipinna*: molecular form, quantity and location. *J. Exp. Zool.* **261,** 414–423.

Dellovade, T. L., Ottinger, M. A. and Rissman, E. F. (1995). Mating alters gonadotropin-releasing hormone cell number and content. *Endocrinology* **136,** 1648–1657.

Dufour, S., Montero, M., Le Belle, N., Bassompierre, M., King, J. A., Millar, R. P., Peter, R. E. and Fontaine, Y. (1993). Differential distribution and response to experimental sexual maturation of two forms of brain gonadotropin-releasing hormone (GnRH) in the European eel, *Anguilla anguilla*. *Fish Physiol. Biochem.* **11,** 99–106.

Goos, H. J. T., Bosma, P. T., Bogerd, J., Tensen, C. P., Li, K. W., Zandbergen, M. A. and Schulz, R. W. (1997). Gonadotropin-releasing hormones in the African catfish: molecular forms, localization, potency and receptors. *Fish Physiol. Biochem.* **17,** 45–51.

Gothilf, Y., Elizur, A., Chow, M., Chen, T. and Zohar, Y. (1995). Molecular cloning and characterization of a novel gonadotropin-releasing hormone from the gilthead seabream (*Sparus aurata*). *Mol. Marine Biol. Biotechnol.* **4,** 27–35.

Gothilf, Y., Munoz-Cueto, J. A., Sagrillo, C. A., Selmanoff, M., Chen, T. T., Kah, O., Elizur, A. and Zohar, Y. (1996). Three forms of Gonadotropin-releasing hormone in a perciform fish (*Sparus aurata*): complementary deoxyribonucleic acid characterization and brain localization. *Biol. Reprod.* **55,** 636–645.

Grober, M. S., Myers, T. R., Marchaterre, M. A., Bass, A. H. and Myers, D. A. (1995). Structure, localization, and molecular phylogeny of a GnRH cDNA from a Paracanthopterygian fish, the plainfin midshipman (*Porichthys notatus*). *Gen. Comp. Endocrinol.* **99,** 85–99.

Kawauchi, H., Suzuki, K., Itoh, H., Swanson, P., Naito, N., Nagahama, Y., Nozaki, M., Nakai, Y. and Itoh, S. (1989). The duality of teleost gonadotropins. *Fish Physiol. Biochem.* **7,** 29–38.

Kim, M. H., Amano, M., Suetake, H., Kobayashi, M. and Aida, K. (1997). GnRH neurons and gonadal maturation in masu salmon and goldfish. *In* "GnRH Neurons-Gene to Behavior" (I. S. Parhar and Y. Sakuma, Eds.), pp. 313–324. Brain Shuppan, Tokyo.

Kim, M. H., Oka, Y., Amano, M., Kobayashi, M., Aida, K. and Kawashima, S. (1993). Immunocytochemical localization of sGnRH and cGnRH-II in the goldfish brain and effects of olfactory tract section on the distribution of GnRH. *Proc. Jpn. Soc. Comp. Endocrinol.* **8,** 11.

Kim, M. H., Oka, Y., Amano, M., Kobayashi, M., Okuzawa, K., Hasegawa, Y., Kawashima, S., Suzuki, Y. and Aida, K. (1995). Immunocytochemical localization of sGnRH and cGnRH-II in the brain of goldfish, *Carassius auratus*. *J. Comp. Neurol.* **356,** 72–82.

King, J. A., Dufour, S., Fontaine, Y. A. and Millar, R. P. (1990). Chromatographic and immunological evidence for mammalian GnRH and chicken GnRH II in eel *Anguilla anguilla* brain and pituitary. *Peptides* **11,** 507–514.

Klungland, H., Lorens, J. B., Andersen, O., Kisen, G. O. and Alestrom, P. (1992). The Atlantic salmon prepro-gonadotropin releasing hormone gene and mRNA. *Mol. Cell. Endocrinol.* **84,** 167–174.

Kobayashi, M., Amano, M., Hasegawa, Y., Okuzawa, K. and Aida, K. (1992). Effects of olfactory tract section on brain GnRH distribution, plasma gonadotropin levels, and gonadal stage in goldfish. *Zool. Sci.* **9,** 765–773.

Kobayashi, M., Amano, M., Kim, M. H., Furukawa, K., Hasegawa, Y. and Aida, K. (1994). Gonadotropin-releasing hormones of terminal nerve origin are not essential to ovarian development and ovulation in goldfish. *Gen. Comp. Endocrinol.* **95,** 192–200.

Kobayashi, M., Amano, M., Kim, M. H., Yoshiura, Y., Sohn, Y. C., Suetake, H. and Aida, K. (1997). Gonadotropin-releasing hormone and gonadotropin in goldfish and masu salmon. *Fish Physiol. Biochem.* **17,** 1–8.

Kudo, H., Hyodo, S., Ueda, H., Hiroi, O., Aida, K., Urano, A. and Yamauchi, K. (1996). Cytophysiology of gonadotropin-releasing-hormone neurons in chum salmon (*Oncorhynchus keta*) forebrain before and after upstream migration. *Cell. Tissue Res.* **284,** 261–267.

Lin, X. W. and Peter, R. E. (1996). Expression of salmon gonadotropin-releasing hormone (GnRH) and chicken GnRH-II precursor messenger ribonucleic acids in the brain and ovary of goldfish. *Gen. Comp. Endocrinol.* **101,** 282–296.

Lin, X. W. and Peter, R. E. (1997). Cloning and expression pattern of a second [$His^5Tryp^7Tyr^8$]gonadotropin-releasing hormone (chicken GnRH-II) mRNA in goldfish: Evidence for two distinct genes. *Gen. Comp. Endocrinol.* **107,** 262–272.

Miyamoto, K., Hasegawa, Y., Nomura, M., Igarashi, M., Kangawa, K. and Matsuo, H. (1984). Identification of the second gonadotropin-releasing hormone in chicken hypothalamus: Evidence that gonadotropin secretion is probably controlled by two distinct gonadotropin-releasing hormones in avian species. *Proc. Natl. Acad. Sci. USA* **81,** 3874–3878.

Montero, M., Lebelle, N., King, J. A., Millar, R. P. and Dufour, S. (1995). Differential regulation of the two forms of gonadotropin-releasing hormone (mGnRH and cGnRH-II) by sex steroids in the European female silver eel (*Anguilla anguilla*). *Neuroendocrinol.* **61,** 525–535.

Montero, M., Vidal, B., King, J. A., Tramu, G., Vandesande, F., Dufour, S. and Kah, O. (1994). Immunocytochemical localization of mammalian GnRH (gonadotropin-releasing hormone) and chicken GnRH-II in the brain of the European silver eel (*Anguilla anguilla* L). *J. Chem. Neuroanat.* **7,** 227–241.

Ngamvongchon, S., Lovejoy, D. A., Fischer, W. H., Craig, A. G., Nahorniak, C. S., Peter, R. E., Rivier, J. E. and Sherwood, N. M. (1992). Primary structure of two forms of gonadotropin-releasing hormone, one distinct and one conserved, from catfish brain. *Mol. Cell. Neurosci.* **3,** 17–22.

Oka, Y. (1997). GnRH neuronal system of fish brain as a model system for the study of peptidergic neuromodulation. *In* "GnRH neurons-Gene to Behavior" (I. S. Parhar and Y. Sakuma, Eds.), pp. 245–276. Brain Shuppan, Tokyo.

Oka, Y. and Matsushima, T. (1993). Gonadotropin-releasing hormone (GnRH)-immunoreactive terminal nerve cells have intrinsic rhythmicity and project widely in the brain. *J. Neurosci.* **13,** 2161–2176.

Okuzawa, K., Amano, M., Aida, K., Hasegawa, Y., Tanaka, H. and Kagawa, H. (1993). Chromatographic and immunological identification of gonadotropin-releasing hormone in five marine teleosts. *Fish Physiol. Biochem.* **12,** 337–345.

Okuzawa, K., Amano, M., Kobayashi, M., Aida, K., Hanyu, I., Hasegawa, Y. and Miyamoto, K. (1990). Differences in salmon GnRH and Chicken GnRH-II contents in discrete brain areas of male and female rainbow trout according to age and stage of maturity. *Gen. Comp. Endocrinol.* **80,** 116–126.

Okuzawa, K., Araki, K., Tanaka, H., Kagawa, H. and Hirose, K. (1994a). Molecular cloning of a cDNA encoding the prepro-salmon gonadotropin-releasing hormone of the red seabream. *Gen. Comp. Endocrinol.* **96,** 234–242.

Okuzawa, K., Furukawa, K., Aida, K. and Hanyu, I. (1994b). The effect of water temperature on gonadotropin-releasing hormone contents in the discrete brain areas and pituitary of male honmoroko *Gnathopogon caerulescens*. *Fish. Sci.* **60,** 155–158.

Okuzawa, K., Granneman, J., Bogerd, J., Goos, H. J. T., Zohar, Y. and Kagawa, H. (1997). Distinct expression of GnRH genes in the red seabream brain. *Fish Physiol. Biochem.* **17,** 71–79.

Parhar, I. S. (1997). GnRH in tilapia: three genes, three origins and their roles. *In* "GnRH neurons-Gene to Behavior" (I. S. Parhar and Y. Sakuma, Eds.), pp. 99–122. Brain Shuppan, Tokyo.

Parhar, I. S., Iwata, M., Pfaff, D. W. and Schwanzel-Fukuda, M. (1995). Embryonic development of gonadotropin-releasing hormone neurons in the sockeye salmon. *J. Comp. Neurol.* **362,** 256–270.

Penlington, M. C., Williams, M. A., Sumpter, J. P., Rand-Weaver, M., Hoole, D. and Arme, C. (1997). Isolation and characterisation of mRNA encoding the salmon- and chicken-II type gonadotrophin-releasing hormones in the teleost fish *Rutilus rutilus* (Cyprinidae). *J. Mol. Endocrinol.* **19,** 337–346.

Peter, R. E., Trudeau, V. L. and Sloley, B. D. (1991). Brain regulation of reproduction in teleosts. *Bull. Inst. Zool., Academia Sinica, Monograph* **16,** 89–118.

Peter, R. E., Yu, K. L., Marchant, T. A. and Rosenblum, P. M. (1990). Direct neural regulation of the teleost adenohypophysis. *J. Exp. Zool.* **4,** 84–89.

Powell, J. F. F., Fischer, W. H., Park, M., Craig, A. G., Rivier, J. E., White, S. A., Francis, R. C., Fernald, R. D., Licht, P., Warby, C. and Sherwood, N. M. (1995). Primary structure of solitary form of gonadotropin-releasing hormone (GnRH) in cichlid pituitary; three forms of GnRH in brain of cichlid and pumpkinseed fish. *Reg. Pept.* **57,** 43–53.

Powell, J. F. F., Krueckl, S. L., Collins, P. M. and Sherwood, N. M. (1996a). Molecular forms of GnRH in three model fishes: rockfish, medaka and zebrafish. *J. Endocrinol.* **150,** 17–23.

Powell, J. F. F., Reska-Skinner, S. M., Prakash, O. M., Fischer, W. H., Park, M., Rivier, J. E., Craig, A. G., Mackie, G. O. and Sherwood, N. M. (1996b). Two forms of gonadotropin-releasing hormone in a protochordate and the evolutionary implications. *Proc. Natl. Acad. Sci. USA* **93,** 10461–10464.

Powell, J. F. F., Standen, E. M., Carolsfeld, J., Borell, M. I., Gazola, R., Fischer, W. H., Park, M., Craig, A. G., Warby, C. M., Rivier, J. E., Val-Sella, M. V. and Sherwood, N. M. (1997). Primary structure of three forms of gonadotropin-releasing hormone (GnRH) from the pacu brain. *Reg. Pept.* **68,** 189–195.

Powell, J. F. F., Zohar, Y., Elizur, A., Park, M., Fischer, W. H., Craig, A. G., Rivier, J. E., Lovejoy, D. A. and Sherwood, N. M. (1994). Three forms of gonadotropin-releasing hormone characterized from brains of one species. *Proc. Natl. Acad. Sci. USA* **91,** 12081–12085.

Prat, F., Sumpter, J. P. and Tyler, C. R. (1996). Validation of radioimmunoassays for two salmon gonadotropins (GTH I and GTH II) and their plasma concentrations throughout the reproductive cycle in male and female rainbow trout (*Oncorhynchus mykiss*). *Biol. Reprod.* **54,** 1375–1382.

Quérat, B. (1995). Structural relationships between "fish" and tetrapod gonadotropins. *In* "Proceedings of the Fifth International Symposium on the Reproductive Physiology of Fish " (F. W. Goetz and P. Thomas, Eds.), pp. 7–9. Fish Symposium 95, Austin.

Schulz, R. W., Bogerd, J., Bosma, P. T., Peute, J., Rebers, F. E. M., Zandbergen, M. A. and Goos, H. J. T. (1995). Physiological, morphological, and molecular aspects of gonadotropins in fish with special reference to the African catfish, *Clarias gariepinus*. *In* "Proceedings of the Fifth International Symposium on the Reproductive Physiology of Fish " (F. W. Goetz and P. Thomas, Eds.), pp. 2–6. Fish Symposium 95, Austin.

Senthilkumaran, B., Okuzawa, K., Gen, K., Ookura, T. and Kagawa, H. (1998). Distribution and seasonal variations in levels of three native GnRHs in the brain and pituitary of perciform Fish. *J. Neuroendocrinol.* In press.

Sherwood, N., Eiden, L., Brownstein, M., Spiess, J., Rivier, J. and Vale, W. (1983). Characterization of a teleost gonadotropin-releasing hormone. *Proc. Natl. Acad. Sci. USA* **80,** 2794–2798.

Sherwood , N. M., Grier, H. J., Warby, C., Peute, J. and Taylor, G. (1993). Gonadotropin-releasing hormones, including a novel form, in snook *Centropomus undecimalis*, in comparison with forms in black sea bass *Centropristis striata*. *Reg. Pept.* **46,** 523–534.

Sherwood, N. M., Schalburg, K. and Lescheid, D. W. (1997). Origin and evolution of GnRH in vertebrates and invertebrates. *In* "GnRH neurons-Gene to Behavior" (I. S. Parhar and Y. Sakuma, Eds.), pp. 3–25. Brain Shuppan, Tokyo.

Somoza, G. M., Stéfano, A., D'Eramo, J. E., Canosa, L. F. and Fridman, O. (1994). Immunoreactive GnRH suggesting a third form of GnRH in addition to cIIGnRH and sGnRH in the brain and pituitary gland of *Prochilodus lineatus* (Characiformes). *Gen. Comp. Endocrinol.* **94**, 44–52.

Stefano, A. V., Canosa, L. F., D'Eramo, J. L., Fridman, O., Affanni, J. M. and Somoza, G. M. (1997). GnRH molecular variants in the brain and pituitary gland of pejerrey, *Odontesthes bonariensis* (Atherniformes). Immunological and chromatographic evidence for the presence of a novel molecular variant. *Comp. Biochem. Physiol.* **118C**, 335–345.

Suetake, H., Kim, M. H., Ashihara, M., Yoshiura, Y., Kikuchi, K., Kobayashi, M. and Aida, K. (1995). Molecular cloning of a cDNA that encodes the precursor to a salmon-type gonadotropin-releasing hormone in the goldfish. *Proc. Jpn. Soc. Comp. Endocrinol.* **10**, 46.

Suetake, H., Yoshiura, Y. and Aida, K. (1997). Structural differences in the 5'- flanking regions of two types of the salmon gonadotropin-releasing hormone genes in goldfish (*Carassius auratus*). *Proc. Jpn. Soc. Comp. Endocrinol.* **12**, 12.

Suzuki, K., Kanamori, A., Nagahama, Y. and Kawauchi, H. (1988). Development of salmon GTH I and GTH II radioimmunoassays. *Gen. Comp. Endocrinol.* **71**, 459–467.

Suzuki, M., Hyodo, S., Kobayashi, M., Aida, K. and Urano, A. (1992). Characterization and localization of mRNA encoding the salmon-type gonadotrophin-releasing hormone precursor of the masu salmon. *J. Mol. Endocrinol.* **9**, 73–82.

Swanson, P. (1991). Salmon gonadotropins: Reconciling old and new idea. *In* "Reproductive Physiology of Fish" (A. P. Scott, J. P. Sumpter, D. E. Kime, and M. S. Rolfe, Eds.), pp. 2–7. FishSymp 91, Sheffield.

Swanson, P., Bernard, M., Nozaki, M., Suzuki, K., Kawauchi, H. and Dickhoff, W. W. (1989). Gonadotropins I and II in juvenile coho salmon. *Fish Physiol. Biochem.* **7**, 169–176.

Tensen, C., Okuzawa, K., Blomenröhr, B., Rebers, F., Leurs, R., Bogerd, J., Schulz, R. and Goos, H. (1997). Distinct efficacies for two endogenous ligands on a single cognate gonadoliberin receptor. *Eur. J. Biochem.* 243, 134–140.

Weber, G. M., Powell, J. F., Park, M., Fischer, W. H., Craig, A. G., Rivier, J. E., Nanakorn, U., Parhar, I. S., Ngamvongchon, S., Grau, E. G. and Sherwood, N. M. (1997). Evidence that gonadotropin-releasing hormone (GnRH) functions as a prolactin-releasing factor in a teleost fish (*Oreochromis mossambicus*) and primary structures for three native GnRH molecules. *J. Endocrinol.* **155**, 121–132.

White, S. A., Kasten, T. L., Bond, C. T., Adelman, J. P. and Fernald, R. D. (1995). Three gonadotropin-releasing hormone genes in one organism suggest novel roles for an ancient peptide. *Proc. Natl. Acad. Sci. USA* **92**, 8363–8367.

Yu, K. L., Sherwood, N. M. and Peter, R. E. (1988). Differential distribution of two molecular forms of gonadotropin-releasing hormone in discrete brain areas of goldfish (*Carassius auratus*). *Peptides* **9**, 625–630.

Zandbergen, M. A., Kah, O., Bogerd, J., Peute, J. and Goos, H. J. T. (1995). Expression and distribution of two gonadotropin-releasing hormones in the catfish brain. *Neuroendocrinol.* **62**, 571–578.

7

GONADOTROPIN-RELEASING HORMONE AS A PARACRINE REGULATOR OF OVARIAN FUNCTION

Hamid R. Habibi

Department of Biological Sciences
University of Calgary
Calgary, Alberta, Canada T2N 1N4

1. INTRODUCTION

Functional diversity of gonadotropin-releasing hormone (GnRH) is well documented in vertebrates. Both GnRH and GnRH receptor (GnRH-R) genes are expressed in a variety of reproductive and nonreproductive tissues including ovarian follicles. Support for the involvement of GnRH as a paracrine/autocrine regulator of ovarian function is provided by observations concerning presence and regulation of GnRH and GnRH-R in the ovary as well as direct GnRH actions on resumption of oocyte meiosis and follicular steroidogenesis in mammals, fish and other vertebrates. This review focuses on the functional significance of GnRH in the ovary, emphasizing the observations made in goldfish and various teleost species. Recent studies carried out in goldfish provides a framework for better understanding of the physiological significance of GnRH peptides in the ovary.

2. STRUCTURAL AND FUNCTIONAL DIVERSITY OF GnRH

GnRH is named for its key regulatory role in vertebrate reproduction and control of gonadotropin (GTH) secretion. Gene duplication and structural changes over the past 500 million years resulted in evolution of multiple molecular forms of GnRH. To date, the primary structure of eleven GnRH variants have been elucidated (Table 1). Two or more GnRH forms are expressed in species from all vertebrate classes (Millar and King, 1994; King and Millar, 1995; Millar et al., 1997) including placental mammals (White et al., 1998; Kasten et al., 1996; Lescheid et al., 1997). The most ubiquitous form of GnRH in vertebrates is [His^5, Trp^7, Tyr^8]-GnRH which was initially identified in the brain of chicken (cGnRH-II) (Table 1). The presence of cGnRH-II has been demonstrated in the

Neural Regulation in the Vertebrate Endocrine System, edited by Prasada Rao and Peter, Kluwer Academic / Plenum Publishers, New York, 1999.

Table 1. Primary structure of native GnRH peptides

	1	2	3	4	5	6	7	8	9	10
	pGlu	His	X	Ser	X	X	X	X	Pro	Gly-NH2
Mammalian GnRH	pGlu	His	Trp	Ser	Tyr	Gly	Leu	Arg	Pro	Gly-NH2
Chicken-I GnRH	pGlu	His	Trp	Ser	Tyr	Gly	Leu	Gln	Pro	Gly-NH2
Chicken-II GnRH	pGlu	His	Trp	Ser	His	Gly	Trp	Tyr	Pro	Gly-NH2
Salmon GnRH	pGlu	His	Trp	Ser	Tyr	Gly	Trp	Leu	Pro	Gly-NH2
Catfish GnRH	pGlu	His	Trp	Ser	His	Gly	Leu	Asn	Pro	Gly-NH2
Dogfish GnRH	pGlu	His	Trp	Ser	His	Gly	Trp	Leu	Pro	Gly-NH2
Lamprey-I GnRH	pGlu	His	Tyr	Ser	Leu	Glu	Trp	Lys	Pro	Gly-NH2
Lamprey-III GnRH	pGlu	His	Trp	Ser	His	Asp	Trp	Lys	Pro	Gly-NH2
Seabream GnRH	pGlu	His	Trp	Ser	Tyr	Gly	Leu	Ser	Pro	Gly-NH2
Tunicate-I GnRH	pGlu	His	Trp	Ser	Asp	Tyr	Phe	Lys	Pro	Gly-NH2
Tunicate-II GnRH	pGlu	His	Trp	Ser	Leu	Cys	His	Ala	Pro	Gly-NH2

brain of many vertebrates along with other form(s) of GnRH (Millar et al., 1994, 1997; King and Millar, 1995). The goldfish brain contains two forms of GnRH, [Trp^7, Leu^8]-GnRH (salmon GnRH; sGnRH) and cGnRH-II (Peter and Yu, 1997; Lin and Peter, 1996). In some species, only one molecular form of GnRH is secreted into the pituitary and has hypophysiotrophic activity and the other form, mostly cGnRH-II, seems to act as a neurotransmitter (Millar et al., 1997). In goldfish both sGnRH and cGnRH-II are secreted into the pituitary and stimulate gonadotropin hormones as well as growth hormone secretion and gene expression (Peter and Yu, 1997; Chang et al., 1997; Mahmoud et al., 1996; Khakoo et al., 1994).

Functional diversity of GnRH has been investigated in fish and other vertebrates, and there is evidence for the presence of both GnRH and GnRH receptors in various peripheral tissues. The existence of specific GnRH binding sites was demonstrated in goldfish ovary (Pati and Habibi, 1993a), testis (Pati and Habibi, 1993b), brain, liver and kidney (Habibi and Pati, 1993b). Further studies demonstrated the presence of a single class of GnRH binding sites in the liver of a number of species including goldfish, trout, frog, turtle, chicken, rat, rabbit (Habibi and Pati, 1993b) and human (Pati and Habibi, 1995; Pati et al., 1996). Subsequent studies also demonstrated expression of GnRH and/or GnRH-R genes in various non-reproductive human tissues such as liver, heart, skeletal muscle, kidney, placenta (Kakar and Jennes, 1995) as well as gonads and tumor cells (see Sealfon et al, 1997; Pati and Habibi, 1995 for review).

3. OVARIAN GnRH

There is now clear evidence for the presence of GnRH in the ovary of different vertebrate species. A large number of earlier studies provided, at best, an indication of the presence of ovarian compounds with GnRH-like activity by demonstrating the ability of ovarian extracts to 1) stimulate GTH production in the pituitary and its inhibition by GnRH antagonists; 2) immunoreact with various GnRH antisera; 3) bind to GnRH receptors in the pituitary and/or ovary; 4) act directly at the level of ovary by altering steroidogenesis and resumption of meiosis in cultured follicle-enclosed oocytes *in vitro* and its inhibition by GnRH antagonists. This information to a large extent has been reviewed in a number of papers dealing with paracrine control of ovarian function (Tsafriri, 1988; Aten, et al., 1986, 1987, 1989; Pati and Habibi, 1998; Habibi et al., 1994; Nabissi, et al., 1997; Hsueh and Jones, 1981, Hsueh and Schaefer, 1985; Knecht, et al., 1985; Habibi and Pati, 1993a; Chieffi et al., 1991). A number of studies attempted to purify the compounds with GnRH-like activity by HPLC for further biochemical characterization, and there is indication of the presence of various GnRH forms in the ovary. In a recent study the primary structure of a GnRH peptide with identical amino acid sequence to that of sGnRH was demonstrated in the goldfish ovary (Pati and Habibi, 1998). This is the first demonstration of an ovarian GnRH, based on amino acid sequencing, as well as clear evidence demonstrating GnRH-like biological activity, receptor binding activity and immunoreactivity of the goldfish ovarian extract (Pati and Habibi, 1998).

Following molecular characterization of complementary DNA (cDNA) sequences encoding the precursor of different forms of brain GnRH (Millar and King, 1994; Millar et al., 1997), a number of investigators have demonstrated the expression of various GnRH genes in the ovary of mammalian and non mammalian species (Peng et al., 1994; Goubau et al., 1992; Oikawa et al., 1990; Sakakibara, 1989; Lin and Peter, 1996). The general structure of GnRH genes is to a large extent conserved among various species, sharing the motif of encoding a signal peptide, a GnRH, and a GnRH-associated peptide (GAP). In terms of sequence, while the GnRH moiety is highly conserved, the GAP and signal peptide regions are not well conserved with less than 50% similarity among different forms.

In goldfish, both sGnRH and cGnRH-II mRNA can be detected in the ovary by northern analysis and PCR (Lin and Peter, 1996; Mirhadi and Habibi, unpublished). However, only sGnRH could be detected in the goldfish ovarian extract, with no evidence of cGnRH-II presence (Pati and Habibi, 1998). This is in accord with the observation that sGnRH gene is expressed more strongly than cGnRH-II in the goldfish ovary (Lin and Peter, 1996; Mirhadi and Habibi, unpublished). Recent studies also demonstrated the expression of three GnRH molecular forms including, salmon, chicken-II and sea bream-GnRH (Table 1) in the seabream ovary (M. Nabissi, L. Soverchia, A. Polzonetti-Magni1 and H.R. Habibi, unpublished). Evidence was provided by these authors for differential splicing of chicken-II and sea bream-GnRH in the seabream ovary.

4. OVARIAN GnRH RECEPTORS

Earlier studies in mammals demonstrated the presence of specific GnRH binding sites in both ovary and testis (for review see Hsueh and Schaefer, 1985, Hsueh and Jones, 1981; Knecht et al., 1985). Similar studies also demonstrated the presence of ovarian GnRH-Rs in a number of fish species including goldfish (Pati and Habibi, 1993a), common carp (Pati and Habibi, 1992), African catfish (Habibi et al., 1994), seabream (Nabissi et al., 1997), and lamprey (Deragon et al., 1997). Goldfish ovary was shown to contain two classes of GnRH binding sites, a high affinity/low capacity site and a low affinity/high capacity sites (Pati and Habibi, 1993a) similar to the characteristic of GnRH-Rs in the goldfish pituitary (Habibi and Peter, 1991; Habibi et al., 1990). However, a single class of high affinity GnRH binding site was observed in the ovary of carp, catfish, seabream, and lamprey (see Pati and Habibi, 1998 for review).

GnRH-Rs have recently been cloned in a number of mammalian species (Sealfon et al., 1997) as well as in African catfish (Tensen et al., 1997) and *Xenopus* (Troskie et al., 1997), resulting in a better understanding of its conformation and molecular structure. In mammals, most GnRH-R cDNAs encode a 328 amino acid protein which contains two to three potential N-linked glycosylation sites and belongs to the seven membrane spanning G-protein coupled receptor (GCRP) family. Like other GPCRs (Probst, 1992), GnRH-Rs are single-chain proteins containing seven hydrophobic alpha-helical membrane-spanning domains (TM1-TM7) connected by extracellular (E1-E3) and intracellular (I1-I3) hydrophilic loops. Other features of GnRH-Rs common to GPCRs include the presence of a pair of conserved cysteine (Cys) residues that mediate the formation of an intracellular disulfide bond between E1 and E2, and several conserved helix breaking proline (Pro) residues in TM2, TM4, TM5, TM6 and TM7 which may play a role in signal propagation (Sealfon et al., 1997). A number of other amino acids conserved among GPCRs are also found in GnRH-Rs. These include Asn^{53} (TM1), Trp^{164} (TM4), Ser^{167} (TM4), Asn^{315} (TM7) and Tyr^{323} (TM7) (Sealfon et al., 1997). Despite high homology with GPCRs, GnRH-Rs detected in mammals have a few unique features including: 1) the absence of a cytoplasmic tail found in all other GRCPs; 2) the reciprocal interchange of the normally highly conserved residues Asn^{87} and Asp^{318}; 3) the change of the Asp-Arg-Tyr motif, a conserved sequence at the intracellular junction of the TM3 region, to Asp-Arg-Ser, which creates a potential phosphorylation site (Millar et al., 1994; Sealfon et al., 1997). Unlike the mammlian GnRH-R, the catfish (Tensen et al., 1997) and *Xenopus* (Troskie et al., 1997) GnRH-R molecules have an intracellular carboxy-terminal tail with an overall similarity of 47% and 38% to human GnRH-R, respectively. While sequence of other nonmammalian GnRH-Rs have not yet been published, preliminary information indicates that GnRH-R molecules in chicken and goldfish also have C-terminal tail with phosphorylation consensus sites (Millar et al., 1997) similar to catfish (Tensen et al., 1997) and *Xenopus* (Troskie et al., 1997).

The cDNA for pituitary GnRH-R was used to isolate ovarian GnRH-R in a number of species. Essentially, the GnRH-R sequence was found to be the same as the pituitary GnRH-R (Fraser et al., 1996; Bauer-Dantoin and Jameson, 1995; Kogo et al., 1995; Minaretzis et al., 1995; Whitelaw et al., 1995; Kakar et al., 1992, 1994a,b; Peng et al., 1994; Moumni et al., 1994). The GnRH-R expression was detected in granulosa cells at varying stages of follicular development and in the corpus luteum (Whitelaw et al., 1995; Leung and Peng, 1996). There is also evidence for the presence of GnRH-R variants in the pituitary, ovary and other peripheral tissues. In this regard, Kakar and co workers (1994b) observed four different GnRH-R transcripts sizes of 5.0, 4.5, 2.5 and 1.3 kb in rat pituitary, ovary and testis by Northern blot analysis, of which the 5.0 kb form was most abundant. No published information is available on ovarian GnRH-R structure in nonmammalian vertebrates.

5. REGULATION OF OVARIAN GnRH PRODUCTION

The regulation of ovarian GnRH production is poorly understood, compared to extensive information available on neuroendocrine control of brain and hypothalamic GnRH synthesis and release in vertebrates (Gore and Roberts, 1997; Peter and Yu, 1997). Most of the studies are based on determination of the steady state mRNA level for ovarian GnRHs because of their low abundance in the ovary, particularly for certain forms of GnRH molecules. Attempts have also been made to investigate immunoreactive GnRH peptides in the ovary (see Pati and Habibi, 1998 for review). However, evidence based on immunoreactivity does not always agree with analysis of gene expression because of low concentrations of ovarian GnRH peptides and antibody cross reactivity with different molecular forms of GnRH. While the problem of antibody crossreactivity can be minimized by analysing the ovarian extracts first by HPLC, the low abundance of GnRH peptides makes this method practically impossible for quantitative determination of GnRH peptides. For example, in the goldfish ovary, both sGnRH and cGnRH-II genes are expressed as determined by Northern analysis and PCR (Lin and Peter, 1996; Mirhadi and Habibi, unpublished). However, no cGnRH-II could be detected by immunoreactivity in combination with HPLC in the extract obtained from approximately 1 Kg of goldfish ovary, while sGnRH was present as confirmed by amino acid analysis and sequencing (Pati and Habibi, 1998). This is the first demonstration that a molecular form of GnRH is transcribed, translated and processed in the same way as hypothalamic GnRH which is an important observation in view of the reports concerning alternative processing of GnRH gene products in the ovary. In the rat ovary, GnRH transcripts were found to differ from those of hypothalamic GnRH since a greater proportion of ovarian GnRH transcripts contained intronic sequences (Goubau et al., 1992). Furthermore, the major transcription start utilized in the ovary was found to be different from that used in the hypothalamus (Goubau et al., 1992). These investigators found low expression of GnRH gene in the rat ovary and were only able to demonstrate fully processed GnRH gene transcripts by RT-PCR and not with Northern blot (Goubau et al., 1992). These problems were not noted by other investigators studying GnRH gene expression in the ovary (Oikawa et al., 1990; Sakakibara, 1989; Leung and Peng, 1996). In the seabream ovary, three forms of GnRH genes coding for sGnRH, cGnRH-II and seabream GnRH (Table 1) are expressed and undergo differential splicing (M. Nabissi, L. Soverchia, A. Polzonetti-Magni1 and H.R. Habibi, unpublished). Similar processes were not noted for GnRH transcripts in the seabream brain (Gothilf et al., 1995, 1996). In goldfish, Pati and Habibi (1998) have established that sGnRH is produced in the ovary and is likely to play a physiological role in the regulation of ovarian function. The physiological significance of cGnRH-II, however, remains to be determined since it could not be detected in the goldfish ovary despite the presence of cGnRH-II transcripts. It is possible, that levels of GnRH peptides available in the ovary are regulated at transcriptional, posttranscriptional, and posttranslational levels as demonstrated to be the case for hypothalamic regulation of GnRH production (Gore and Roberts, 1997). In cultured human granulosa-luteal cells, GnRH treatment at 1 nM level was found to increase GnRH mRNA level, while treatment with hCG (1 IU/ml) was without effect (Peng, et al., 1994). In cultured goldfish ovarian follicles, treatment with hCG was found to stimulate both sGnRH and cGnRH-II transcripts with different sensitivity; low doses stimulated sGnRH and higher doses stimulated cGnRH-II (Mirhadi and Habibi, unpublished). Administration of 17ß -estradiol, *in vitro,* was also found to stimulate sGnRH transcript in the goldfish follicles (Mirhadi and Habibi, unpublished). These findings demonstrate that GnRH peptides are produced in the ovary at low concentrations, and their production is regulated by pituitary and ovarian hormones.

6. CONTROL OF OVARIAN GnRH RECEPTORS

There are a number of reports concerning regulation of GnRH-R in the ovary of various mammalian species. These studies have used *in situ* hybridization and quantitative PCR for measurement of ovarian GnRH-R mRNA levels. In general, the level of GnRH-R transcripts was found to be significantly lower in the ovary than pituitary (Kakar et al., 1994a,b; Fraser et al., 1996; Minaretzis et al., 1995). Significant variation in GnRH-R mRNA levels were observed in the rat ovary with respect to the degree of follicular development as well as estrus cycle stage. Higher levels of GnRH-R mRNA were found at proestrus and estrus stages compared to metestrus (Bauer-Dantoin and Jameson, 1995). Furthermore, GnRH-R gene expression was found to be greatest in the granulosa cells of Graafian and atretic follicles, with lower levels of expression present in preantral and small antral follicles and corpora lutea in rat (Bauer-Dantoin and Jameson, 1995). Moreover, GnRH-R mRNA level was found to undergo substantial variation across the 4-day rat estrus cycle, with mRNA levels increasing on the day of proestrus coincident with the preovulatory gonadotropin surge (Bauer-Dantoin and Jameson, 1995). Administration of FSH or human menopausal gonadotropin to hypophysectomized rats was found to alter the distribution of ovarian GnRH-R expression. Higher levels of GnRH-R transcripts were detected in medium-sized follicles not expressing LH receptor mRNA, and those showing signs of follicular atresia (Whitelaw et al., 1995). These investigators also found expression of GnRH-R mRNA in healthy preovulatory rat follicles (Whitelaw et al., 1995). Higher expression of GnRH-R gene in atretic follicles was also observed by Kogo et al. (1995). These investigators found no signals in the granulosa cells of healthy small, preantral, or early antral follicles. High levels of GnRH-R transcripts were detected in the Graafian and preovulatory granulosa cells with no expression of GnRH-R in the cumulus oophorus cells (Kogo et al., 1995). Pretreatment with different combinations of gonadotropins yielded different ovarian histology, but had no influence on the localization of hybridization signals (Kogo et al., 1995). In cultured rat granulosa cells, treatment with LH reduced GnRH-R mRNA levels in a dose-related fashion, with maximal inhibition seen at 100–1000 ng, whereas FSH was without effect (Olofsson et al., 1995). In GnRH-treated granulosa cells, a dose-dependent increase in GnRH-R mRNA levels was observed, reaching maximum level of 2.4-fold at 10(-6) M (Olofsson et al., 1995). In human, measurable levels of GnRH-R mRNA were found by ribonuclease protection assay in 2 of 10 ovarian follicular cells obtained from women whose ovarian hyperstimulation involved a GnRH agonist (Minaretzis et al., 1995).

Similar studies on GnRH-R transcripts have not been carried out in the ovary of nonmammalian vertebrates. However, evidence based on receptor binding studies suggest that GnRH-R might undergo hormonally-, and developmentally-dependent regulation in the fish ovary. In this regard, mature and medium size goldfish ovarian follicles contain two classes of GnRH binding sites, a high affinity/low capacity site and a low affinity/high capacity site. A single class of low affinity GnRH binding site was observed in the immature goldfish ovary which lack high affinity GnRH-Rs (Pati and Habibi, 1993a).

These findings demonstrate that the expression and localization of ovarian GnRH-R change during the ovulatory cycle in mammals, and support the view that GnRH might be involved in the induction of follicular atresia, as well as follicular development, ovulation, and luteinization. Furthermore, the observed regulation of GnRH-R gene expression in granulosa cells provides support for the physiological importance of ovarian GnRH peptides during the periovulatory period.

7. GnRH ACTIVITY IN THE OVARY

The demonstration of ovarian GnRH and GnRH-R gene expression and their regulation by various factors in the ovary provide clear evidence that GnRH peptides are important autocrine/paracrine factors involved in the regulation of ovarian function. In mammals, GnRH peptides have been shown to have various effects on follicular steroidogenesis and resumption of meiosis. Treatment with GnRH was shown to stimulate basal steroidogenesis, and also to inhibit gonadotropin induced cAMP and steroid production in a maturational stage-dependant manner (for review see Hsueh and Jones, 1981; Hsueh and Schaefer, 1985; Knecht et al., 1985; Leung and Steele, 1992). The actions of GnRH on ovary are through activation of specific GnRH receptors coupled to Gq and phospholipase C (Leung and Steele, 1992). Recent studies on isolated human granulosa cells demonstrated that GnRH actions on steroidogenesis are significantly enhanced in the presence of prostaglandin PGF2alpha (Vaananen et al., 1997)

Direct ovarian action of GnRH has also been investigated in fish and amphibians (for review see Habibi and Pati, 1993a; Chieffi et al., 1991). In goldfish, GnRH exerts both stimulatory and inhibitory actions on oocyte meiosis and follicular steroidogenesis, depending upon the presence or absence of GTH. In follicle-enclosed goldfish oocytes, an agonist analog of sGnRH (sGnRH-A) was found to inhibit GTH-induced resumption of meiosis as indicated by germinal vesicle breakdown (GVBD) by approximately 30–40%, whereas sGnRH by itself stimulated GVBD (Habibi et al., 1988). Treatment with sGnRH-A was also found to impair GTH-induced testosterone production in goldfish follicles (Habibi et al., 1989). Administration of a number of GnRH variants including sGnRH, cGnRH-II, mammalian GnRH, chicken GnRH-I and lamprey GnRH-I were found to significantly stimulate GVBD in the follicle-enclosed goldfish oocytes (Habibi and Pati, 1993a; Pati, 1995). In the presence of gonadotropin hormone, sGnRH treatment significantly inhibited GTH-induced GVBD response, whereas cGnRH-II treatment was without an inhibitory effect (Habibi and Pati, 1993a; Pati, 1995). This suggests that goldfish ovary might contain GnRH receptor subtypes or mechanisms through which sGnRH and cGnRH-II could interact with the same receptor molecule in different ways leading to activation of alternate post-receptor mechanisms. Similar suggestions were also made based on the actions of sGnRH and cGnRH-II on the GTH-II release and transcription in the goldfish pituitary (Khakoo et al., 1994). This is consistent with a suggestion that rat luteal cells contain GnRH receptor subtypes coupled independently to phospholipase C or A_2 (Watanabe et al., 1990).

8. HYPOTHESIS UNDERLYING THE PHYSIOLOGICAL SIGNIFICANCE OF GnRH IN THE GOLDFISH OVARY

In view of the observed levels of sGnRH and cGnRH-II in the goldfish ovary (Pati and Habibi, 1998), it is likely that sGnRH might be the main form of ovarian GnRH peptide in this species (Pati and Habibi, 1998). One postulate is that ovarian sGnRH increases during mid to late recrudescence in goldfish approaching preovulatory stages, characterized by elevated levels of estrogen and increasing concentrations of maturational GTH (GTH-II) (Kobayashi et al., 1986). These conditions were shown to be stimulatory on the sGnRH transcript level in the goldfish ovary (Mirhadi and Habibi, unpublished). The presence of sGnRH at that stage could block GTH-induced meiosis in larger follicles and prevent premature ovulation. This is important in a species like goldfish which is a batch spawner and depend on mechanisms for maintaining an ovary containing large number of preovulatory follicles. Higher concentrations of GTH-II and lower 17ß-estradiol levels which occur at the late stages of ovarian recrudescence could potentially result in lower expression of sGnRH in the goldfish ovary (Mirhadi and Habibi, unpublished) and allow synchronized ovulation, triggered by the ovulatory surge of GTH-II. This is a hypothesis to be tested by further investigation of sGnRH and cGnRH-II gene expression and their regulation in the goldfish ovary.

9. CONCLUDING REMARKS

There is clear and unambiguous evidence for the presence of GnRH and GnRH-R in the ovary of a number of vertebrates. The general structure of GnRH genes appears to be conserved among various species and there is evidence for the first time in goldfish that a molecular form of GnRH is transcribed, translated and processed in the same way as hypothalamic GnRH. However, levels of GnRH transcript in the ovary are significantly lower than that in the hypothalamus and are regulated by both hypophyseal and gonadal hormones. Similarly, the sequence of ovarian GnRH-R was found to be the same as the pituitary GnRH-R, and its transcript level in ovary is significantly lower than the pituitary. The GnRH-R is expressed in granulosa and luteal cells, and there is a change in its level and localization during the ovulatory cycle. Results of a number of studies in mammals indicate that GnRH may play a role in the control of follicular development, ovulation, and luteinization as well as induction of follicular atresia.

The demonstration of GnRH production, activity and GnRH-R gene regulation in the ovary strongly supports' the hypothesis that GnRH is an important physiological factor involved in the regulation of ovarian function during the preovulatory and postovulatory periods. A key to better understanding of the physiological significance of ovarian GnRH would be the elucidation of the mechanisms controlling GnRH and GnRH-R production in the ovary.

REFERENCES

Aten, R.F., Williams, A.T., Behrman, H.R., and Wolin, D.L. (1986). Ovarian gonadotropin-releasing hormone-like protein(s): Demonstration and characterization. *Endocrinol.* **118,** 961–967.

Aten, R.F., Ireland, J.J., Weems, C.W., and Behrman, H.R. (1987). Presence of gonadotropin-releasing hormone-like protein in bovine and ovine ovaries. Endocrinology 120, 1727–1733.

Aten, R.F., and Behrman, H.R. (1989). A gonadotropin-releasing hormone inhibitor from bovine ovaries: purification and identification as histone H2A. *J. Biol. Chem.* **264,** 11065–11071.

Bauer-Dantoin, A.C. and Jameson, J.L. (1995) Gonadotropin-releasing hormone receptor messenger ribonucleic acid expression in the ovary during the rat estrous cycle. *Endocrinol.* **136,** 4432–4438.

Chang, J.P., Van Goor, F., Lo, A., Johnson, J.D., Jobin, R.M. and Goldberg, J.I. (1997) Signal transduction in gonadotropin (GTH)-II secretion in goldfish pituitary cells. *In* "Proceedings of XIII Congress of Comparative Endocrinology, Yokohama Japan", (S. Kawashima and S. Kikuyama, Eds.), pp. 29–33. Monduzzi Editore, Bologna (Italy).

Chieffi, G., Pierantoni, R. and Fasano, S. (1991) Immunoreactive GnRH in hypothalamic and extrahypothalamic areas. *Int. Rev. Cytology* **127,** 1–55.

Deragon, K.L., Gazourian, L., Pati, D., Habibi, H.R. and Sower, S.A. (1997) Characteristics of GnRH binding in the gonads and Effects of lamprey GnRH-I and -III on reproduction in the adult seal lamprey. *Gen. Comp. Endocrinol.* **108,** 327–339.

Fraser, H.M., Sellar, R.E., Illingworth, P.J. and Eidne, K.A. (1996) GnRH receptor mRNA expression by in-situ hybridization in the primate pituitary and ovary. *Mol. Human Reprod.* **2,** 117–121.

Gore, A.C., and Roberts, J.L. (1997) Regulation of gonadotropin-releasing hormone gene expression in vivo and in vitro. *Frontiers in Neuroendocrinol.* **18,** 209–245.

Gothilf, Y., Munoz-Cueto, J.A., Sagrillo, C.A., Selmanoff, M., Chen, T.T., Kah, O., Elizur, A. and Zohar, Y. (1996) Three forms of gonadotropin-releasing hormone in a perciform fish (*Sparus aurata*): complementary deoxyribonucleic acid characterization and brain localization. *Biol. Reprod.* **55,** 636–645.

Gothilf, Y., Elizur, A., Chow, M., Chen, T.T. and Zohar, Y. (1995) Molecular cloning and characterization of a novel gonadotropin-releasing hormone from the gilthead seabream (*Sparus aurata*). *Molec. Marine Biol. Biotech.* **4,** 27–35.

Goubau, S., Bond, C.T., Adelman, J.P., Misra, V., Hynes, M.F., and Murphy, B.D. (1992). Partial characterization of gonadotropin-releasing hormone (GnRH) gene transcript in the rat ovary. *Endocrinol.* **130,** 3098–4000.

Habibi, H.R., Van Der Kraak, G., Bulanski, E. and Peter, R.E. (1988). Effects of teleost GnRH on reinitiation of oocyte meiosis in goldfish in vitro. *Amer. J. Physiol.* **255,** R268-R273.

Habibi, H.R., Van Der Kraak, G., Fraser, R., and Peter, R.E. (1989). Effect of a teleost GnRH analog on steroidogenesis by the follicle-enclosed goldfish oocytes, in vitro. *Gen. Comp. Endocrionol.* **76,** 95–105.

Habibi, H.R., Peter, R.E. and Hazum E. (1990) Photoaffinity labeling of pituitary gonadotropin-releasing hormone (GnRH) receptors in goldfish (*Carassius auratus*). *Biol. Reprod.* **43,** 1006–1011.

Habibi, H.R. and Peter R.E. (1991) Gonadotropin-releasing hormone (GnRH) receptors in teleosts. *In* "Reproductive Physiology of Fish", (Scott A.P., Sumpter J.P., Kime D.E. and Rolfe M., eds), pp,109–114. FishSymp 91, Sheffield.

Habibi, H.R., and Pati, D.P. (1993a). Endocrine and paracrine control of ovarian function: role of compounds with GnRH-like activity in goldfish. *In* "Cellular Communication in Reproduction", (F. Facchinetti, I. W. Henderson, R. Pierantoni, and A. M. Polzonetti-Magni, eds.), pp. 59–70. J. of Endocrinology Ltd, Bristol, U.K.

Habibi H.R. and Pati D. (1993b) Extrapituitary gonadotropin-releasing hormone (GnRH) binding sites in goldfish. *Fish Physiol Biochem.* **11,** 43–49.

Habibi, H.R., Pati, D., Ouwens, M., and Goos, H.J.Th. (1994). Presence of gonadotropin-releasing hormone (GnRH) binding sites and compounds with GnRH-like activity in the ovary of African catfish, *Clarias gariepinus*. *Biol. Reprod.* **50,** 643–652.

Hsueh, A.J.W., and Jones P.B.C. (1981) Extrapituitary actions of gonadotropin-releasing hormone. *Endocr. Rev.* **2,** 437–461.

Hsueh, A.J.W., and Schaefer J.M. (1985) Gonadotropin-releasing hormone as a paracrine hormone and neurotransmitter in extra-pituitary sites. *J. Steroid Biochem.* **23,** 757–764.

Kakar, S.S., and Jennes, L. (1995) Expression of gonadotropin-releasing hormone and gonadotropin-releasing hormone receptor mRNAs in various non-reproductive human tissues. *Cancer Letters,* **98,** 57–62,

Kakar, S.S., Musgrove, L.C., Devor, D.C., Sellers, J.C. and Neill, J.D. (1992) Cloning, sequencing, and expression of human gonadotropin releasing hormone (GnRH) receptor. *Bioch. Biophys. Res. Commu.* **189,** 289–295.

Kakar, S.S., Grantham, K., Musgrove, L.C., Devor, D., Sellers, J.C. and Neill, J.D. (1994a) Rat gonadotropin-releasing hormone (GnRH) receptor: tissue expression and hormonal regulation of its mRNA. *Mol. Cell. Endocrinol.* **101,** 151–157.

Kakar, S.S., Grizzle, W.E. and Neill, J.D. (1994b) The nucleotide sequences of human GnRH receptors in breast and ovarian tumors are identical with that found in pituitary. *Mol. Cell. Endocrinol.* **106,** 145–149.

Kasten, T.L., White, S.A., Norton, T.T., Bond, C.T., Adelman, J.P. and Fernald R.D. (1996) Characterization of two new preproGnRH mRNAs in the tree shrew: first direct evidence for mesencephalic GnRH gene expression in a placental mammal. *Gen. Comp. Endocrinol.* **104,** 7–19.

Khakoo, Z., Bhatia, A., Gedamu, L., and Habibi, H.R. (1994). Functional specificity for salmon gonadotropin-releasing hormone (GnRH) and chicken GnRH-II coupled to gonadotropin release and subunit messenger ribonucleic acid level in the goldfish pituitary. *Endocrinol.* **134,** 838–847.

King, J.A., and Millar, R.P. (1995) Evolutionary aspects of gonadotropin-releasing hormone and its receptor. *Cell. Mol. Neurobiol.* **15,** 5–23.

Knecht, M., Ranta, T., Feng, P., Shinohara, O. and Catt, K.J. (1985) Gonadotropin-releasing hormone as a modulator of ovarian function. *J. Steroid Biochem.* **23,** 771–777.

Kobayashi, M., Aida, K., Hanyu, I. (1986) Annual changes in plasma levels of gonadotropin and steroid hormones in goldfish. *Bull Jap Soc Sci Fish.* **52,** 1153–1158.

Kogo, H., Kudo, A,. Park, M.K., Mori, T. and Kawashima, S. (1995) In situ detection of gonadotropin-releasing hormone (GnRH) receptor mRNA expression in the rat ovarian follicles. *J. Exp. Zool.* **272,** 62–68.

Lescheid, D.W., Terasawa, E., Abler, L.A., Urbanski, H.F., Warby, C.M., Millar, R.P. and Sherwood, NM (1997) A second form of gonadotropin-releasing hormone (GnRH) with characteristics of chicken GnRH-II is present in the primate brain. *Endocrinol.* **138,** 5618–5629.

Leung, P.C.K., and Steele, G.L. (1992). Intracellular signaling in gonads. *Endocr. Rev.* **13,** 476–498.

Leung, P.C. and Peng, C. (1996) Gonadotropin-releasing hormone receptor: gene structure, expression and regulation. *Biol. Signals.* **5,** 63–69.

Lin, X.W., and Peter, R.E. (1996) Expression of salmon gonadotropin-releasing hormone (GnRH) and chicken GnRH-II precursor messenger ribonucleic acids in the brain and ovary of goldfish. *Gen Comp Endocrinol.* **101,** 282–296.

Mahmoud, S.S., Moloney, M.M. and Habibi, H.R. (1996) Cloning and Sequencing of the goldfish growth hormone cDNA. *Gen. Comp. Endocrinol.* **101,** 139–144.

Millar, R.P. and King, J.A. (1994) Plasticity and conservation in gonadotropin-releasing hormone structure and function. *In* "Perspectives in Comparative Endocrinology", (K.G. Davey, R.E. Peter, and S.S. Tobe, Eds)., pp, 129–136., National Research Council of Canada, Ottawa.

Millar, R.P., Troskie, B., Sun, Y-M., Ott, T., Wakefield, I., Myburch, D., Pawson, A., Davidson, J.S., Flanagan, C., Katz, A., Hapgood, J., Sealfon, S.C., Peter, R.E., Terasawa, E. and King, J.A. (1997) Plasticity in the structural and functional evolution of GnRH: a peptide for all seasons. *In* "Proceedings of XIII Congress of

Comparative Endocrinology, Yokohama Japan", (S. Kawashima and S. Kikuyama, Eds.), pp, 15–28. Monduzzi Editore, Bologna (Italy).

Minaretzis, D., Jakubowski, M., Mortola, J.F. and Pavlou, S.N. (1995) Gonadotropin-releasing hormone receptor gene expression in human ovary and granulosa-lutein cells. *J. Clin. Endocrinol. Metab.* **80,** 430–434.

Moumni, M., Kottler, M.L. and Counis, R. (1994) Nucleotide sequence analysis of mRNAs predicts that rat pituitary and gonadal gonadotropin-releasing hormone receptor proteins have identical primary structure. *Biochem. Biophys. Res. Comm.* **200,** 1359–1366.

Nabissi, M., Pati, D., Polzonetti-Magni, A., and Habibi H.R. (1997) Presence and activity of compounds with GnRH-like Activity in the Ovary of Seabream, *Sparus aurata. Am. J. Physiol.* **272,** R111-R117.

Oikawa, M., Dargan, C., Ny, T., and Hsueh, A.J.W. (1990) Expression of the gonadotropin-releasing hormone and prothymosin-a messenger ribonucleic acid in the ovary. *Endocrinol.* **127,** 2350–2356.

Olofsson, J.I., Conti, C.C. and Leung, P.C. (1995) Homologous and heterologous regulation of gonadotropin-releasing hormone receptor gene expression in preovulatory rat granulosa cells. *Endocrinol.* **136,** 974–980.

Pati, D. (1995). Role of GnRH and compounds with GnRH-like activity in the control of ovarian function in goldfish (*Carassius auratus*). Ph.D. dissertation, University of Calgary, Alberta, Canada.

Pati, D., and Habibi, H.R. (1998) Presence of salmon GnRH and compounds with GnRH-like activity in the ovary of goldfish. *Endocrinol.* **139,** 2015–2024.

Pati, D., Habibi, H.R. and Gedamu L. (1996) Inhibition of zinc-induced production of metallothionein mRNA by gonadotropin-releasing hormone in human hepatocarcinoma cell line HepG2. *Europ. J. Biochem.* **242,** 36–40.

Pati, D. and Habibi, H.R. (1995). Inhibition of human hepatocarcinoma cell proliferation by mammalian and fish gonadotropin-releasing hormones. *Endocrinol.* **136,** 75–84.

Pati, D. and Habibi, H.R. (1992) Characterization of gonadotropin-releasing hormone(GnRH) receptors in the ovary of common carp (*Cyprinus carpio*). *Can. J. Physiol. Pharmacol.* **70,** 268–274.

Pati, D. and Habibi, H.R. (1993a) Characterization of gonadotropin-releasing hormone receptors in goldfish ovary: Variation during follicular development. *Am. J. Physiol.* **264,** R227-R234.

Pati, D. and Habibi, H.R. (1993b) Gonadotropin-releasing hormone (GnRH) binding characteristics in the testis of goldfish (*Carassius auratus*). *J. Exp. Zool.* **267,** 155–163.

Pati, D. and Habibi, H.R. (1992) Characterization of gonadotropin-releasing hormone(GnRH) receptors in the ovary of common carp (Cyprinus carpio). *Can. J. Physiol. Pharmacol.* **70,** 268–274.

Peter, R.E. and Yu, K.L. (1997) Neuroendocrine regulation of ovulation in fishes: basic and applied aspects. *Rev. Fish. Biol. Fisheries.* **7,** 173–197.

Peng, C., Fan, N.C., Ligier, M., Vaananen, J. and Leung, P.C. (1994) Expression and regulation of gonadotropin-releasing hormone (GnRH) and GnRH receptor messenger ribonucleic acids in human granulosa-luteal cells. *Endocrinol.* **135,** 1740–1746.

Probst, W.C. (1992) Sequence alignment of the G-protein coupled receptor superfamily. *DNA and Cell Biol.* **11,** 1–20.

Sakakibara, H. (1989) Gene expression of gonadotropin-releasing hormone in the rat ovary. *Acta. Obst. Gynapc. Jpn.* **41,** 1817–1822.

Sealfon, S.C., Weinstein, H. and Millar, R.P. (1997) Molecular Mechanisms of ligand interaction with the gonadotropin-releasing hormone receptor. *Endoc. Rev.* **18,** 180–205.

Tensen, C., Okuzawa, K., Blomenrohr, M. Rebers, F., Leurs, R., Bogerd, J., Schulz, R. and Goos, H. (1997) Distinct efficacies for two endogenous ligands on a single cognate gonadoliberin receptor. *Europ. J. Biochem.* **243,** 134–140.

Troskie, B., Hapgood, J., Millar, R. and Illing, N. (1997) The Xenopus laevis gonadotropin-releasing hormone receptor: coloning and characterisation. *In* "Proceedings of XIII Congress of Comparative Endocrinology, Yokohama Japan", (S. Kawashima and S. Kikuyama, Eds.), pp, 747–751. Monduzzi Editore, Bologna (Italy).

Tsafriri, A.(1988). Local nonsteroidal regulators of ovarian function. *In* "The Physiology of Reproduction", (E. Knobil and J. D. Neil et al., eds.), pp, 527–565. Raven Press, Ltd, New York.

Vaananen, J.E., Tong, B.L., Vaananen, C.M., Chan, I.H., Yuen, B.H. and Leung, P.C. (1997) Interaction of prostaglandin F2alpha and gonadotropin-releasing hormone on progesterone and estradiol production in human granulosa-luteal cells. *Biol. Reprod.* **57,** 1346–1353.

Watanabe, H., Tanaka, S., Akino, T., Hasegawa-Sasaki, H. (1990) Evidence for coupling of different receptors for gonadotropin-releasing hormone to phospholipases C and A_2 in cultured rat luteal cells. *Biochem. Biophys. Res. Comm.* **168,** 328–334.

White, R.B,. Eisen, J.A., Kasten, T.L. and Fernald, R.D. (1998) Second gene for gonadotropin-releasing hormone in humans. *P.N.A.S., US.* **95,** 305–309.

Whitelaw, P.F., Eidne, K.A., Sellar, R., Smyth, C.D. and Hillier, S.G. (1995) Gonadotropin-releasing hormone receptor messenger ribonucleic acid expression in rat ovary. *Endocrinol.* **136,** 172–179.

8

ROLE OF CENTRAL MONOAMINES IN REGULATION OF GONADOTROPIN-II SECRETION

Catfish Model

K. P. Joy

Department of Zoology
Banaras Hindu University
Varanasi 221 005, India

1. INTRODUCTION

Two types of gonadotropin: gonadotropin I (GTH-I) and gonadotropin II (GTH-II) have been demonstrated in many teleosts (Kawauchi *et al.*, 1989, Swanson and Dittman, 1997). Prat *et al.* (1996) and Swanson and Dittman (1997) suggested designating them as follicle stimulating hormone (FSH) and luteinizing hormone (LH), respectively in line with the tetrapod nomenclature. However, in African catfish (*Clarias gariepinus*), only a single GTH corresponding to GTH-II or LH has been identified (Koide *et.al.*, 1992, Schulz *et al.*, 1995). The latter authors have suggested three possibilities to explain the absence of FSH in the catfish. They are, 1) FSH is synthesized from the same gonadotropes as does LH and is biochemically, immunologically and biologically (steriodogenic activity) indistinguishable from LH, 2) FSH is devoid of any steroidogenic activity and is synthesized from a cell type that does not show ultrastructural features of the gonadotropes, and 3) the gene encoding for β-subunit of FSH, if present, is "silent". Most studies on the regulation of GTH in teleosts are related to GTH-II secretion and demonstrate the participation of multiple brain regulatory/modulatory systems, as has been detailed in goldfish (Peter *et al.*, 1986, 1991, Trudeau and Peter, 1995, Peter and Yu, 1997) and African catfish (de Leeuw *et al.*, 1987). The dominant brain mechanism controlling GTH-II is stimulatory by gonadotropin-releasing hormone (GnRH). The activity of GnRH (synthesis and release) is influenced by a number of factors originating from both intrinsic and extrinsic environments. The well known variables, among them, are the circulating levels of sex steroids, seasonally varying day length, temperature, rainfall, sex pheromones, etc. Deciphering the mechanisms by which these factors influence the GnRH activity and GTH-II

Neural Regulation in the Vertebrate Endocrine System, edited by Prasada Rao and Peter, Kluwer Academic / Plenum Publishers, New York, 1999.

secretion is an exciting task. To date, a number of neurohormones have been demonstrated to modulate positively (stimulatory) or negatively (inhibitory) GnRH activity and consequently GTH-II secretion. They include a wide range of substances like monoamines, neuropeptides, amino acids, etc. (see Trudeau and Peter, 1995 for details).

Central monoamines (MAs) play a key role in the regulation of GTH-II secretion by interacting with the GnRH system. The dominant inhibitory hormone is dopamine (DA) which inhibits basal and/or GnRH-induced GTH-II release (Peter *et al.*, 1991). The DA inhibition is mediated through a high affinity DA_2 receptor mechanism on gonadotropes (Chang *et al.*, 1990, Van Asselt *et al.*, 1990). The DA_2 receptor-mediated blockade of GnRH-induced GTH-II release has been demonstrated in other species (Tharakan and Joy, 1996, for references). This has led to the development of Ovaprim—a formulation of domperidone (DA_2 antagonist) and salmon GnRH analogue, and Spawnrite (unstated GnRH-analogue and DA antagonist) for induced spawning of fish under captivity (Yaron, 1995, Peter and Yu, 1997). In Atlantic croaker, however, the DA inhibition of GTH-II release is not evident (Copeland and Thomas, 1989) which means that this mechanism is not common for all species. In fact, species with strong, moderate or little DA-ergic inhibitory tone on GTH-II release has been recognized (Yaron, 1995). Other monoamines, noradrenaline (NA) and serotonin (5-HT), in contrast, stimulate GTH-II secretion. In goldfish (Somoza *et al.*, 1988), 5-HT stimulates GTH release with high responsiveness in prespawning fish. On the other hand, in Atlantic croaker, 5-HT elicits a potentiating effect on GnRH analogue-induced GTH-II release (Khan and Thomas, 1992). The effect of 5-HT on GTH-II release seems to be mediated through a 5-HT_2 receptor. A stimulatory effect of NA on GTH-II release was demonstrated in goldfish (Chang and Peter, 1984). This was attributed to the stimulation of GnRH release (Yu and Peter, 1992). The stimulatory effect of NA varies with the reproductive status of the fish, NA being most sensitive in sexually regressed or regressing fish. The effect is mediated through an α_1-adrenergic receptor. To date, there is no report of adrenaline (A) involved in GTH-II secretion in teleosts.

While the role of individual monoamines and to a great extent, their interactive role on GnRH-induced GTH-II release are elucidated, recent studies indicate that they serve as the main integrators of environmental information. In the present article, a brief review of the hypothalamic MA-ergic system, seasonal variations in MAs, and their integrating role in GTH-II secretion is presented, keeping the catfish model in mind.

2. ANATOMICAL DISTRIBUTION OF HYPOTHALAMIC MONOAMINES

Monoaminergic system forms an integral part of the hypothalamo-hypophysial neurosecretory system of vertebrates. In teleosts, the organization of the MA-ergic system has been demonstrated by various methods:histochemical/fluorescence methods for monoamine oxidase (MAO) and monoamines, immunocytochemical methods using antibodies against various monoamines and MA-synthesizing enzymes (see Joy , 1993 for references). In catfish hypothalamus (*Mystus vittatus*: Prasada Rao *et al.*, 1987, *Clarias gariepinus*: Corio *et al.*, 1991, *Clarias batrachus*: Joy *et.al.*, 1992), the majority of MA-containing neurons are located in the nucleus recessus lateralis and posterioris (NRL and NRP), paraventricular organ pars anterior, nucleus ventro-medialis thalami, nucleus posterior periventricularis, in one component of the nucleus lateralis tuberis (NLT) and in preoptic (nucleus preopticus, NPO) region. Furthermore, a high density of MA-ergic fibres is noticed throughout the hypothalamus, particularly in the NPO and NLT. Immunocytochemical studies reveal that the monoamines involved are DA and 5-HT. In the pituitary, MA-ergic fibres innervate the gland cells. In *C. gariepinus*, the DA-immunoreactivity was found associated with type B neurosecretory fibres, especially those which innervate the gonadotropes (Peute *et al.*, 1987). A unique feature of the MA-organization in catfishes is the occurrence of perikarya in one component of the NLT and in the pituitary (see review, Joy, 1993).

3. SEASONAL VARIATIONS IN HYPOTHALAMIC MONOAMINE ACTIVITY

Hypothalamus, the major neuroendocrine centre of brain records the highest concentration of catecholamines (CA) and 5-HT (Juorio, 1973, Crim *et.al.*, 1989, Gurrero *et al.*, 1990, Senthilkumaran and Joy, 1993). The amines show significant annual/seasonal variations in their contents and turnovers (activity), as reported in winter flounder (Crim *et.al.*, 1989), and subtropical species like *Channa punctatus* (murrel, Khan and Joy, 1990a), *Clarias batrachus* (Manickam and Joy, 1990), and *Heteropneustes fossilis* (Senthilkumaran and Joy, 1993, 1995a). The data from *H. fossilis* show clearly a marked increase in the contents and turnovers of NA and A, and enzymes, DA-β-hydroxylase (DβH), phenylethanolamine-n-methyltransferase (PNMT), catechol-o-methyltransferase (COMT) and MAO during the progress of ovarian recrudescence, and a corresponding decrease in DA content and turnover (Table 1). This pattern holds true for murrel and *C. batrachus* hypothalamic amines. DA is progressively hydrolyzed to NA as is evident from the peak activity of both NA and DβH in prespawning phase. In spawning phase, NA is methylated to A as indicated by the maximal level and turnover of A and PNMT activity. Thus, during the reproductive cycle, CAs show differential dominance in each phase which is reflected also in similar seasonal differences of their inhibition by α-methyl-paratyrosine (α-MPT, a tyrosine hydroxylase inhibitor, Senthilkumaran and Joy, 1995a). Concerning the functional implication of the differential CA dominance, correlations can be drawn with seasonal cyclic secretion of GTH-II. In the resting phase when the inhibitory DA activity predominates, plasma GTH-II level is not detectable (< 1 ng/ml) and, with the progressive decline of DA and concomitant increase of NA in preparatory phase and later, it appears in circulation and increases steadily. The peak GTH-II level coincides with the peak NA activity. The significance of peak A activity in the spawning phase to the secretion of GTH-II is debatable at present. In contrast to the situation in the murrel and catfishes,

Table 1. Seasonal changes in hypothalamic monoamines and enzymes (DβH, PNMT, COMT and MAO) in *Heteropneustes fossilis* : a summary[a]

	Ovarian cycle				
	Resting (Nov - Jan)	Preparatory (Feb - April)	Prespawning (May - June)	Spawning (July - Aug)	Postspawning (Sept - Oct)
Dopamine	++++	+	+	+	+
Noradrenaline	+	++++	++++	+++	+
Adrenaline	-	+	+	++++	+
Serotonin	+	++++	+++	+++	+
DβH	+	+++	++++	+	not measured
PNMT	+	+	+++	++++	+
COMT	+	+	++++	+++	+
MAO	+	+	++++	++++	+++

[a]++++ : maximal; +: minimal; - : absent.

the seasonal patterns of CAs in goldfish and rainbow trout hypothalamus are totally different (Saligaut *et al.*, 1992a, Trudeau *et.al.*, 1993a,b). In these species, DA is the dominant amine in recrudescing or recrudescent fish. The species differences can be attributed to various factors, such as habit and habitat, photoperiod, temperature, domestication, if any, of fish, stress, etc. While goldfish and trout hypothalamus belongs to the DA dominating type, in catfish and murrel it belongs to the NA/A dominating type during recrudescence.

Hypothalamic 5-HT content and turnover also show significant seasonal/annual variations, as has been demonstrated in goldfish (Olcese *et al.*, 1981), murrel (Khan and Joy, 1990a) and *H. fossilis* (Senthilkumaran and Joy, 1993). The highest 5-HT activity is present during early gonad recrudescence (preparatory phase) and it declines soon after spawning (Table 1).

While studying the annual 5-HT dynamics, we came across an important phenomenon that has a bearing on timing the breeding cycle. Both its content and activity, together with MAO, show significant diurnal variations exclusively in gonadal preparatory phase (February and March). The diurnal pattern is restricted to the hypothalamus and telencephalon (among the brain regions), the centres associated with GTH-II regulation (Khan and Joy, 1990a, Senthilkumaran and Joy, 1993). The day-time amplitude of 5-HT activity peaks in March which is the early recrudescent phase but the content peaks in late preparatory-prespawning phases. Detailed experimental studies in the murrel and catfish indicate that the rhythm is driven by the prevailing photoperiod in February and March (day length varying from 11.00 to 12.36 hr) as evident from the following evidence (Khan and Joy, 1990a,b, Senthilkumaran and Joy, 1993, 1994a) : 1) short photoperiod, total darkness and continuous light abolish the diurnal patterns, 2) reversal of LD cycle causes a corresponding shift in the diurnal patterns, 3) the pattern can be artificially induced in the resting phase by exposing the fish to long photoperiod, 4) long photoperiod intensifies 5-HT activity, and 5) the diurnal patterns can be abolished by a combination of pinealectomy and blinding, implying the involvement of photoreceptor organs. Both the timing and region-specificity of the rhythm coincide with initiation of gonadal recrudescence after the winter photorefractoriness. The hypothalamic serotonergic system may be the ultimate link in the neural mechanism(s) involved in the time measurement of day length difference and transmission of that information to the centres controlling GTH-II secretion. Therefore, the diurnal pattern can be viewed as an important neural component signaling the hypothalamo-hypophysial system of the arrival of suitable environmental (photoperiod) condition to initiate gonadal recrudescence. Coincident with the rhythm is the appearance of a detectable level of GTH-II in plasma after the long duration of its absence (below the detection range) during November, December and January (unpublished data).

From the seasonal study it is clear that in recrudescing fish, the stimulatory NA and 5-HT mechanisms prevail on the GTH-II regulatory centres while the inhibitory DA mechanism is suppressed. The reverse is true in nonbreeding fish. The two mechanisms alternate cyclically during the annual reproductive cycle.

4. MONOAMINE-AFFECTING DRUGS MODIFY GTH-II SECRETION

MA function can be modified by interventions at different enzymatically controlled metabolic steps (Figure 1) and at receptor levels using specific inhibitors. Such a pharmacological approach was followed in *H. fossilis* to demonstrate the functional nature of MAs on GTH-II secretion.

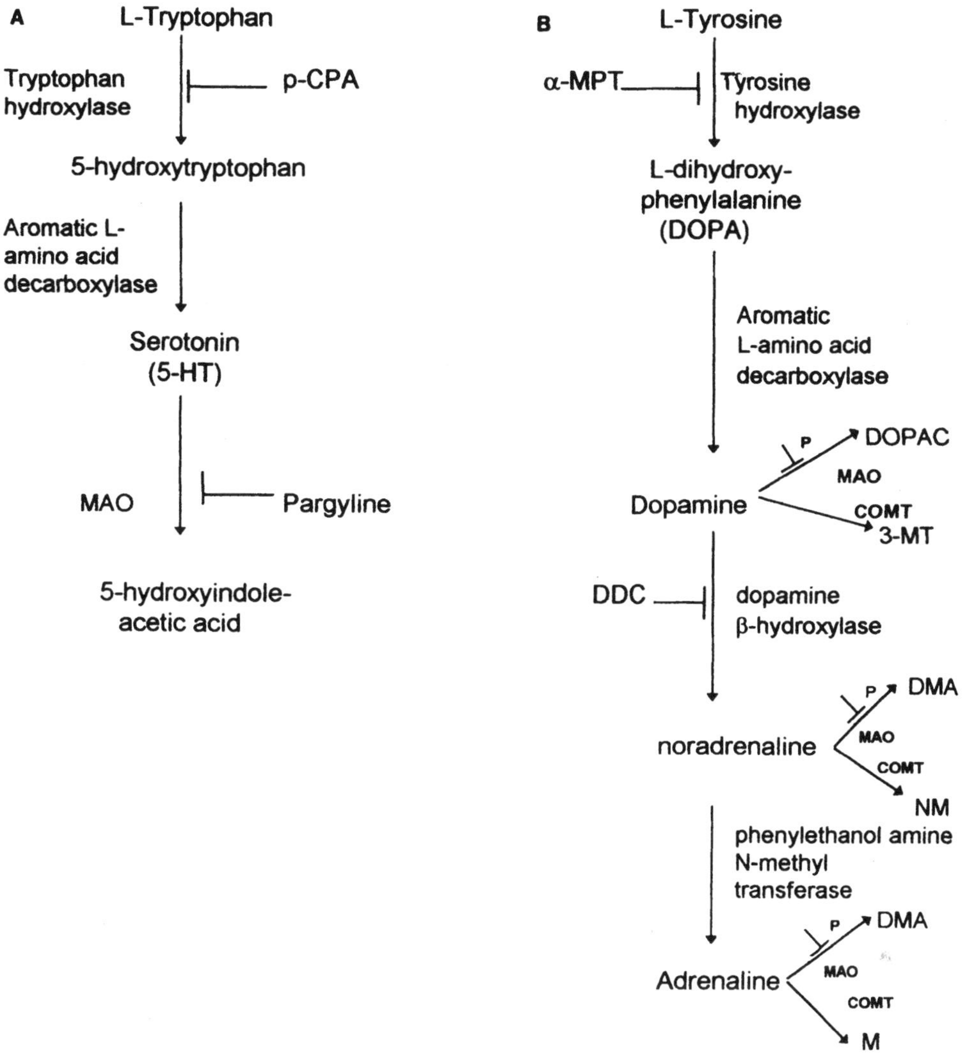

Figure 1. Serotonin (A) and catecholamine (B) biosynthesis and degradation pathways showing specific sites of blockade by some drugs. P - Pargyline, α-MPT - α-methylparatyrosine, pCPA - parachlorophenylalanine, DOPAC - 3,4-dihydroxyphenylacetic acid, 3-MT - 3-methoxytyramine, DMA - 3,4-dihydroxymandellic acid, NM - normetanephrine, M - metanephrine, MAO - monoamine oxidase, COMT-catechol-o-methyltransferase, DDC-diethyl-dithiocarbamate.

4.1. Catecholamines

CA synthesis can be blocked by inhibiting tyrosine hydroxylase, the rate limiting enzyme, using α-MPT. In goldfish, it was reported that α-MPT caused DA inhibition and resulted in rise of GTH-II level (Chang *et al.*, 1983, Trudeau *et al.*, 1993a). In the catfish, α-MPT produced differential inhibition of CAs depending on the season (Senthilkumaran and Joy, 1995a). In the prespawning phase, NA was maximally inhibited, followed by A and DA. In intact and 3-week ovariectomized catfish, it reduced plasma GTH-II level significantly (Senthilkumaran and Joy, 1995a,b; 1996). This is in contrast to the observation in the goldfish and the difference can be ascribed to whether the stimulatory or inhibitory mechanism was maximally inhibited. Supplementation with NA counteracted the effect of α-MPT, indicating a stimulatory role of the amine on GTH-II. DOPA supplementation elevated the GTH-II level but significantly below the level of the ovariectomized fish. This can be attributed to the elevated levels of both NA and DA acting in opposite directions. Thus, the degree of DA tone seems to influence the stimulatory effect of NA. In another experiment, DDC, a DβH inhibitor, was used to block the conversion of DA to NA. It was found that GTH-II level dropped significantly, clearly showing the inhibitory effect of DA. A further study showed that the DA inhibitory effect was mediated through a DA_2 receptor mechanism. Pimozide inhibited the GnRHa- induced GTH-II release, but not basal release, in prespawning and spawning phases. This effect was not seen in preparatory phase, perhaps due to pituitary insensitivity to GnRHa treatment (Tharakan and Joy, 1996). The combination resulted in induced ovulation of the fish as has been reported in other catfishes (de Leeuw et al, 1985, Manickam and Joy, 1989a). However, pimozide treatment is not obligatory since GnRHa alone can induce the fish to spawn (Tharakan and Joy, 1996, Joy et al, 1998).

4.2. Serotonin

Serotonin stimulates GTH-II secretion in goldfish and Atlantic croaker (see Introduction). In catfish also a stimulatory effect of 5-HT was demonstrated. Inhibition of 5-HT synthesis by p-CPA, a tryptophan hydroxylase inhibitor, lowered GTH-II level in *H. fossilis* held in normal or long photoperiods in the preparatory phase and 5-HTP supplementation reversed the effect of p-CPA on 5-HT and GTH-II (Senthilkumaran and Joy, 1995a, Joy and Senthilkumaran, 1995). In 3-week ovariectomized catfish also similar effects were reported after p-CPA treatment and 5-HTP replacement (Senthilkumaran and Joy, 1996). The stimulatory effect of 5-HT is mediated through a $5\text{-}HT_2$ receptor, as shown in other species. Ketanserin, a $5\text{-}HT_2$ receptor blocker inhibited GTH-II in the preparatory phase, but not in the prespawning and spawning phases (unpublished data).

Simultaneous inhibition of both CA and 5-HT by α-MPT and p-CPA has led to a very drastic reduction of GTH-II in 3-week ovariectomized fish almost to the basal level, along with reductions of both CA and 5-HT. This strongly supports that MAs have a primary role in the modulation of GnRH and GTH-II activity.

4.3. MAO Inhibition

Both CAs and 5-HT are catabolized by MAO and oxidative deamination is, therefore, important for the availability of MAs for GTH-II regulation. In goldfish, this is the exclusive pathway of DA degradation; MAO inhibition by pargyline and other inhibitors decreased GTH-II level consequent upon increased DA activity (Sloley *et al.*, 1992). In an extensive study in *H. fossilis*, it has been shown that the effect of MAO inhibition by pargyline varied with the steroid environment (Senthilkumaran and Joy, 1995c). Pargyline inhibits MAO in a time-dependent manner causing 80% inhibition at 6 hr and by 21 days the steady state level is returned. Pargyline injection elevated plasma GTH-II level, unlike in goldfish, at 3 and 6 hr, and 7 days in sham ovariectomized and ovariectomized fish, the effect being maximal at maximal MAO inhibition (6 hr). The GTH-II rise could be correlated with changes in hypothalamic MAs which increased differentially with the percentage increase of 5-HT and NA being about 2 times more than that of DA (Table 2). A low dose of E_2 (0.1 μg/g BW) restored the GTH-II level to that of sham control group, but pargyline

Table 2. Effects of pargyline (P, 75mg/Kg BW; one injection) on plasma gonadotropin (GTH-II) and hypothalamic monoamine contents at maximal MAO inhibition (80%; 6 hr)[a]

Groups	Plasma GTH-II	5-HT	DA	NA
3-OVX + P	+ 41	+ 42	+ 17	+ 42
3-OVX+0.1μg E_2 + P	+ 35	+ 45	+ 33	+ 42
3-OVX+1.0μg E_2 + P	- 32	+ 27	+ 49	+ 17

[a]Values show percentage of increase (+) or inhibition (-) over respective control.
OVX : Ovariectomy; Sham : Sham ovariectomy.

treatment elevated it which could be correlated with a similar pattern of aminergic changes. In contrast, a high dose of E_2 (1.0 μg/g BW) decreased the GTH-II level at 3 and 6 hr, and 7, 14 and 21 days. Pargyline treatment decreased it further at 3 and 6 hr only, the effect being more pronounced at 6 hr; concomitantly, DA increased more than 5-HT and NA.

5. MONOAMINE ACTIVITY CHANGES WITH ALTERED GTH-II SECRETION PATTERN

GTH-II secretion can be altered by various procedures such as exposure to long photoperiod and high temperature, ovariectomy, E_2 replacement, GnRH analogue, methallibure, etc. In this section, effects of such treatments on GTH-II secretion are correlated with changes in MA-ergic activity.

5.1. Influences of Photoperiod and Temperature

It is well established in many fishes including *H. fossilis* that gonads undergo recrudescence when environmental photoperiod and temperature are increased (Sundararaj, 1981). On this reckoning, exposure of the fish to long photoperiods and high temperatures causes precocious maturity of gonads. Several studies have shown that such treatments elevate plasma/serum GTH and sex steroid levels (Fostier *et al.*, 1983). Exposure to high temperatures has a more pronounced effect on these variables than to long photoperiods as has been illustrated in the catfish (Sundararaj, 1981). In the catfish plasma GTH-II level increased significantly in fish exposed to high temperature or high temperature + long photoperiod regimes in the early preparatory phase. However, the neural mechanism(s) underlying the activation of GnRH and GTH-II secretion is not known. In the murrel (Khan and Joy, 1990b), exposure to long photoperiods (16L) acted differently on 5-HT and CAs: the long photoperiod exposure stimulated 5-HT activity but did not influence CA activity. In the catfish, exposure to short photoperiod and total darkness depressed, and that to long photoperiod and continuous light stimulated the 5-HT activity (Senthilkumaran and Joy, 1994a). The importance of 5-HT stimulation to GTH-II secretion was demonstrated in catfish treated with p-CPA. The treatment decreased significantly the content and turnover of 5-HT concurrently with inhibition of plasma levels of GTH-II, E_2, testosterone and GSI in fish held under normal (control) or long photoperiods (Joy and Senthilkumaran, 1995). These changes were reversed when 5-hydroxytryptophan (the immediate precursor of 5-HT) was supplemented along with p-CPA. These results conclusively show that 5-HT is involved in the photoperiodic stimulation of the pituitary-ovarian axis.

Table 3. A comparison of effects of melatonin and p-CPA on hypothalamic monoamines, plasma levels of GTH-II and E_2, and gonadosomatic index (GSI) in *Heteropneustes fossilis*

Parameters	Melatonin	p-CPA
5-HT content	decrease	decrease
5-HT turnover	decrease	decrease
CA (DA, NA & A) contents	No effect	No effect
CA(DA, NA & A) turnovers	No effect	No effect
Plasma GTH II	decrease	decrease
Plasma E_2	decrease	decrease
GSI	decrease	decrease

Melatonin, the pineal hormone has been demonstrated to inhibit ovarian recrudescence in the catfish, murrel and other species held in normal or long photoperiods (Joy and Khan, 1991). This inhibitory effect of melatonin on pituitary-ovarian axis has been shown to be mediated through 5-HT and shows close similarities with the effects of p-CPA that impairs 5-HT activity (Joy and Khan, 1991, Senthilkumaran and Joy, 1995b), as shown in Table 3. These results underscore the functional intactness of the 5-HT system as essential to photoperiodic stimulation of the pituitary-ovarian axis. The schematic diagram (Figure 2) explains the possible functional linkage between the retinal light pathway, pineal melatonin and the 5-HT system. The pineal organ is a direct photosensory organ and functions as a "dosimeter" of incident light via melatonin secretion which exerts an inhibitory "brake" on the 5-HT activity. As day length increases, pineal melatonin secretion is gradually and progressively diminished in which state 5-HT responds positively to light received via the retinal pathway (Joy and Khan, 1991).

Both photoperiod and temperature influence CA-ergic activity but the temperature effects are more pronounced, as shown in the murrel (Khan and Joy, 1990b) and catfish (Senthilkumaran and Joy, 1995a). In the catfish, exposure to a high temperature regime (28 $\pm$2°C) inhibited more significantly the DA content and turnover than the long photoperiod (16L) exposure alone. In contrast, the contents and turnovers of both NA and A were more significantly elevated in the high temperature or high temperature + long photoperiod groups. The mechanisms underlying the changes in MA correlates were examined by monitoring simultaneously enzyme activities. It was found that the high temperature and long photoperiod exposures caused significant rise of DβH and PNMT activities resulting in stepwise conversion of DA to NA to A (Figure 2). The effects were highly pronounced in the high temperature groups. Thus, the seasonal dominance and differential activity of CAs in the recrudescent phase of the annual reproductive cycle of catfish can be related to the cycles of natural photoperiod and temperature. The stimulatory effect of these variables on GTH-II release is mediated through the stimulation of NA and 5-HT activity concurrently with inhibition of DA activity.

5.2. Influences of Gonadal Steroids

Among the intrinsic factors, gonadal steroids play a crucial feedback control of GTH-II secretion. Physiological studies involving gonadectomy and steroid replacement, and *in vitro* or *in vivo* steroid treatment indicate the existence of a negative or positive feedback as assessed by changes in gonadotropic morphology and ultrastructure, and pituitary/blood GTH level. However, recent GTH gene expression studies have shown that the feedback effects are more complex and their classification into positive and negative

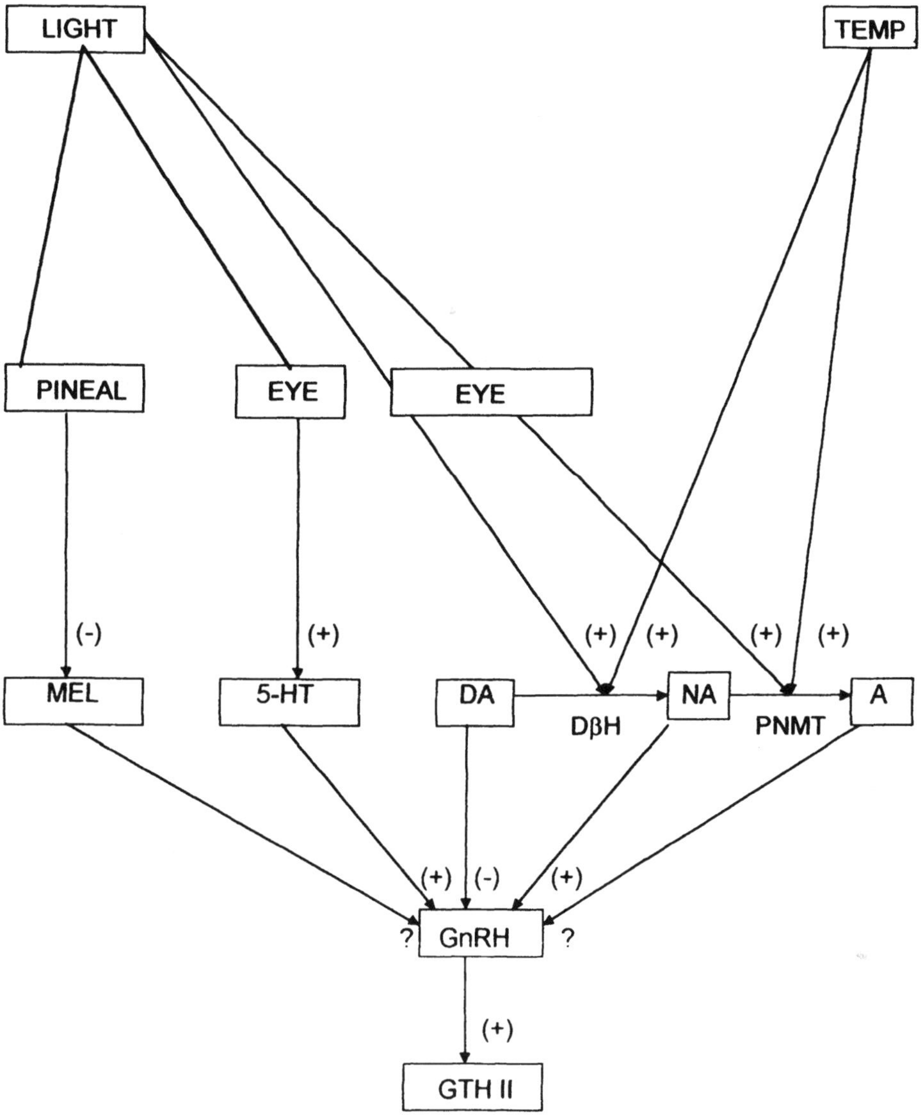

Figure 2. A schematic diagram showing the influences of light and temperature (temp) on monoaminergic activity in *Heteropneustes fossilis*. (+) stimulation, (–) inhibition, ? - not known, MEL - melatonin.

types needs reconsideration, as steroids can affect several aspects of gonadotropic function in a positive or negative manner concurrently (Goos *et al.*, 1998).

In teleosts, as in other vertebrates, gonadal steroids modulate the neurohormonal systems that control GTH-II secretion. Hypothalamic MA-ergic system is one such target. A strong support for steroid-monoamine interaction is evident from the demonstration of anatomical proximity or overlapping distribution of the two systems. In rainbow trout, double immunocytochemical studies have documented that most estrogen receptor-positive neurons are tyrosine hydroxylase-positive (DA-ergic) in the anterior preoptic region (NPO-pars anteroventralis), implying these CA neurons as E_2 feedback sites (Linard *et al.*, 1996a). Physiological studies employing measurements of MA contents, turnovers or concentrations of MA-related enzymes have been used to understand the nature of influence of steroids on MA-ergic activity. The seasonal variations in MA content and turnover and enzyme activity have been attributed partially to the changing level of circulating sex steroids (Manickam and Joy, 1989b, 1990, Khan and Joy, 1991, Saligaut *et al.*, 1992a,b, Trudeau *et al.*, 1993a,b, Senthilkumaran and Joy, 1993, 1994b, 1995a). Due to variations in species, season, and dose and mode of administration of steroids, it is difficult to arrive at a uniform conclusion about steroid influences on amine metabolism. A few examples are presented here to show the diversity of effects. In gonad intact goldfish, E_2 or testosterone treatment does not affect DA turnover in the hypothalamus but enhances pituitary DA turnover and DA inhibition of GTH-II release (Trudeau *et al.*, 1993a). E_2 decreases the NA turnover in the hypothalamus of sexually regressed fish and increases it in sexually recrudescent phase (Trudeau et al, 1993b). In rainbow trout, ovariectomy does not modify brain aminergic function, E_2 replacement increases DA turnover and NA contents in the hypothalamus but 5-HT function is not affected (Saligaut *et al.*, 1992b; Linard *et al.*, 1995). In catfishes, ovariectomy and E_2 replacement modify aminergic activity in the recrudescent phase but not in quiescent phase (Senthilkumaran and Joy, 1994b, 1995a). In male tilapia (*Oreochromis mossambicus*), moderate doses of E_2 increase 5-HT concentration and high doses increase the 5-HIAA/5-HT ratio in hypothalamic incubates (Tsai and Wang, 1997).

Ovariectomy (OVX) followed by E_2 replacement is a dynamic model system for demonstrating the relationship between GTH-II secretion and steroid feedback notwithstanding the usual criticism against the procedure that it removes several bioactive substances other than steroids and, therefore, the reliability of drawing conclusions is weak. E_2 exerts a negative feedback on GTH-II secretion during vitellogenesis which is a major and prolonged event in oogenesis and involves sustained activation of pituitary-ovarian axis. OVX elevates GTH-II level and a low dose of 0.1 μg/g BW of E_2 administered in 3-week ovariectomized catfish restores it to that of sham ovariectomized + vehicle treated fish (restorative or physiological dose). We used this model system to show the functional relationship between E_2 feedback and MA-ergic activity on GTH-II secretion (Senthilkumaran and Joy, 1994b, 1995a). Ovariectomy produced a significantly high level of GTH-II release over time (up to 6 weeks). Based on the responses, two patterns (biphasic) emerge. In the first phase of OVX (2,3 and 4 weeks, short duration), the amplitudes of GTH-II increased steadily to the peak at week 4. In the second phase (5 and 6 weeks, long duration), the amplitudes decreased. These can be attributed to hypersecretory activity followed by exhausion, vacuolation and atrophy of GTH cells, OVX evoked biphasic responses on amines. In the short duration, the increase in the amplitude of GTH-II level was matched by a corresponding increase of the stimulatory aminergic mechanisms (5-HT and NA) with a concomitant suppression of the DA inhibitory mechanism. In the long duration, the reverse changes were noticed. The results of the E_2 replacement further strengthen these observations. The restoration of GTH-II level to the control range by the 0.1 μg/g E_2 dose involves an increase of DA activity, and suppression of NA activity and 5-HT (content only). The higher doses of E_2 (0.5–5 μg/g) increased further the DA activity and suppressed the NA and A activities in a dose-related manner to inhibit further the GTH-II level. Surprisingly, 5-HT content and turnover also increased with E_2 doses.

Regarding the relative roles of photoperiod and temperature, and E_2 on maintaining the seasonal variations of MA, it is reasoned that in nonbreeding fish which experiences short photoperiod and low temperature and lacks E_2 feedback, DA is the only amine dominant in the hypothalamus in the absence of its conversion into NA. As photoperiod and temperature increase, fish undergoes recrudescence and E_2 feedback is active, DA is metabolized into NA and A which dominate sequentially. 5-HT also dominates this time. E_2

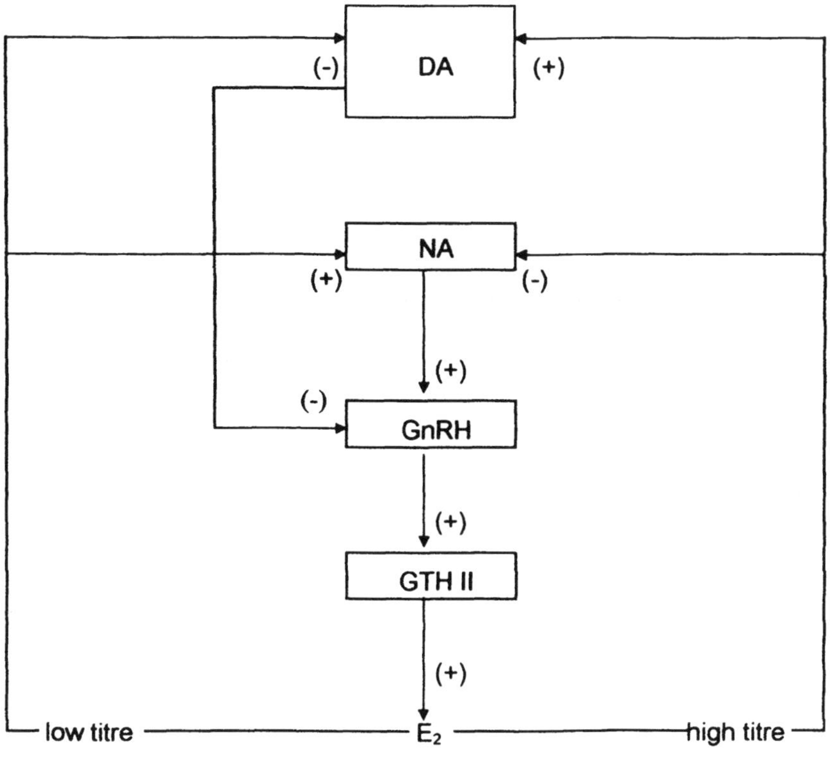

Figure 3. A schematic diagram showing the influence of low and high titers of E_2 feedback on DA and NA activity in *Heteropneustes fossilis*. (+) stimulation, (–) inhibition.

fine-tunes the secretion of GTH-II, a fall in its level activates the stimulatory control mechanism and suppresses the inhibitory mechanism and a rise in its titre activates the inhibitory mechanisms and suppresses the stimulatory mechanism (Figure 3).

The mechanism(s) by which E_2 modulates the steady-state level of monoamines requires elucidation. Investigations in the catfish have shown that all enzymically controlled steps in CA metabolism are susceptible to E_2 actions and the effect at one step can lead to cascading effects on other amines/metabolities. This was examined in several species by measuring enzyme concentrations (Manickam and Joy, 1989b, Khan and Joy, 1991, Senthilkumaran and Joy, 1994b, 1995a,c) or various metabolites (Saligaut *et al.*, 1992a,b, Sloley, *et al.*, 1992, Hernandez-Rauda, *et al.*, 1996, Linard *et al.*, 1996b, Tsai and Wang, 1997). In the catfish, ovariectomy and E_2 replacement produced biphasic effects on DβH, PNMT, MAO and COMT only in the recrudescing fish (Senthilkumaran and Joy, 1994b, 1995a; Joy and Senthilkumaran, 1998). The differential effects of E_2 feedback on these enzymes can, in turn, alter differentially the availability of CAs for GTH-II regulation. The mechanism underlying the biphasic effects of E_2 on MAO turnover was elucidated in the catfish (Senthilkumaran and Joy, 1995c). This study shows that ovariectomy increased the half-life (19.04 days) of the enzyme by decreasing both degradation and synthesis rates. Replacement with a low dose (0.1 μg/g) enhanced the half-life (19.63 days) by elevating its synthesis rate. The high dose of E_2 (1.0 μg/g) decreased the half-life (13.33 days) by enhancing the degradation rate and reducing the synthesis rate simultaneously. Similar mechanisms may operate in the case of other enzymes showing biphasic responses to E_2.

5.3. Periovulatory Changes

The terminal events in oogenesis are oocyte maturation and ovulation, and the hormonal and neuroendocrine changes associated with them. In catfishes and other species the arrival of South-West Monsoon in July-August (at Varanasi) precipitates spawning activity. Flooding and accompanying changes in water seem to activate the neuroendocrine mechanisms and trigger the surgic release of GTH-II. In the absence of these signals fishes do not spawn, as in captivity. DA is considered to be the neuroendocrine mechanism responsible for this refractoriness and its inhibition by using a combination of a DA_2 receptor antagonist and GnRH analogue can induce successful spawning in many species held in confined water. Recent studies indicate that the low titre of E_2, coinciding with the periovulatory period can suppress DA activity directly through an E_2 receptor-mediated mechanism at the level of the preoptic region and facilitate preovulatory surge in GTH-II (Saligaut *et al.,* 1992a,b; Linard *et al.*, 1996a). The involvement of an endogenous mechanism (E_2 feedback) altering the aminergic metabolism differentially during GnRH analogue-induced ovulation in catfish was recently demonstrated (Joy *et al.*, 1998). The GnRH analogue treatment lowered E_2 levels parallel to the GTH-II rise and the surge at 8 hr of injection and elicited feedback effects on hypothalamic and pituitary amines. At 8 hr, DA level was lowered with a concurrent rise in NA and 5-HT. After egg stripping at 16 hr, the amine levels were restored to the 0 hr level except the pituitary DA which remained low. The decrease in DA was due to its conversion into NA since DβH activity was high at 8 hr. Thus, the periovulatory release of GTH-II also involved changes in amine metabolism, as in the case of tonic release of GTH-II in vitellogenic fish, both triggered by E_2 feedback. The catfish study also indicates that the exogenously administered GnRH triggers the endogenous mechanism involving lowered E_2 titre, DA inhibition, and NA and 5-HT activation to stimulate the release of native GnRH, both together may account for the prolonged and significant rise of GTH-II for about 16 h. The differential effects on the amines may explain why GnRH analogue alone is able to induce ovulation in the catfish.

5.4. Methallibure Effects

Methallibure (AIMAX) is a known non-steroidal GTH inhibitor used extensively to elucidate pituitary-gonad relationship in a variety of vertebrates. In teleosts also, the antigonadotropic effects of methallibure are documented in several species. However, there are few studies explaining its mode of action. Kitay and Westfall (1977) have reported an inhibition of uptake of tritiated DA and NA in hypothalamic slices incubated with the drug. In catfish, the methallibure-induced inhibition of GTH-II secretion can be correlated to the stimulation of hypothalamic DA and inhibition of 5-HT and NA mechanisms (Senthilkumaran and Joy, 1998). Methallibure shows chemical similarities with DDC, a DβH inhibitor (Chang *et al.*, 1995) and shares similarities in its actions on GTH-II release and aminergic activity (Table 4).

6. CONCLUSIONS

Monoamines exercise a primary role in the regulation of GnRH-GTH-axis by mediating and integrating cues originating from both external (photoperiod, temperature, etc.) and internal (E_2 feedback, etc.) environments. As the main integrating system, their differential tones can switch on (stimulate) and switch off (inhibit) GTH-II secretion both tem-

Table 4. A comparison of effects of methallibure and diethyldithiocarbamate (DDC) on GTH-II, E_2, hypothalamic monoamines and enzymes

	Methallibure (a GTH inhibitor)	DDC (a DβH inhibitor)
Plasma GTH-II	inhibition	inhibition
Plasma E_2	inhibition	inhibition
Dopamine	elevation	elevation
Noradrenaline	inhibition	inhibition
Adrenaline	inhibition	inhibition
Serotonin	inhibition	no effect
DβH	inhibition	inhibition
PNMT	inhibition	not measured
COMT	elevation	not measured
MAO	elevation	not measured

porally and tonically to synchronize reproductive activity. The diurnal rhythm in hypothalamic 5-HT forms an important component of photoperiodic stimulation of reproduction, the occurrence of which in gonadal preparatory phase signals the approach of favourable season. Recent studies also indicate that reproductive pheromones exert their neuroendocrine effects by modulating DA activity (Dulka *et al.*, 1992; Zheng and Stacey, 1997) or by differentially modulating DA, NA and 5-HT tones (unpublished data). Central monoamines are closely linked to the control of all aspects of reproduction.

REFERENCES

Chang, J.P., and Peter, R.E. (1984). Influences of norepinephrine and α-adrenergic mechanisms on gonadotropin secretion in female goldfish, *Carassius auratus*. *Gen. Comp. Endocrinol.* **55,** 89–95.

Chang, J.P., Cook, A.F., and Peter, R.E. (1983). Influence of catecholamines on gonadotropin secretion in goldfish, *Carassius auratus*. *Gen. Comp. Endocrinol.* **49,** 22–31.

Chang, J.P., Yu, K.L., Wong, A.O.L., and Peter, R.E. (1990). Differential actions of dopamine receptor subtypes on gonadotropin and growth hormone release *in vitro* in goldfish. *Neuroendocrinology* **51,** 664–674.

Chang, W.J., Barb, R., Kraeling, R.R., Rampacek, G.B., and Wright, J.T. (1995). Biogenic amines in the hypothalamus of rats after diethyldithiocarbamate or AIMAX treatment, an alternative for norepinephrine depletion. *J. Anim. Sci.* **73,** 1147–1151.

Copeland, P.A., and Thomas, P. (1989). Control of gonadotropin release in the Atlantic croaker (*Micropogonias undulatus*) , Evidence for lack of dopaminergic inhibition. *Gen. Comp. Endocrinol.* **74,** 474–483.

Corio, M., Peute, J., and Steinbusch, H.W.M. (1991). Distribution of serotonin- and dopamine-immunoreactivity in the brain of the teleost *Clarias gariepinus*. *J. Chem. Neuroanat.* **4,** 79–95.

Crim, L.W., Evans, D.M., and Moreland, K. (1989). Monoaminergic substances in the teleost brain , catecholamine levels in male and female winter flounder, *Pseudopleuronectes americanus* Walbaum associated with gonadal recrudescence. *Fish Physiol. Biochem.* **7,** 95–100.

de Leew, R., Goos, H.J.Th., Richter, C.J.J., and Eding, E.H. (1985). Pimozide-LHRHa induced breeding of the catfish, *Clarias gariepinus* (Burchell). *Aquaculture* **44,** 295–302.

de Leeuw, R., Goos, H.J. Th., and van Oordt P.G.W.J. (1987). The regulation of gonadotropin release by neurohormones and gonadal steroids in the African catfish, *Clarias gariepinus*. *Aquaculture* **63,** 43–58.

Dulka, J.G., Sloley, B.D., Stacey, N.E., and Peter, R.E. (1992). A reduction in pituitary dopamine turnover is associated with sex pheromone-induced gonadotropin secretion in male goldfish. *Gen. Comp. Endocrinol.* **86,** 496–505.

Fostier, A., Jalabert, B., Billard, B., Breton, B., and Zohar, Y. (1983). The gonadal steroids. *In* "Fish Physiology" (W.S. Hoar, D.J. Randall, E.M. and Donaldson, Eds.), Vol. IX B pp. 277–372. Academic Press, New York.

Goos, H.J. Th., Senthilkumaran, B., and Joy, K.P. (1998). Neuroendocrine integrative mechanisms in the control of gonadotropin secretion in teleosts. *In* "Comparative Endocrinology and Reproduction" (K.P. Joy, A. Krishna, and C. Haldar, Eds.), pp. 114–137, Narosa, New Delhi.

Guerrero, H.Y., Caceres, G., Paia, C.L., and Marcano, D. (1990). Hypothalamic and telencephalic catecholamine content in the brain of the teleost fish, *Pygocentrus notatus*, duing the annual reproductive cycle. *Gen. Comp. Endocrinol.* **80,** 257–263.

Hernandez-Rauda, R., Otero, J., Rey, P, Rozas, G., and Aldegunde, M. (1996). Dopamine and serotonin in the trout (*Onchorhynchus mykiss*) pituitary:main metabolites and changes during gonadal recrudescence. *Gen. Comp. Endocrinol.* **103,** 13–23.

Joy, K.P., and Khan, I.A. (1991). Pineal-gonadal relationship in the teleost *Channa punctatus* (Bloch):Evidence for possible involvement of hypothalamic serotonergic system. *J. Pineal Res.* **11,** 12–22.

Joy, K.P., Manickam, P., and Sathyanesan, A.G. (1992). Paraformaldehyde and glyoxylic acid histofluorescence localization of monoamines in the hypothalamo-hypophysial system of the catfish, *Clarias batrachus* (L.). *Biol. Struct. Morphogen.* **4,** 58–67.

Joy, K.P. (1993). Hypothalamic monoamines and gonadotropin regulation in teleosts. *Ind. J. Life Sci.* **13,** 83–119.

Joy, K.P., and Senthilkumaran, B. (1995). A serotonergic control of pituitary-gonadal activity in the female catfish, *Heteropneustes fossilis*, under normal and long photoperiods. *In* "Reproductive Physiology of Fish" (F.W. Goetz and P. Thomas, Eds.), pp. 67–69, Fish Symposium 95, Austin, USA.

Joy, K.P., and Senthilkumaran, B. (1998). Annual and diurnal variations in, and effects of altered photoperiod and temperature, ovariectomy, and estradiol-17β replacement on catechol-O-methyltransferase level in brain regions of the catfish, *Heteropneustes fossilis. Comp. Biochem. Physiol.* **119C,** 37–44.

Joy, K.P., Senthilkumaran, B., and Sudhakumari, C.C. (1998). Periovulatory changes in hypothalamic and pituitary monoamines following GnRH analogue treatment in the catfish *Heteropneustes fossilis* : a study correlating changes in plasma hormone profiles. *J. Endocrinol.* **156,** 365–372.

Juorio, A.V. (1973). The distribution of catecholamines in the hypothalamus and other brain areas of some lower vertebrates. *J. Neurochem.* **20,** 641–645.

Kawauchi, H., Suzuki, K., Itoh, H., Swanson, P., Naito, N., Nagahama, Y., Nozaki, M., Nakai, Y., and Itoh, S. (1989). The duality of teleost gonadotropins. *Fish Physiol. Biochem.* **7,** 29–38.

Khan, I.A., and Joy, K.P. (1990a). Effects of season, pinealectomy, and blinding, alone and in combination on hypothalamic monoaminergic activity in the teleost *Channa punctatus* (Bloch). *J. Pineal Res.* **8,** 277–287.

Khan, I.A., and Joy, K.P. (1990b). Differential effects of photoperiod and temperature on hypothalamic monoaminergic activity in the teleost *Channa punctatus* (Bloch). *Fish Physiol. Biochem.* **8,** 291–297.

Khan, I.A., and Joy, K.P. (1991). Changes in hypothalamic monoamine oxidase activity in relation to 17β - estradiol and clomiphene citrate treatments in the teleost *Channa punctatus* during three seasons. *Gen. Comp. Endocrinol.* **81,** 426–432.

Khan, I.A., and Thomas, P. (1992). Stimulatory effects of serotonin on maturational gonadotropin release in the Atlantic croaker, *Micropogonias undulatus. Gen. Comp. Endocrinol.* **88,** 388–396.

Kitay, D.S., and Westfall, T.C. (1977). Dose related effect of methallibure (ICI 33828), *in vitro*, upon hypothalamic catecholamine uptake in the teleost, *Poecilia latipinna. Gen. Comp. Endocrinol.* **31,** 402–406.

Koide, Y., Noso, T., Schouten, G., Peute, J., Zandbergen, M.A., Bogerd, J., Schulz, R.W., Kawauchi, H., and Goos, H.J. Th. (1992). Maturational gonadotropin from the African catfish, *Clarias gariepinus* : Purification, characterization, localization and biological activity. *Gen. Comp. Endocrinol.* **87,** 327–341.

Linard, B., Bennani, S., and Saligaut, C. (1995). Involvement of estradiol in a catecholamine inhibitory tone of gonadotropin release in the rainbow trout (*Oncorhynchus mykiss*). *Gen. Comp. Endocrinol.* **99,** 192–196.

Linard, B., Anglade, I., Corio, M., Navas, J.M., Pakdel, F., Saligaut, C., and Kah, O. (1996a). Estrogen receptors are expressed in a subset of tyrosine hydoxylase-positive neurons of the anterior preoptic region in the rainbow trout. *Neuroendocrinology* **63,** 156–165.

Linard, B., Bennani, S., Jego, P., and Saligaut, C. (1996b). Tyrosine hydroxylase activity and dopamine turnover of rainbow trout (*Oncorhynchus mykiss*) brain : the special status of the hypothalamus. *Fish Physiol. Biochem.* **15,** 41–48.

Manickam, P., and Joy, K.P. (1989a). Induction of maturation and ovulation by pimozide-LHRH analogue treatment and resulting high quality egg production in the Asian catfish, *Clarias batrachus* (L.). *Aquaculture* **83,** 193–199.

Manickam, P., and Joy, K.P.(1989b). Changes in hypothalamic monoamine oxidase activity in relation to season, ovariectomy, and 17β-estradiol administration in intact and ovariectomized catfish, *Clarias batrachus* (L). *Gen. Comp. Endocrinol.* **75,** 437–445.

Manickam, P., and Joy, K.P. (1990) Changes in hypothalamic catecholamine levels in relation to season, ovariectomy, and 17β-estradiol replacement in the catfish, *Clarias batrachus* (L). *Gen. Comp. Endocrinol.* **80,** 167–174.

Olcese, J., Hall, T.R., Figueroa, H.R., and De Vlaming, V.L. (1981). Pinealectomy and melatonin effects on daily variations of the hypothalamic serotonergic system in the goldfish. *Comp. Biochem. Physiol.* **70A,** 69–72.

Peter, R.E., Chang, J.P., Nahorniak, C.S., Omeljaniuk, R.J., Sokolowska, M., Shih, S.H., and Billard, R. (1986). Interactions of catecholamines and GnRH in regulation of gonadotropin secretion in teleost fish. *Rec. Prog. Horm. Res.* **42,** 513–548.

Peter, R.E., Trudeau, V.L., and Sloley, B.D. (1991). Brain regulation of reproduction in teleosts. *Bull. Inst. Zool. Acad. Sinica,* Monogr. **16,** 89–118.

Peter, R.E., and Yu, K.L. (1997). Neuroendocrine control of ovulation in fishes : basic and applied aspects. *Rev. Fish Biol. Fisheries* **7,** 173–197.

Peute, J, Schild, R.G., Schild, V.A., Buijs, R.M., van Asselt, L.A.C., and van Oordt, P.G.W.J. (1987). Immunocytochemical evidence for peptidergic (GnRH) and dopaminergic innvervation of the gonadotropic cells in the pituitary of the African catfish, *Clarias gariepinus. Gen. Comp. Endocrinol.* **67,** 303–310.

Prasada Rao, P.D., Rajendra Babu, and Jadhao, A.G. (1987). Distribution of serotonin-like immunoreactivity in the hypothalamus and pineal organ of the catfish, *Mystus vittatus* (Bl). *In* "Proceedings of First Congress of Asia and Oceania Society for Comparative Endocrinology" (E. Ohinishi, Y. Nagahama, and H. Ishizaki, Eds.), pp. 115–116, University of Nagoya Printing Corporation, Nagoya.

Prat, F., Sumpter, J.P., and Tyler, C.R. (1996). Validation of radioimmunoassays for two salmon gonadotropins (GTH I and GTH-II) and their plasma concentration throughout the reproductive cycle in male and female rainbow trout (*Oncorhynchus mykiss*). *Biol. Reprod.* **54,** 1375–1382.

Saligaut, C., Salbert, G., Bailhache, T., Bennani, S., and Jego, P. (1992a). Serotonin and dopamine turnover in the female rainbow trout (*Oncorhynchus mykiss*) brain and pituitary : Changes during the annual reproductive cycle. *Gen. Comp. Endocrinol.* **85,** 261–268.

Saligaut, C, Garnier, D., Bennai, S, Salbert, G, Bailhache, T., and Jego, P. (1992b) Effects of estradiol on brain aminergic turnover of the female rainbow trout (*Oncorhynchus mykiss*) at the beginning of vitellogenesis. *Gen. Comp. Endocrinol.* **88,** 209–216.

Schulz, R.W., Bogerd, J., Bosma, P.T., Peute, J., Robers, F.E.M., Zandbergen, M.A., and Goos, H.J. Th. (1995). Physiological, morphological, and molecular aspects of gonadotropins in fish with special reference to the African catfish, *Clarias gariepinus. In* "Reproductive Physiology of Fish" (F. Goetz and P. Thomas, Eds.), 2–6, Fish Symposium 95, Austin.

Senthilkumaran, B., and Joy, K.P. (1993). Annual variations in hypothalamic serotonin and monoamine oxidase in the catfish *Heteropneustes fossilis*, with a note on brain regional differences of day-night variations in gonadal preparatory phase. *Gen. Comp. Endocrinol.* **90,** 372–382.

Senthilkumaran, B., and Joy, K.P. (1994a). Effects of photoperiod alterations on day-night variations in hypothalamic serotonin content and turnover, and monoamine oxidase activity in the female catfish, *Heteropneustes fossilis* (Bloch). *Fish Physiol. Biochem.* **13,** 301–307.

Senthilkumaran, B., and Joy, K.P. (1994b). Effects of ovariectomy and oestradiol replacement on hypothalamic serotonergic and monoamine oxidase activity in the catfish, *Heteropneustes fossilis* : a study correlating plasma oestradiol and gonadotrophin levels. *J. Endocrinol.* **142,** 193–203.

Senthilkumaran, B., and Joy, K.P. (1995a). Changes in hypothalamic catecholamines, dopamine-β-hydroxylase and phenylethanolamine-N-methyl-transferase in the catfish, *Heteropneustes fossilis* in relation to season, raised photoperiod and temperature, ovariectomy, and estradiol-17β replacement. *Gen. Comp. Endocrinol.* **97,** 121–134.

Senthilkumaran, B., and Joy, K.P. (1995b). Effects of melatonin, p-chloro-phenylalanine, and α-methylparatyrosine on plasma gonadotropin level and ovarian activity in the catfish, *Heteropneustes fossilis* : A study correlating changes in hypothalamic monoamines. *Fish Physiol. Biochem.* **14,** 471–480.

Senthilkumaran, B., and Joy, K.P. (1995c). A turnover study of hypothalamic monoamine oxidase (MAO) and effects of MAO inhibition on gonadotropin secretion in the female catfish, *Heteropneustes fossilis. Gen. Comp. Endocrinol.* **97,** 1–12.

Senthilkumaran, B., and Joy, K.P. (1996). Effects of administration of some monoamine-synthesis blockers and precursors on ovariectomy-induced rise in plasma gonadotropin II in the catfish *Heteropneustes fossilis. Gen. Comp. Endocrinol.* **101,** 220–226.

Senthilkumaran, B., and Joy, K.P. (1998). Methallibure-induced inhibition of hypothalamo-hypophysial-ovarian activity in the catfish *Heteropneustes fossilis* involves changes in hypothalamic monoamine activity. *Fish Physiol. Biochem.* (in press).

Sloley, B.D, Trudeau, V.L., and Peter, R.E. (1992). Dopamine catabolism in goldfish (*Carassius auratus*) brain and pituitary : Lack of influence of catecholestrogens on dopamine catabolism and gonadotropin secretion. *J. Exp. Zool.* **263,** 398–405.

Somoza, G., Yu, K.L., and Peter, R.E. (1988). Serotonin stimulates gonadotropin release in female and male goldfish, *Carassius auratus L. Gen. Comp. Endocrinol.* **72,** 374–382.

Sundararaj, B.I. (1981). Reproductive physiology of teleost fishes: A review of present knowledge and needs for future research. *ADCP/REP/81/16*, pp. 1–82, FAO, Rome.

Swanson, P., and Dittman, A. (1997). Pituitary gonadotropins and their receptors in fish. *In* "Advances in Comparative Endocrinology" (S. Kawashima and S. Kikuyama, Eds.), pp. 841–846, Mondruzzi Editore, Bologna.

Tharakan, B., and Joy, K.P. (1996). Effects of mammalian gonadotropin-releasing hormone analogue, pimozide, and the combination on plama gonadotropin levels in different seasons and induction of ovulation in female catfish. *J. Fish Biol.* **48,** 623–632.

Trudeau, V.L., Sloley, B.D., Wong, A.O.L., and Peter, R.E. (1993a). Interactions of gonadal steroids with brain dopamine and gonadotropin-releasing hormone in the control of gonadotropin-II secretion in the goldfish. *Gen. Comp. Endocrinol.* **89,** 39–50.

Trudeau, V.L., Sloley, B.D., and Peter, R.E. (1993b). Norepinephrine turnover in the goldfish brain is modulated by sex steroids and GABA. *Brain Res.* **624,** 29–34.

Trudeau, V.L., and Peter, R.E. (1995). Functional interactions between neuroendocrine systems regulating GTH-II release. *In* "Reproductive Physiology of Fish" (F.W. Goetz and P. Thomas, Eds.), pp. 44–48, Fish Symposium 95, Austin.

Tsai, C.L., and Wang, L.H. (1997). Effects of estradiol on the serotonin secretion and turnover in the hypothalamus of male tilapia *Oreochromis mossambicus, in vitro. Gen. Comp. Endocrinol.* **106,** 175–180.

Van Asselt, L.A.C., Goos, H.J. Th., de Leeuw, R., Peter, R.E., Hol, E.M., Wassenberg, F.P., and van Oordt, P.G.W.J. (1990). Characterization of dopamine D_2 receptors in the pituitary of the African catfish, *Clarias gariepinus. Gen. Comp. Endocrinol.* **80,** 107–115.

Yaron, Z. (1995). Endocrine control of gametogenesis and spawning induction in the carp. *Aquaculture* **129,** 49–73.

Yu, K.L., and Peter, R.E. (1992). Adrenergic and dopaminergic regulation of gonadotropin-releasing hormone release from goldfish preoptic-anterior hypothalamus and pituitary *in vitro. Gen. Comp. Endocrinol.* **85,** 138–146.

Zheng, W., and Stacey, N.E. (1997). A steroidal pheromone and spawning stimuli act via different neuroendocrine mechanisms to increase gonadotropin and milt volume in male goldfish *Carassius auratus. Gen. Comp. Endocrinol.* **105,** 228–238.

9

OPIOIDERGIC REGULATION OF LUTEINIZING HORMONE SECRETION AND GONADAL STEROIDOGENESIS IN MALE RAT

R. N. Saxena and Poonam A. Kant

Department of Zoology
University of Delhi
Delhi 110007, India

1. INTRODUCTION

Endogenous Opioid Peptides (EOPs) are a class of neuropeptides which were initially recognized as analgesic agents with actions similar to morphine. Ever since their discovery in mid-seventies, they have aroused special interest not only because of their key roles in different domains of physiology, but because they have belied all norms and fundamentals of our knowledge of hormones and their action. They have compelled endocrinologists to rethink on the manner in which they view hormones. First, EOPs do not have a specific site of production and second, they lack a specific target organ as well. They are ubiquitously produced in the body and have actions of varied kinds at multiple sites. These observations have paved the way for deeper insight and research into the nature and mode of action of other classical hormones as well. The knowledge of opioids may well hallmark a new concept in the understanding of physiological mechanisms.

Notwithstanding the unorthodox nature of endogenous opioids, the following text is an attempt to review the role of these peptides in reproduction in the male rat with reference to other animals, wherever relevant.

2. DISCOVERY

In 1975, Hughes and co-workers isolated two pentapeptides with morphine-like activity from pig brain. These they named met-enkephalin and leu-enkephalin. Simantov and Snyder (1976) isolated similar peptides from calf brain. Many other peptides, all containing the basic amino-acid sequence of enkephalins were later found to have morphine-like activity. The most potent of these is beta-endorphin, a 31aa peptide. Beta-endorphin and

Neural Regulation in the Vertebrate Endocrine System, edited by Prasada Rao and Peter, Kluwer Academic / Plenum Publishers, New York, 1999.

various other endorphins have now been synthesized. Sequences of human, ox, sheep, pig, camel and salmon beta-endorphins have been determined; the mammalian peptides are very similar, differing only at three residues.

3. BIOSYNTHESIS

EOPs are derived from three different precursors each coded by distinct genes and expressed in distinct neuron systems.

3.1. Proenkephalin A

This 241aa sequence yields four molecules of met-enkephalin, one molecule each of leu-enkephalin, enkephalin heptapeptide and enkephalin octapeptide.

3.2. Prodynorphin or Proenkephalin B

This is a 234aa peptide chain which generates alpha- and beta-neoendorphins, dynorphins A and B, and possibly leu-enkephalin.

3.3. Proopiomelanocortin (POMC)

It has 241aa and gives rise to beta-endorphin. It also forms ACTH (Adrenocorticotropic hormone), beta-LPH (beta-Lipotropic Hormone), beta-MSH (beta- Melanocyte Stimulating Hormone) and CLIP (Corticotropin-like Intermediary Peptide) in the pituitary (Mains *et al.*, 1977).

EOPs have been localized in the brain, hypothalamus, pituitary, blood, gastrointestinal tract, pancreas, adrenal, skin, cells of immune system, male and female genital tracts and spermatozoa in many species of mammals, birds, amphibians and fishes (Olson, 1993; Vethamani-Globus, 1989; Cuello, 1983).

4. RECEPTORS

Opioids as well as EOPs exert their action via interaction with cell surface receptors. Opioid receptors have been pharmacologically classified into three major categories: mu-, delta- and kappa-, each containing at least two subtypes. In addition, an opioid receptor-like protein called nociceptin (Meunier *et al.*, 1995) and orphain FQ (Reinscheid *et al.*, 1995) has also been shown to bind endogenous opioids. All of these receptors are members of the G-protein coupled receptor superfamily and, thus, are predicted to contain seven transmembrane domains. Activation of opioid receptors results, through a pertussis-toxin sensitive mechanism, in the inhibition of adenylate cyclase and the modulation of Ca^{2+} and K^{+} conductances (Barnard and Simon, 1993). The enkephalins possess a preferential affinity for delta-receptors, dynorphins and alpha-neoendorphins bind preferentially to kappa-receptors; while beta-endorphin has a high affinity for all three types of receptors (Pfeiffer and Herz, 1982; Pfeiffer *et al.*, 1982). Kappa-receptors are abundantly present in areas which may be involved in the endocrine effects of opiates such as anterior hypothalamus, amygdala and hippocampus. The mu-opioid receptor mediates the analgesic and addictive effects of morphine (Matthes *et al.*, 1996).

5. BIOLOGICAL ACTIONS

Many actions of enkephalins, endorphins and related peptides are similar to those of morphine such as induction of analgesia, hypothermia, salivation, inhibition of peristalsis and catatonia in humans and rats. In rats, beta-endorphin also causes abnormal sexual behaviour, excessive grooming, stimulation of food intake, wet dog shakes and a decrease in general motor activity (Olson *et al.*, 1993; Dalayeun *et al.*, 1993). EOPs also markedly increase prolactin and growth hormone release and inhibit secretion of LH (luteinizing hormone), FSH (follicle stimulating hormone) and TSH (thyroid stimulating hormone) (Bruni *et al.*, 1977; Blank *et al.*, 1980; Kalra, 1983).

6. ROLE IN REPRODUCTION

The role of EOPs in reproduction is now well recognized and based on ample evidence. As early as 1955, Barraclough and Sawyer reported blockade of ovulation in rats by administration of morphine prior to the spontaneous discharge of ovulatory hormone on proestrus. It is only about two decades ago that there has been renewed interest in opiate participation in reproductive function. In general, it has been found that EOPs inhibit release of gonadotropin-releasing hormone (GnRH), gonadotropins (mainly LH) and steroids.

6.1. Effect on GnRH

EOPs exert an inhibitory effect on the release of GnRH. Endogenous opioid peptides-containing neurons are found in close association with GnRH-secreting neurons in the hypothalamus (Bloom *et al.*, 1978; Finley *et al.*, 1981a, b; Watson *et al.*, 1982). Direct axo-axonic link between GnRH and EOP neurons have also been demonstrated in the arcuate nucleus of the hypothalamus (Thind and Goldsmith, 1988). Opioids inhibit the firing frequency and pulse generation by GnRH releasing neurons (Ito *et al.*, 1993) and decrease the frequency of GnRH pulses in the hypophyseal portal blood (Gilbeau *et al.*, 1985). GnRH-secreting nuclei in the hypothalamus possess kappa-type opiate receptors (Pfeiffer *et al.*, 1982). EOPs also block GnRH release from MBH fragments *in vitro* (Rotsztejn *et al.*, 1978).

Hypothalamic opiate activity changes with change in steroid milieu. Gonadectomy decreases opiate receptors in hypothalamus and replacement regimes replenish these receptors (Wardlaw *et al.*, 1982a; Weiland and Wise, 1990). Some reports, however, deny the role of steroids in the opioid activity of hypothalamus. It has, therefore, been suggested that there exist both steroid dependent and steroid independent opioidergic pathways in the hypothalamus (Kimura *et al.*,1991; Babu *et al.*, 1987).

In addition to a direct opioid action in GnRH neurons, a large body of evidence also suggests indirect effects mediated via monoaminergic neurons (Khatchaturian and Watson, 1982; Kalra and Kalra, 1984). Alpha- adrenergic blockers and inhibitors of noradrenaline synthesis prevent the rise in plasma LH in response to naloxone (an opioid receptor antagonist) injection suggesting that endogenous opioids may also suppress release of noradrenaline (Kalra and Simpkins, 1981; Dyer *et al.*, 1985). Opioid receptors are located on the axons and terminals of the adrenergic neurons close to the GnRH neurons in the anterior hypothalamic area (Henderson, 1983). However, adrenergic modulation of these neurons is probably not the only source of input influenced by opioids. Serotoninergic projections, for instance, show a similar close anatomical relationship with GnRH neurons (Jennes *et al.*, 1982) and EOPs appear to stimulate serotonin turnover in the preoptic region (Johnson and Crowley, 1984). It has also been suggested that GnRH secretory terminals themselves may possess opioid receptors since deafferentation of the medio-basal hypothalamus results in a fall in the content of the receptors (Rostene *et al.*, 1982).

6.2. Effect on Gonadotropins

In general, opioids inhibit the synthesis and secretion of gonadotropins in males and females of several species. The action is more pronounced on LH than on FSH. Opiate receptor antagonists (e.g. naloxone) are able to elevate plasma levels of LH in many species including humans (Delitala *et al.*, 1981; Van Vugt *et al.*, 1982). Similar effects have been observed with antisera of beta-endorphin and dynorphin (Schulz *et al.*, 1981). This indicates that endogenous opiates act as tonic inhibitors of LH secretion. Administration of beta-endorphin reduces LH pulsatility in conscious castrated rats (Van Vugt *et al.*, 1982; Kinashita *et al.*, 1980). On the other hand, administration of naloxone increases LH pulse frequency in normal men as well as women (Delitala *et al.*, 1981; Ellingboe *et al.*, 1982; Baranowska *et al.*, 1984; Rupert *et al.*, 1981). Naloxone also reverses the sex-steroid induced decrease in LH pulse frequency in men (Veldhius *et al.*, 1984).

The role of EOPs in regulation of LH secretion appears highly dependent upon the endocrine condition of experimental animals. Naloxone-induced elevation of mean LH levels is extremely reduced or absent in gonadectomized rats (Bhanot and Wilkinson, 1983). It can be restored by pretreatment with testosterone or estrogens and further enhanced by progesterone (Gabriel *et al.*, 1983). Estrogens decrease hypothalamic concentration of beta-endorphin and increase its release into the hypophyseal portal blood (Wardlaw *et al.*, 1982a, b; Sarkar and Yen, 1985). It is reported that sex steroids affect the number of opiate binding sites in the brain (Wilkinson *et al.*, 1985; Weiland and Wise 1990). Contrary to this, several reports indicate the effectiveness of naloxone in increasing LH secretion in rats gonadectomized for considerably long periods (Funabashi *et al.*, 1990). Naloxone has been shown to induce increase in pulsatile LH secretion between estrus and diestrus I in the rat estrus cycle, a time when ovarian steroid negative feedback is possibly absent (Babu *et al.*, 1987). Thus, as with GnRH, there are both steroid dependent and steroid independent EOP suppressions of LH secretion.

Most evidence suggests that EOPs act centrally via hypothalamic mechanisms to alter LH secretion. There exists, however, a possibility of their action directly on the pituitary cells. Opiate receptors have been found in the pituitary (Atwen and Kuhar,1983; Lightman *et al.*, 1983). Direct effect of met-enkephalin, on *in vitro* LH and FSH release has been demonstrated in rats (Cacicedo and Franco,1986), cattle (Chao *et al.*, 1986) and pig (Barb *et al.*, 1990). Simpkins *et al.* (1991) have reported that a potent narcotic antagonist, nalmefene methyliodide, which does not cross the blood-brain barrier, blocks morphine induced suppression of LH release.

We have examined the action of opioids on adult male rat pituitary in terms of LH release using both *in vivo* and *in vitro* systems. Our data show that beta-endorphin inhibits and naloxone increases LH levels in intact male rats. Since naloxone is an opiate receptor antagonist, it acts by blocking the inhibitory effect of endogenous opioids. Naloxone is effective within 20 minutes, while beta-endorphin does not depress LH levels until 60 minutes later. This time difference is probably due to the fact that in the normal adult male rat, the hypothalamic-pituitary axis is already continuously inhibited by androgen-dependent feedback systems. Any further inhibition is difficult to detect and takes a longer time to become apparent. Pituitaries excised from rats acutely treated with naloxone or beta-endorphin exhibit a significant alteration in their capacity to release LH. Pituitaries of nalox-

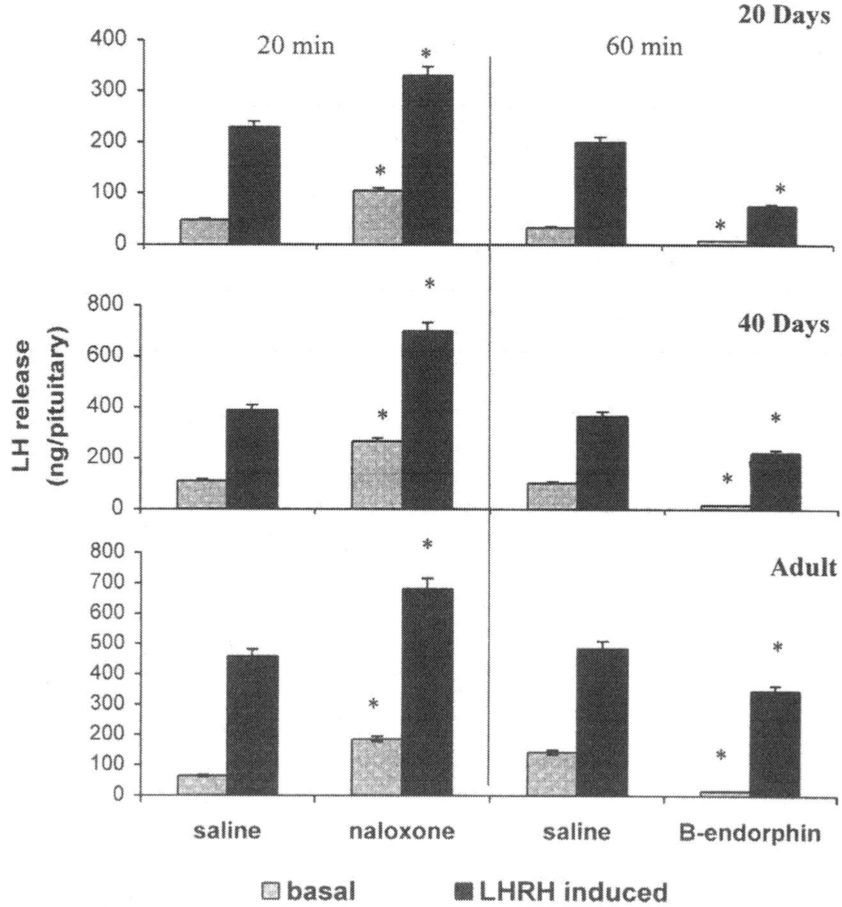

Figure 1. Effect of acute treatment with naloxone (s.c., 2mg/kgbw; autopsy after 20min) and beta-endorphin (s.c., 10μg/kgbw, autopsy after 60 min) on *in vitro* pituitary responsiveness to GnRH in male rats of different age groups. Values represent mean ± s.e.m. (n = 4), *p<0.01 compared with saline.

one-treated rats showed enhanced basal and GnRH induced LH release when compared to those from saline treated rats, whereas treatment with beta-endorphin depressed both (Figure 1).

These observations demonstrate a definite change in pituitary responsiveness following opioid treatment and raise the possibility of a direct opioid action on pituitary cells. To examine this possibility, isolated cells of intact male rat pituitaries were incubated with beta-endorphin and naloxone in the presence or absence of GnRH. It was observed that both had no effect on basal LH release, but GnRH-stimulated LH release was significantly augmented by naloxone, whereas beta-endorphin suppressed it. Naloxone not only augmented the effect of GnRH on LH release but also countered the inhibitory effect of beta-endorphin on GnRH-stimulated LH release. This indicates the presence of a local inhibitory opioid tone in the pituitary which is mediated via opiate receptors. Since receptor sites for opioids and GnRH are separate, one can expect to see an augmentation of the release of LH after simultaneous addition of GnRH plus naloxone. If isolated anterior pituitary cells are challenged with naloxone and beta-endorphin in the presence of different

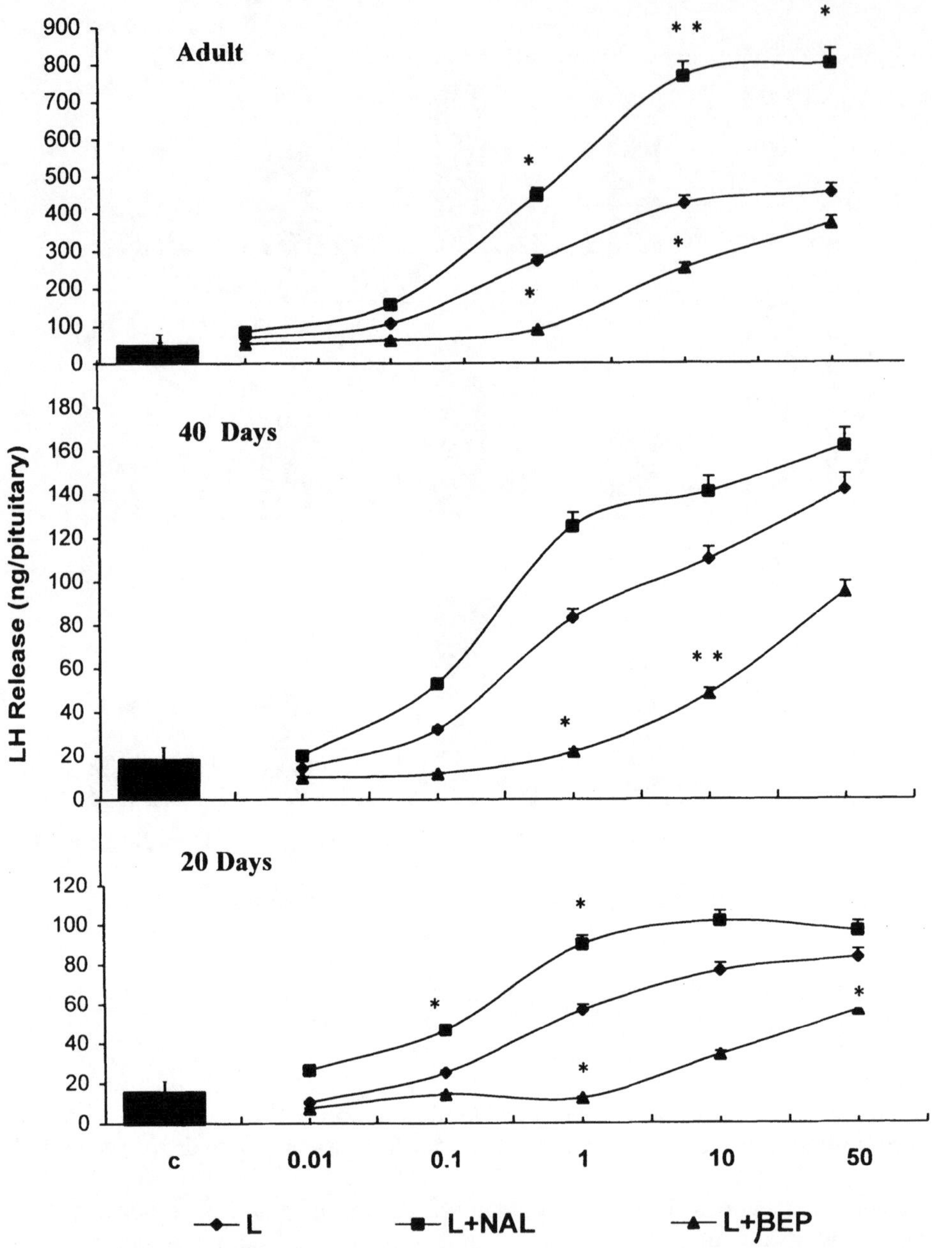

Figure 2. Effect of β-endorphin (βEP) and naloxone (NAL) on LH release from isolated pituitary cells in the presence of different concentrations of GnRH (L). Values represent mean ± s.e.m. (n = 4); C, Control; p values; *<0.01 Vs L; **<0.001 Vs L.

concentrations of GnRH, naloxone augments LH release at almost all stimulatory doses of GnRH. Therefore, naloxone enhanced pituitary responsiveness but not sensitivity to GnRH in adult rats.

Beta-endorphin, however, reduced the sensitivity of the cells i.e., higher concentrations of GnRH were required to elicit LH release (Figure 2).

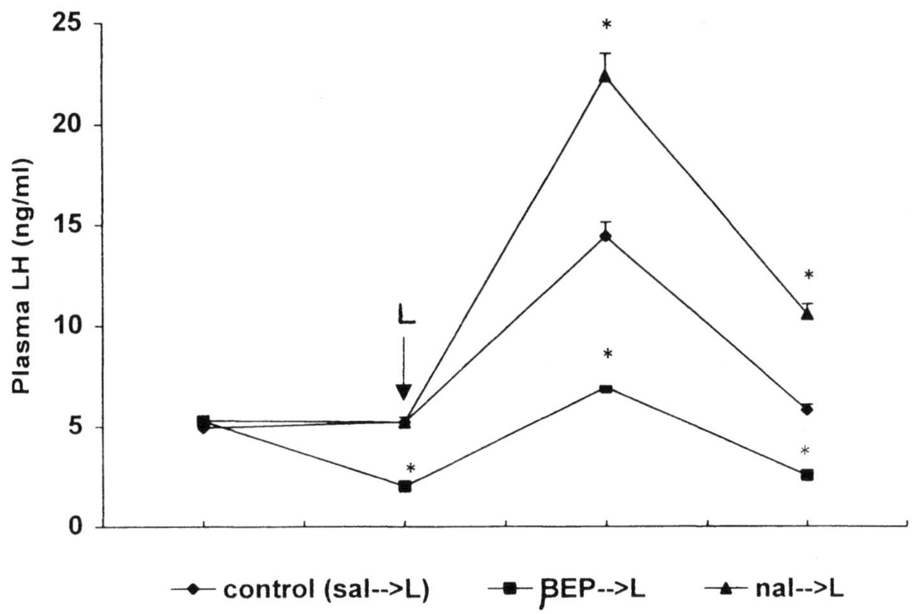

Figure 3. Effect of priming (60min) with naloxone (nal; s.c., 2mg/kgbw) and beta- endorphin (bEP;s.c., 10μg/kgbw) on *in vivo* GnRH (L;s.c.,25ng) induced LH release in adult male rats. Values are mean ± s.e.m. (n = 4); *p<0.01 Vs Control (sal →L).

At high doses, GnRH could override the inhibition of LH release by beta-endorphin. If anterior pituitary cells were preincubated with beta-endorphin their response to subsequent treatment with GnRH was greatly diminished. No effect of pretreatment with naloxone was observed. Similarly, priming of rats with beta-endorphin greatly diminished the response of the animals to GnRH, whereas treatment with naloxone caused a significant increase in the responsiveness (Figure 3).

All these results indicate that beta-endorphin exerts a direct effect on the anterior pituitary cells and suppresses their response to GnRH. The responsiveness of the gonadotropes is known to depend on the GnRH receptor numbers and their binding potency (Katt *et al.*, 1985). Opioid effects on pituitary gonadotropes, therefore, may be through the regulation of either receptors or some post-receptor mechanism.

There is ample evidence to suggest that opioids and steroids interact with each other in alteration of gonadotropin secretion via hypothalamic mechanisms. It is also known that steroids regulate LH secretion not only at the level of the hypothalamus, but also at hypophyseal level by altering pituitary responsiveness to GnRH (Drouin and Labrie, 1981). It is, therefore, possible that opioids and steroids may interact at the level of pituitary to regulate LH secretion. *In vitro* experiments with isolated anterior pituitary cells have shown that beta-endorphin intensifies the inhibitory effect of testosterone but reduces the stimulatory effect of estradiol on GnRH-induced LH release (Kant, 1995).

6.3. Effect on Testicular Steroidogenesis

POMC-derived peptides have been found to be present in the gonads and in the accessory reproductive organs as chemical signals working in a paracrine fashion (Polzonetti-Magni *et al.*, 1995). Relatively high concentrations have been measured in human semen

(Sharp and Pekary, 1981); Leydig cells (Tsong *et al.*, 1982a; Bardin *et al.*, 1987; Chen *et al.*, 1984), Sertoli cells, spermatogenic cells (Kilpatrick *et al.*, 1987) and even spermatozoa (Carnevali *et al.*, 1997). The amount of beta-endorphin in the testis seems to be under direct control of the gonadotropins and other substances including steroids, GnRH, and glucocorticoids (Fabbri and Dufau, 1988). Treatment with human chorionic gonadotropin (hCG) increases the concentration of radioimmunoassayable beta-endorphin in the interstitial fluid and POMC-like mRNA concentration in the Leydig cells (Valenca and Negro-Vilar, 1986; Chen and Madigan, 1987). There is some doubt over the functional role of beta-endorphin in the testis, though increasing evidence suggests that it regulates the production of testosterone. Intratesticular administration of beta-endorphin decreases testosterone in interstitial fluid (Chandrashekhar and Bartke,1992), whereas naloxone increases serum testosterone levels without any apparent changes in LH. It has been reported that testosterone response to LH treatment in hypophysectomized rats that had received intratesticular injections of beta-endorphin was significantly reduced (Cicero *et al.*, 1989).

Experiments performed by the authors have shown that bilateral intratesticular administration of naloxone resulted in a rise in plasma testosterone in adult rats at 60 and 90 min. after injection. This effect of locally administered naloxone indicates that the adult rat testes have an opioid tone similar to the one existing at the hypothalamic and pituitary level. Experimental increase in beta-endorphin levels by intratesticular administration suppressed testosterone secretion in adult rats at 90 min post injection. No change occurred in plasma LH levels following these treatments. This however, does not rule out the role of LH in the opioid effects on testosterone secretion. It has been suggested that for such effects the presence of LH in circulation is required. In hypophysectomized animals, naloxone, even at high doses, does not cause a major change in testosterone secretion (Cicero *et al.*, 1989). Thus it is possible that gonadotropin milieu is necessary for paracrine opiate effects.

Pretreatment with naloxone (intratesticular)-enhanced hCG-induced increase in testosterone secretion in adult rats (Figure 4).

This may represent an additive effect of dual stimulation by naloxone and hCG or a potentiation of the effects of hCG by naloxone through blockade of EOP tone. In beta-endrophin treated rats, the hCG-induced testosterone release was lower than in controls. The mechanism by which EOPs regulate steroidogenesis is a matter of discussion and necessitates further work. There is a possibility of a direct autocrine effect on the Leydig cells or an action mediated by other cell types of the testis. Opiates have been shown to specifically bind to Sertoli cells and inhibit FSH-stimulated production of inhibin by Sertoli cells in culture (Fabbri *et al.*, 1985). Since inhibin amplifies LH-induced testosterone secretion, opioids may indirectly affect Leydig cell steroidogenesis by altering inhibin production. The results of our *in vitro* experiments using isolated interstitial cells of both adult and immature rats, however, show that opioids can exert their effect even on isolated cells and therefore rule out the involvement of any tubular elements or factors. Naloxone enhanced hCG-induced testosterone release but had no effect on basal testosterone release. Beta-endorphin, on the other hand, suppressed both basal and hCG-induced testosterone release. The decrease in basal levels is indicative of a direct inhibitory effect of EOPs on Leydig cell's synthesis and release of testosterone (Kant and Saxena, 1995).

7. ROLE IN PUBERTY

The developmental changes that occur in the CNS prior to sexual maturation culminate in the appearance of pulsatile secretion of GnRH. It is asssumed that excitatory

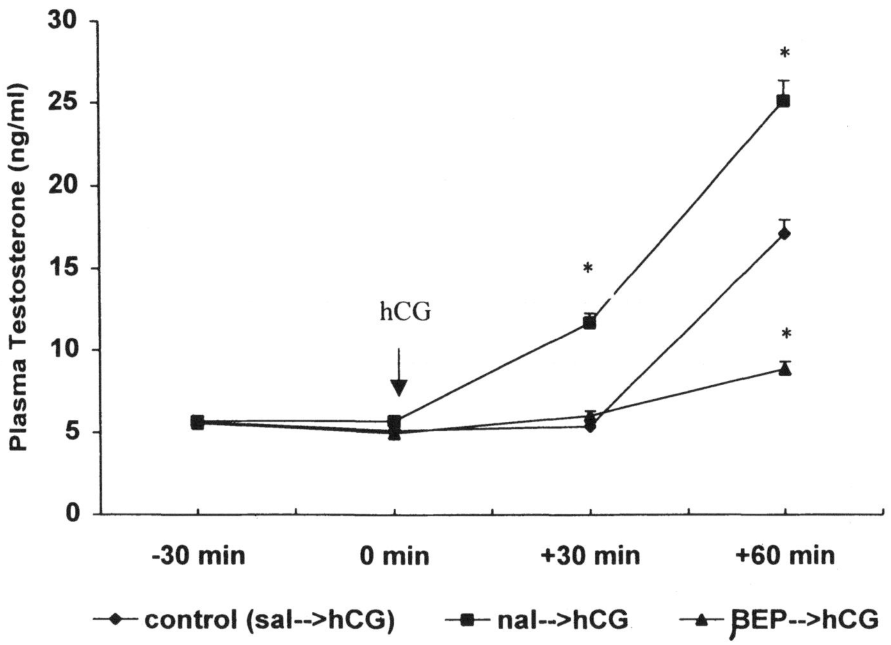

Figure 4. Effect of intratesticular administration of beta-endorphin (100ng/kgbw/testis) and naloxone (50mg/kgbw/testis) on plasma testosterone in adult male rats. Values are mean ± s.e.m. (n = 4), *p<0.01 compared with saline.

neuronal inputs provide the drive for increased pubertal secretion of GnRH. An alternative hypothesis suggests that the maturing brain somehow becomes less sensitive to the negative feedback influence of gonadal steroids (Gonadostat hypothesis). Because of the link between steroid feedback and opioids, it has been proposed that the resetting of the gonadostat occurs by reduction in opioid sensitivity. Corroborative evidence for a reduction in opioid tone at the time of puberty can be gleaned from several studies (Foster *et al.*, 1987; Martensz, 1985; Cicero, 1986). In the male rat, it is generally assumed that naloxone has no effect until atleast 30 days after birth (Marini *et al.*, 1984; Cicero, 1986; Valenca and Negro-Vilar, 1986; Mauras *et al.*, 1986; Kant, 1995).

The reason for this may be associated with a delay in the development of opioid neuron receptor complex and/or due to the existence of other more powerful inhibitory drives in the GnRH neurons. This marked difference between the response of prepubertal and peripubertal or adult rats does not appear to be due to lack of functional receptors since no major differences have been found between opioid binding in the young or adult rat hypothalamus (Leposavic *et al.*, 1991; Petrillo *et al.*, 1987).

As in adults, pituitaries excised from immature rats acutely treated with naloxone or beta-endorphin exhibit a significant alteration in their capacity to release LH (Figure 1). Pituitaries of naloxone treated rats show enhanced basal and GnRH induced LH release as compared to those from saline treated rats. Treatment with beta-endorphin depresses both. The uniformity of response in all age groups indicates that the cause of the differential *in vivo* response of adult and prepubertal rats to naloxone may not lie in the pituitary. Alternatively, an inhibitory input or lack of stimulatory effect at the gonadotropes which prevents the secretory response to naloxone in the intact prepubertal male rat *in vivo* may not be operative under *in vitro* conditions.

The direct *in vitro* effects of naloxone and beta-endorphin on isolated pituitary cells are also similar in prepubertal and adult rats. Beta-endorphin seemed to have a stronger inhibitory effect on GnRH stimulation of LH release in prepubertal rats since very high doses of GnRH are not able to override the inhibitory effect of beta-endorphin (Figure 2).

Differences have been observed in the response of testes of adult and immature rats to opioids. Intratesticular administration of naloxone has no effect in immature rats implying thereby that either the opioid tone is absent in these animals and develops only on attainment of sexual maturity or other inhibitory factors in the immature testis prevent the rise in testosterone in response to naloxone. Bardin *et al.* (1984) have shown that the content of beta-endorphin in the testis is low in immature animals, rises with age and is maximum in adults. It may, therefore, be that the levels of EOPs in immature rat testes are not enough to cause an inhibitory effect. With the rise in their levels towards maturity, the 'tone' may become operative and only then naloxone becomes effective in blocking the inhibitory effect. *In vitro* experiments using isolated interstitial cells indicate similar effects of beta-endorphin in adult as well as immature rats.

Based on the above gamut of information it may be proposed that EOPs, atleast in the male rat, have multiple sites of action of their inhibitory effect on the reproductive system which may include all the three levels of the hypothalamo-hypophyseal-testicular axis. The significance of direct hypophyseal and testicular effects is still unclear and it is not readily apparent how independent actions on the testis and pituitary contribute to their well established effects on central nervous system, or which of these mechanisms predominate under normal physiological conditions.

REFERENCES

Atwen, S.F. and Kuhar, M.J. (1983). Distribution and physiological significance of opioid receptors in the brain. *Brit. Med. Bull.* **39** ; 47–52.

Babu, G.N., Marco, J., Bona-Gallo, A. and Gallo, R.V. (1987). Steroid independent endogenous opioid peptide suppression of pulsatile Luteinizing Hormone release between estrus and diestrus 1 in the rat. *Brain Res.* **416**; 235–242.

Barb, C.R., Barrett, J.B., Wright, J.T., Kraeling, R.R. and Rampacek, G.B. (1990). Opioid modulation of Luteinizing Hormone secretion by pig pituitary cells *in vitro*. *J. Reprod. Fertil.* **90** ; 213–219.

Bardin, C.W., Shaha, C., Mather, J., Salomon, Y., Margioris, A.N., Liotta, A.S., Gerendai, I., Chen, C.L. and Krieger, D.T. (1984). Identification and possible function of pro-opiomelanocortin derived peptides in the testis. *Ann. NY Acad. Sci.* **438,** 346–364.

Bardin, C.W., Chen, C.L.C., Morris, P.L., Boitani, C., Liotta, A.S., Margioris, A. and Krieger, D.T. (1987). Proopiomelanocortin-derived peptides in testis, ovary and tissues of reproduction. *Rec. Prog. Horm. Res.* **43,** 1–28.

Barnard, E.A. & Simon, J. (1993). Opoid receptors. *In*: Neurotransmitter Receptors, (F. Hucho, Ed.), pp. 297–323. Elsevier, Amsterdam.

Barraclough, C.A. and Sawyer, C.H. (1955). Inhibition of the release of pituitary ovulatory hormone in the rat by morphine. *Endocrinology* **57**, 329–336.

Bhanot, R. and Wilkinson, M. (1984). Opiatergic control of Luteinizing Hormone secretion is eliminated by gonadectomy. *Endocrinology* **112**, 399–401.

Blank, M.S., Panerai, A.E. and Friesen, H.G. (1980). Effects of naloxone on luteinizing hormone and prolactin in serum of rats. *J. Endocrinology* **85**, 307–315.

Bloom, F., Battenberg, E., Rossier, J., Ling, N. and Guillemin, R. (1978). Neurons containing beta-endorphin in rat brain exist separately from those containing enkephalin: immunocytochemical studies. *Proc. Natl. Acad. Sci.* **75,** 1591–1594.

Bruni, J.F., Van Vugt, D., Marshall, S. and Meites, J. (1977). Effects of naloxone, morphine and met-enkephalin on serum prolactin, luteinizing hormone, follicle stimulating hormone, thyroid stimulating hormone and growth hormone. *Life Sci.* **21,** 461–466.

Cacicedo, L. and Franco, S.F. (1986). Direct Action of Opioid Peptides and naloxone on gonadotropin secretion by cultured rat anterior pituitary cells. *Life Sci* **38**, 617–625.

Carnaveli, O., Mosconi, G., Degli Innoceti, S., Polzonetti-Magni, A.M. (1997). POMC-derived peptide, b-Endorphin in stallion spermatozoa. *In* "Advances in Comparative Endocrinology" (S. Kawashima and S. Kikuyama, Eds.), pp. 1019–1022. Mondruzzi Editore, Bologna.

Chandrashekar, V. and Bartke, A. (1992). The influence of b-endorphin on testicular endocrine function in adult rats. *Biol. Reprod.* **47**, 1–5.

Chao, C.C., Moss, G.E. and Malven, P.V. (1986). Direct opioid regulation of pituitary release of bovine luteinizing hormone. *Life Sci.* **39**, 527–534.

Chen, C.L.C., Marber, J.P., Morris, P.L. and Bardin, C.W. (1984). Expression of proopiomelanocortin-like gene in the testis and epididymis. *Proc. Natl. Acad. Sci.* U.S.A. **81**, 5672–5675.

Chen,C.L.C. and Madigan, M.B. (1987). Regulation of testicular proopiomelanocort in gene expression. *Endocrinology* **121**, 590–596.

Cicero, T.J. (1986). Ontogeny of the opioid-mediated control of reproductive endorinology in the male and female rat. *J. Pharmacol. Exptl. Therap.* **236**, 627–633.

Cicero, T.J., Adams, M.L., O'Connor, L.H. and Nock, B. (1989). In vivo evidence for a direct effect of naloxone on steroidogenesis in the male rat. *Endocrinology* **125**, 957–964.

Cuello, A.C. (1983). Central distribution of opioid peptides. *Brit. Med. Bull.* **39**, 11–16.

Dalayeun, J.F., Nore's. J.M. and Bergal, S. (1993). Physiology of beta-endorphins. A close-up view and a review of the literature. *Biomed. Pharmacother*, **47**, 311–320.

Delitala, G., Devilla, L. and Arata, L. (1981). Opiate receptors and anterior pituitary hormone secretion in man. *Acta Endocrinology/* (Copenh.) **97**, 150–156.

Drouin, J. and Labrie, F. (1981). Interactions between 17b-estradiol and progesterone in the control of LH and FSH release in rat anterior pituitary cells in culture. *Endocrinology* **108**, 52–57.

Dyer, R.G., Manfield, S., Corbet, H. and Dean, A.D.P. (1985). Fasting impairs LH secretion in female rats by activating an inhibitory opioid pathway. *J. Endocrinology* **105**, 91–97.

Ellingboe, J., Veldhius, J.D., Mendelson, J.H., Kuehule, J.C. and Mello, N.K. (1982). Effect of endogenous opioid blockade on the amplitude and frequency of pulsatile luteinizing hormone secretion in normal men. *J. Clin. Endocrinology* Metab. **54**, 854–859.

Fabbri, A., Tsai-Morris, C.H., Luna, S., Fraioli, F. and Dufau, M.L. (1985). Opiate receptors are present in the rat testis. Identification and localization in Sertoli cells. *Endocrinology* **117**, 2544–2550.

Fabbri, A. and Dufau, M.L. (1988). Hormonal regulation of beta-endorphin in the testis. *J. Steroid. Biochem.* **29**, 475–482.

Finley, J.W., Maderdrut, J. and Petrusz, P. (1981a). The immunocytochemical localization of enkephalin in the central nervous system of the rat. *J. Comp. Neurol.* **198**, 541–546.

Finley, J.C.W., Lindshon, P. and Petrusz, P. (1981b). Immunocytochemical localization of b-endrophin containing neurons in the rat brain. *Neuroendocrinol.* **33**, 28–42.

Foster, D.L., Karsch, F.J., Olster, D.H., Ryan, K.D. and Yellon, S.M. (1987). Determinants of puberty in a seasonal breeder. *Rec. Prog. Horm. Res.* **42**, 331–368.

Funabashi, T., Kato, A. and Kimura, F. (1990). Naloxone affects the LH secretory pattern in the short and long term ovariectomized rats. *Neuroendocrinology* **52**, 35–41.

Gabriel, S.M., Simpkins, J.W. and Kalra, S.P. (1983). Modulation of endogenous opioid influence on LH secretion by Progresterone and Estrogen. *Endocrinology* **113**, 1806–1811.

Gilbeau, P.H., Almirez, R.G., Holaday, J.W. and Smith, C.G. (1985). Opioid effects on plasma concentrations of luteinizing hormone and prolactin in the adult rhesus monkey. *J. Clin. Endocrinology Metab.* **60**, 299–305.

Henderson, G. (1983). Electrophysiological analysis of opioid actions in the central nervous system. *Br. Med. Bull.* **39**, 59–64.

Hughes, J., Smith, T.W., Kosterlitz, H.W., Fothergill, L., Morgan, B.A. and Morris, H.R. (1975). Identification of two related pentapeptides from the brain with potent opiate-agonist activity. *Nature/London* **258**, 577–580.

Ito, K., Tanaka, T. and Mori, Y. (1993). Opioid peptidergic modulation of GnRH pulse generation activity in the ovariectomized goat. *Neuroendocrinology* **57**, 634–639.

Jennes, L., Beckman, W.C., Stumpf, W.E. and Grzanna, R. (1982). Anatomical relationships of serotoninergic and noradrenergic projections with the GnRH system in septum and hypothalamus. *Exptl. Brain Res.* **46**, 331–338,

Johnson, M.D. and Crowley, W.R. (1984). Effects of opiate antagonists on serotonin turnover and on luteinizing hormone and prolactin secretion in estrogen or morphine-treated rats. *Neuroendocrinology* **38**, 322–327.

Kalra, S.P. and Simpkins, J.W. (1981). Evidence for noradrenergic mediation of opioid effects on LH secretion. *Endocrinology* **109**, 776–782.

Kalra, S.P. (1983). Opioid peptides - inhibitory neuronal system in regulation of gonadotropin secretion *In* "Role of Peptides and Proteins in Control of Reproduction" (S.M. McCann and D.S. Dhindsa, Eds.), pp. 63–128. Elsevier Biomedical, New York.

Kalra, S.P. and Kalra, P.S. (1984). Opioid adrenergic steroid connection in regulation of luteinizing hormone secretion in the rat. *Neuroendocrinology* **38**, 418–426.

Kant, P.A. (1995). Ph.D. Thesis, Studies on the Opioidergic Regulation of Luteinizing hormone secretion and Testicular Steroidogenesis in the male albino rat. University of Delhi.

Kant, P.A. and Saxena, R.N. (1995). Effect of beta-endorphin and naloxone on rat testicular steroidogenesis. *Ind. J. Exp. Biol.* **33,** 165–168.

Katt, J.A., Duncan, J.A., Herbon, L., Barkan, A. and Marshall, J.C. (1985). The frequency of GnRH stimulation determines the number of pituitary GnRH receptor. *Endocrinology* **116,**2113–2115.

Kew, D. and Kilpatrick, D.I. (1989). Expression and regulation of the proenkephalin gene in rat Sertoli cells. *Mol. Endocrinology* **3**, 179–185

Kimura, F., Nishihara, M., Hiruma, H. and Funabashi, T. (1991). Naloxone increases the frequency of the electrical activity of LHRH pulse generator in long-term overiectomized rats. *Neuroendocrinol.* **53**, 97–102.

Kinashita, F., Nakai, I., Katakami, H., Kato, G., Yajima, H. and Imura, H. (1980). Effect of b-endorphin on pulsatile luteinizing hormone release in conscious castrated rats. *Life Sci.* **27**, 843–847.

Leposavic, G., Cover, P.O. and Buckingham, J.C. (1991). In vivo and in vitro studies on the Opioidergic control of the secretion of Gonadotrophin-Releasing Hormone and LH in sexually immature and adult male rats. Neuroendocrinology 53, 579–588.

Lightman, S.L., Ninkovic, M., Hunt, S.P. and Iversen, L.L. (1983). Evidence for opiate receptors on pituicytes. *Nature* **305,** 235–236.

Mains, R.E., Eipper, B. and Ling, N, (1977). Common precursor to corticotropin and endorphins. *Proc. Natl. Acad. Sci.* USA **74,** 3014–3023.

Marini, R., Attanasio, A., Burg, P. and Gupta, D. (1984). Modulation of LHRH and gonadotropin secretion by naloxone in male rats during development. *In* "Central and Peripheral Endorphins: Basic and Clinical Aspects, (E.E. Muller and A.R. Genazzani, Eds.), pp. 175–180. Raven Press, New York.

Martensz, N.D. (1985). Changes in the processing of b-endorphin in the hypothalamus and pituitary gland of female rats during sexual maturation. *Neuroscience* **16**, 625–640.

Matthes, H.W.D., Maldonado, R., Simonin, F., Valverde, O., Slowe, S., Kitchen, L., Befort, K., Dierich, A., Le Meur, M., Dolle, P., Tzavara, E., Hanoune. J. Roques, B.P. & Kieffer, B.L. (1996). Loss of morphine-induced analgesia, reward effect amd withdrawal symptoms in mice lacking the mu-opioid receptor gene. *Nature* (London) **383**, 819–823.

Mauras, N., Veldhius, J.D. and Rogol, A.D. (1986). Role of endogenous opiates in pubertal maturation: opposing actions of naltrexone in prepubertal and late pubertal boys. *J. Clin. Endocrinology Metab.* **62**, 1256–1263.

Meunier, J.C., Mollereau, C., Toll, L., Suaudeau, C., Moisand, C., Alvinerie, P., Butour, J.L., Guillemot. J.C., Ferrara, P., Monsarrat, B., Mazarguil, H., Vassart, G., Parmentier, M. & Costentin, J. (1995). Isolation and structure of the endogenous agonist of opioid receptor like peptide *Nature* (London) **377**, 532–538.

Olson, G.A., Olson, R.D. and Kastin, A.J. (1993). Endogenous Opiates: 1992, *Peptides* **14**, 1339–1378.

Petrillo, P., Tavani, A., Verotta, D., Robson, L.K. and Kosterlitz, H.W. (1987). Different post-natal development of mu, delta and kappa-opioid binding sites in rat brain. *Dev. Brain Res.* **31**, 53–58.

Pfeiffer, A. and Herz, A. (1982). Discrimination of three opiate receptor binding sites with the use of a computerized curve fitting technique. *Molecular Pharmacology* **21**, 266–271.

Pfeiffer, A., Pasi, A., Mehraein, P. and Herz, A. (1982). Opiate receptor binding sites in human brain. *Brain Res.* **248**, 87–96.

Polzonetti - Magni, A.M., Mosconi, G., Gallinelli, A., Fachinetti, F. (1995). Ovarian Opioid system in vertebrates. *Adv. Contra. Deliv. Syst.* **11,** 273–281.

Reinscheid, R.K., Nothacker, H.P., Bourson, A., Ardati, A., Henningsen, R.A., Bunzow, J.R., Grandy, D.K., Langen, H., Monsma, F.J., Jr. & Civelli, O. (1995). Orphain FQ: A neuropeptide that activates an opioid-like G-protein coupled receptor. *Science* **270**, 792–794.

Rostene, W.H., Drouva, S.V., Pollard, H., Sokoloff, P., Pattou, E. and Kordon, C. (1982). Further evidence for the existence of opiate binding sites on neurosecretory LHRH mediobasal hypothalamic terminals. *Eur. J. Pharmacol.* **80,** 139–141.

Rotsztejn, W.H., Drouva, S.V., Pattou, E. and Kordon, C. (1978b). Effect of morphine on the basal and dopamine-induced release of LHRH from mediobasal hypothalamic fragments in vitro. *Eur. J. Pharmacol.* **50,** 285–286.

Rupert, J.F., Quigley, M.E. and Yen, S.S.C. (1981). Endogenous opiates modulate pulsatile luteinizing hormone release in humans. *J. Clin. Endocrinology Metab.* **52**, 583–87.

Sarkar, D.H. and Yen, S.S. (1985). Changes in b-endorphin-like immunoreactivity in pituitary Portal blood during the estrous cycle and after ovariectomy in rats. *Endocrinology* **116**, 2075–2079.

Schulz, R., Wilhelm, A., Pirke, K.M., Gramsch, C. and Herz, A. (1981). b-Endorphin and dynorphin control serum LH level in immature rats. *Nature* **294**, 757–759.

Sharp, B. and Pekary, A.B. (1981). b-Endorphin 61–91 and other b-endorphin immuno-reactive peptides in human semen. *J. Clin. Endocrinology Metab.* **52**, 586–590.

Simantov, R. and Snyder, S.H. (1977). Morphine-like peptides in mammalian brain. Isolation, structure elucidation and interaction with the opiate receptor. *Proc. Natl. Acad. Sci.* USA **73**, 2515–2515.

Simpkins, J.W., Suvager, D. and Millard, W.J. (1991). Evaluation of the sites of opioid influence on anterior pituitary hormone secretion using a quaternary opiate antagonist. *Neuroendocrinology* **54**, 384–390.

Thind, K.K. and Goldsmith, P.C. (1988). Infundibular gonadotropin-releasing hormone neurons are inhibited by direct opioid and autoregulatory synapses in juvenile monkeys. *Neuroendocrinology* **47**, 203–216.

Tsong, S.D., Phillips, D.M., Halmi, N., Krieger, D.T. and Bardin, C.W. (1982a). b-endorphin is present in the male reproductive tract of five species. *Biol. Reprod.* **27**, 755–759.

Valenca, M.M. and Negro-Vilar, A. (1986). Proopiomelanocortin-derived peptides in testicular interstitial fluid: characterization and changes in secretion after hCG or LHRH analog treatment. *Endocrinology* **118**, 32–41.

Van Vugt, D.A., Sylvester, P.W., Aylsorth, C.F. and Meites, J. (1982). Counteraction of gonadal steroid inhibition of LH by naloxone. *Neuroendocrinology* **34**, 274–280.

Veldhius, J.D., Rogol, A.D., Samojik, E. and Ertel, N.H. (1984). Role of endogenous opiates in the expression of negative feedback actions of androgen & estrogen on pulsatile properties of luteinizing hormone secretion in man. *J. Clin. Invest.* **74**, 47–55.

Vethamani-Globus, S. (1989). Immunohistochemical localization of b-endorphin like material in the urodele and anuran amphibian tissue. *Gen. Comp. Endo.* **75**, 271–279.

Wardlaw, S.L., Thoron, L. and Frank, A.G. (1982a). Effects of sex steroids on brain b-endorphin. *Brain Res.* **245**, 327–331.

Wardlaw, S.L., Wehrenberg, W.B., Ferin, M., Antunes, J.L. and Frank, A.G. (1982b). Effect of sex steroids on b-endorphin in hypohyseal portal blood. *J. Clin. Endocrinology Metab.* **55**, 877–882.

Watson, S.J., Khatchaturian, H., Akil, H., Coy, D.H. and Goldstein, A. (1982). Comparison of the distribution of dynorphin system and enkephalin system in brain. *Science* **218**, 1132–1135.

Weiland, N.G. and Wise, P.M. (1990). Estrogen and progesterone regulate opiate receptor densities in multiple brain regions. *Endocrinology* **126** ; 804–808.

Wilkinson, M., Brawer, J.R. and Wilkinson, D.A. (1985). Gonadal steroid-induced modification of opiate-binding sites on the anterior hypothalamus of female rats. *Biol. Reprod.* **32**, 501–506.

10

PHOTOPERIODIC CONTROL OF GONADOTROPHIN-RELEASING HORMONE SECRETION IN SEASONALLY BREEDING BIRDS

Alistair Dawson

Institute of Terrestrial Ecology
Monks Wood, Abbots Ripton, Huntingdon, Cambridgeshire
PE17 2LS, United Kingdom

1. INTRODUCTION

Birds are highly photoperiodic. By controlling the time of gonadal maturation and the time of gonadal regression, photoperiod defines the period of the year during which birds are physiologically able to breed. The fine tuning of exactly when within this period that birds actually do breed may be controlled by other factors. The activity of the neurons that synthesise and secrete gonadotrophin-releasing hormone (GnRH) is central to the control of breeding activity. The activity of these cells is regulated by neural signals integrated from external (e.g., photoperiod) and internal (e.g., circadian clock) information. In response they secrete GnRH. The physiological, anatomical and behavioural changes necessary for reproduction are all dependent on the rate of secretion of GnRH. This chapter reviews the seasonal changes in activity of these cells, their neural control and the possible influence of other hormone systems.

2. BREEDING SEASONS

2.1. The Ultimate Factors which Determine Breeding Seasons

In seasonally changing environments, natural selection will favour the genotypes whose young are reared at the most auspicious time of year. Attempts to breed earlier or later than the optimal breeding season will incur disproportionate mortality of young, and possibly also of adults. The factors which directly or indirectly affect survival and which therefore ultimately lead to the evolution of particular breeding seasons, are termed the ul-

Neural Regulation in the Vertebrate Endocrine System, edited by Prasada Rao and Peter, Kluwer Academic / Plenum Publishers, New York, 1999.

timate factors (Baker, 1938). Reproduction is energetically demanding. Consequently, food supply is often the most significant factor.

The importance of food availability as an ultimate factor is possibly more acute in birds than in mammals—they do not have the flexibility that feeding young on milk permits. In mammals, adequate milk production requires only that the food on which adults feed is sufficiently plentiful. In most birds, the supply of food specifically suitable for the young has to be abundant. In altricial species (those which raise dependent young in a nest), parents need to provide young with adequate amounts of high quality food to ensure rapid growth rates, in addition to satisfying their own needs. In precocial species (those whose young hatch in a sufficiently advanced state of development to feed independently), food has to be abundant enough for the young themselves to find ample amounts, in spite of their inexperience. The exception to this is the Columbiformes (pigeons and doves) where young are fed on crop milk. It may be significant that these species tend to have particularly long breeding seasons (Murton and Westwood, 1977).

Breeding is not the only energetically demanding aspect of a bird's annual cycle. Renewal of the feathers (moult) is also energetically demanding and is vital for survival. Birds normally moult immediately after breeding: there is little or no overlap between the two activities. Both require comparatively abundant food supply. So breeding seasons have not evolved simply to occupy the whole period of food abundance; there has to be a compromise between the requirements of breeding and the need to moult. Birds that breed late will suffer the penalty of less time to moult, resulting in poorer quality feathers. This will increase the risk of mortality and reduce the chances of breeding successfully in the following year. Migration may impose an additional constraint. Migratory species need a period of hyperphagia to lay down sufficient energy reserves for their migration. This too must coincide with a period of comparative food abundance. All of these energetic demands act as ultimate factors. So may other aspects, such as weather, predation and the availability of nest sites or nesting material. Breeding seasons evolve as the optimal compromise between all of the factors that ultimately affect survival.

2.2. Proximate Control of Breeding Seasons

Ultimate factors determine why birds breed when they do. However, they may have little or no value as predictive information enabling individuals to time their reproductive effort correctly. For example, in Great tits, the dominant ultimate factor, as with most species, is the availability of food on which the young are fed—in this case caterpillars. Breeding is timed so that nestlings are as near as possible 9 days old (when their demand for food is greatest) when the availability of caterpillars is maximal (Perrins and Birkhead, 1983). The appearance of caterpillars cannot itself be used by the parent as a cue to time egg-laying. For young to hatch just prior to maximum food availability requires that a cascade of physiological and behavioural processes (acquisition of territory, mate selection, gonadal maturation, nest building, copulation, ovulation, egg laying and incubation) is set in train many weeks beforehand. Rather, timing of this sequence is brought about by proximate factors (Baker, 1938). These are reliable environmental cues that the individual can perceive and use as predictive information. They may themselves have little or no effect on survival. An ideal proximate factor is one which changes predictably during the year and reliably between years, so that it can act as a calendar. Outside the tropics, the seasonal change in photoperiod meets these criteria. In comparison, other environmental variables, such as temperature, rainfall and food supply, are uncertain and unreliable. Consequently, the most important proximate factor outside the tropics is the seasonal change in photoperiod.

The first scientific demonstration that increasing photoperiod during spring was the stimulus to breeding was by Rowan (1926). He found that Juncos would breed in the middle of the Canadian winter if they were provided with an extended period of daily illumination. It is probably true to say that all temperate zone birds are highly photoperiodic. This is even true of species such as crossbills which do not have a fixed breeding season because of the unpredictable nature of their food resource (Hahn, 1995). Tropical species too may be photoperiodic even though such birds would naturally experience little change in photoperiod (Hau *et al.*, 1998).

2.3. Photostimulation and Photorefractoriness

In most species of birds from temperate and high latitudes, the breeding season begins during spring when gonadal maturation is stimulated by the vernal increase in photoperiod. The length of the breeding season varies widely between different species, to suit their particular food resource, but in most species, breeding ends and the gonads regress while photoperiod is still longer than that which earlier stimulated gonadal maturation. So breeding seasons are not symmetrical with respect to the annual change in photoperiod, and therefore it follows that the control of gonadal regression is not the reverse of effects during spring. For example, European starlings breed during April. Gonadal maturation before then progresses at a gradually increasing rate as photoperiod increases (Figure 1). But after the breeding season gonadal regression occurs rapidly, and this happens while photoperiod is still increasing. So the annual cycle in gonadal maturity is not symmetrical with respect to photoperiod either as regards when during the year the gonads are fully mature, i.e. the breeding season, or as regards the rate of gonadal maturation compared to the rate of gonadal regression.

Gonadal regression is caused, not directly by a decrease in photoperiod, but by the development of a state of photorefractoriness, an insensitivity to the stimulatory effects of long photoperiods. In many, if not most, species of birds outside the tropics, prolonged exposure to long photoperiods leads to spontaneous gonadal regression, after which any further increase in photoperiod fails to induce renewed gonadal maturation. This situation is referred to as absolute photorefractoriness (for reviews see Nicholls *et al.*, 1988; Wilson

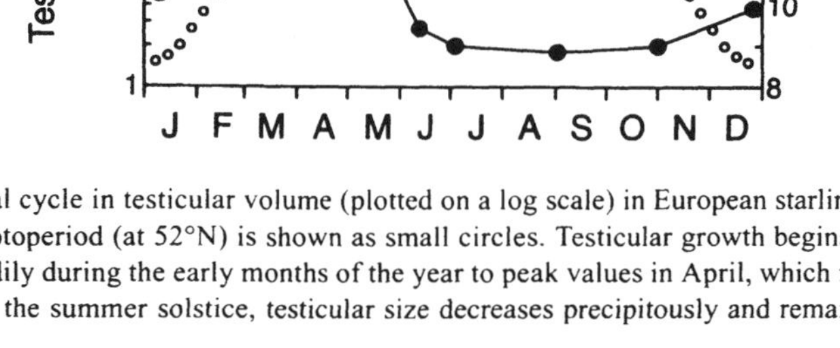

Figure 1. The annual cycle in testicular volume (plotted on a log scale) in European starlings held in outdoor aviaries. Change in photoperiod (at 52°N) is shown as small circles. Testicular growth begins before the winter solstice, increases steadily during the early months of the year to peak values in April, which is when starlings breed. During May, before the summer solstice, testicular size decreases precipitously and remains low during summer months.

and Donham, 1988; Cockrem, 1995; Sharp, 1996). In contrast, when Japanese and European quail are exposed to long photoperiods, there is no spontaneous gonadal regression; the gonads remain mature indefinitely. The induction of regression requires a decrease in photoperiod (and possibly also a decrease in temperature (Wada, 1993)). However, the natural breeding season of these species is still slightly asymmetric with respect to change in photoperiod because the photoperiod when gonadal regression occurs is slightly longer than that which initiated gonadal maturation during spring. This condition is referred to as relative photorefractoriness (Robinson and Follett, 1982; Boswell *et al.*, 1993). The length of the photoperiod which induces gonadal regression is dependent on the length and duration of preceding photoperiods. An increase in photoperiod after gonadal regression always induces renewed gonadal maturation. Although the characteristics of relative photorefractoriness have been defined clearly in these two species of quail, a similar situation has yet to be demonstrated in other species.

Many species of birds have quite short breeding seasons, similar to that of starlings, which end before the summer solstice, and clearly such species must become absolutely photorefractory. Perhaps the majority of temperate latitude species have somewhat longer breeding seasons which end soon after the summer solstice e.g., the House finch (Hamner, 1968). Gonadal regression therefore occurs during long but decreasing photoperiods. Studies on one such species, the House sparrow, show that this species does become absolutely photorefractory, but also that a decrease in photoperiod accelerates gonadal regression (Dawson, 1991; Hahn and Ball, 1995; Dawson, 1998). Whether this means that this species shows characteristics of both relative and absolute photorefractoriness, or whether a decrease in photoperiod simply accelerates gonadal regression after absolute photorefractoriness has already been set in train by long photoperiods, remains far from clear.

Photorefractoriness is induced by photoperiods in excess of a critical photoperiod. However, photorefractoriness does not become apparent immediately the critical photoperiod is exceeded. It may be several weeks later that gonadal regression begins. In experimental situations, where birds are exposed to constant photoperiods in excess of the critical photoperiod, the time taken before gonadal regression begins decreases as photoperiod increases (Hamner, 1971; Dawson and Goldsmith, 1983). In free-living starlings, and probably most species which become absolutely photorefractory, breeding occurs after the critical photoperiod for the development of photorefractoriness has been exceeded. So breeding actually happens as birds are becoming photorefractory.

By controlling the time of gonadal maturation and the time of gonadal regression, photoperiod defines the window within which birds are physiologically able to breed. This may still leave a certain amount of flexibility. The fine tuning of exactly when individuals actually do breed may involve a variety of extrinsic and intrinsic factors which need be less predictive than photoperiod (Wingfield, 1980; Ball, 1993; Hahn, 1995; Wingfield *et al.*, 1997).

Termination of photorefractoriness (the recovery of photosensitivity) occurs during short photoperiods. In starlings, if birds are held indefinitely on long photoperiods, they remain photorefractory indefinitely. However, in some absolutely photorefractory species there is evidence that spontaneous gonadal recrudescence can occur during prolonged exposure to long photoperiods e.g., ducks (Storey and Nicholls, 1978), House finches (Hamner, 1968) and House sparrows (unpublished results). In relatively photorefractory species, the recovery of photosensitivity presumably begins at the same time as gonadal regression starts, since both are caused by a decrease in photoperiod. Some care must be exercised when using the term photorefractoriness. To mammalian biologists it means the spontaneous gonadal recrudescence which occurs in many small mammals some time after short day-induced gonadal regression. This may be analogous to the *termination* of photorefractoriness in birds.

3. SEASONAL CHANGES IN THE ACTIVITY OF GnRH NEURONS

Secretion of gonadotrophins and hence gonadal maturity is directly related to the rate of secretion of GnRH from the median eminence. A key question to be addressed is: how does symmetrical change in photoperiod during the year lead to an asymmetrical pattern of GnRH secretion?

3.1. Avian GnRH

In birds there are two forms of GnRH, chicken GnRH-I (cGnRH-I; [Gln^8]-GnRH) (King and Millar, 1982; Miyamoto *et al.*, 1982) and cGnRH-II ([His^5-Trp^7-Tyr^8]-GnRH) (Miyamoto *et al.*, 1984). Experimental treatment with both cGnRH-I and cGnRH-II stimulates gonadotrophin release (Hattori *et al.*, 1986). Of the two, GnRH-II is the more potent (Chou *et al.*, 1985; Millar *et al.*, 1986; Sharp *et al.*, 1986; Sharp *et al.*, 1987). However, only cGnRH-I is present in the median eminence and so can have a physiological gonadotrophin releasing function (Mikami *et al.*, 1988; Sharp *et al.*, 1990). In contrast, cGnRH-II is found in extra-hypothalamic areas of the chicken brain and may act as a neuromodulator or neurotransmitter. Indeed, cGnRH-II is ubiquitous in vertebrates, and is even found in tunicates (Millar *et al.*, 1997). Consequently the description "chicken" is inappropriate and it is now referred to as Type II.

Avian GnRH-I is synthesised as a precursor peptide, pro-GnRH-I-GAP. As in mammals, the precursor peptide comprises a 23 amino-acid signal peptide, the decapeptide hormone and a Gly-Lys-Arg cleavage site, followed by a 56 amino-acid GnRH-associated peptide (GAP). This precursor peptide undergoes post-translational processing to form GnRH-I and GAP (Seeburg and Adelman, 1984).

3.2. Seasonal Changes

Figure 1 shows how testicular size varies in starlings during the year. Starlings become photorefractory during May, and return to being photosensitive in October. So they are photorefractory for 6 months of the year and photosensitive for the other 6 months. The hypothalamo-pituitary-gonad axis differs considerably between the photosensitive and photorefractory states. In photorefractory starlings, and other species, the amount of GnRH in the hypothalamus is low, as judged by radioimmunoassay and immunocytochemistry (Dawson *et al.*, 1985; Dawson *et al.*, 1986; Foster *et al.*, 1987; Bluhm *et al.*, 1991; Saldanha *et al.*, 1994; Hahn and Ball, 1995; Cho *et al.*, 1998). Figure 2 shows that GnRH-I in photorefractory starlings is 30-fold lower than in photosensitive birds. Plasma gonadotrophin concentrations are low and they are unaffected by gonadal steroid feedback (Goldsmith and Nicholls, 1984; Wilson, 1985) or an increase in photoperiod. In photosensitive birds, hypothalamic GnRH-I is high. Although plasma gonadotrophin concentrations remain low under short photoperiods, they are sensitive to steroid feedback; gonadectomy results in an increase. An increase in photoperiod stimulates an increase in gonadotrophin secretion and consequent gonadal maturation.

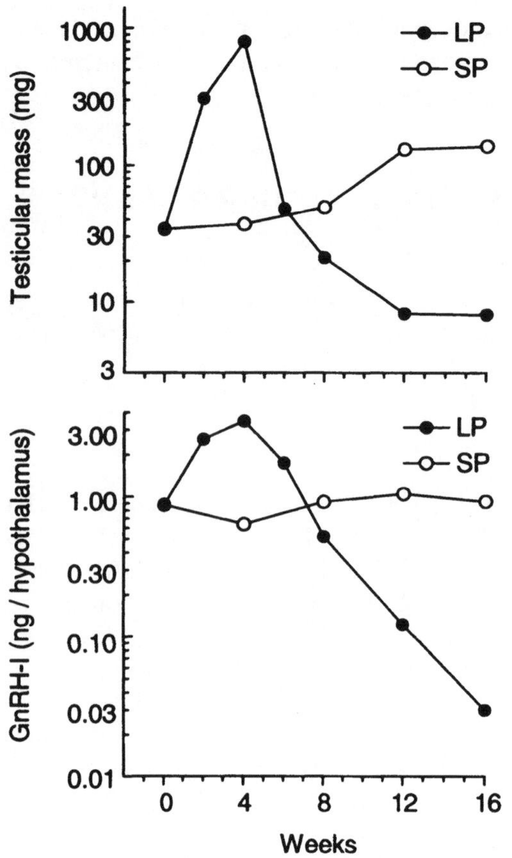

Figure 2. The upper panel shows changes in combined testicular mass in photosensitive starlings transferred from a short photoperiod (8h of light per day) to a long photoperiod (18h) at week 0 (solid circles - LP) or kept on a short photoperiod (open circles - SP). In response to photostimulation, testicular mass increased to a peak 4 weeks later, after which it decreased, as birds became photorefractory, to levels lower than initial values. In SP birds there was a slow increase. At 16 weeks testicular mass was 15-fold greater in SP birds than in LP birds at the same time. The lower panel shows changes in hypothalamic content of GnRH-I in the same birds, as measured by radioimmunoassay. Photostimulation induced a 3.5-fold increase in GnRH-I to a peak at 4 weeks, after which values decreased. In SP birds, there was no change in GnRH-I. Values after 16 weeks were 30-fold higher in SP photosensitive birds than in LP photorefractory birds at the same time. Based on unpublished data from Dawson, Talbot and Sharp.

Long photoperiods have two effects. One is gonado-stimulatory, inducing gonadal maturation in photosensitive birds, the other is gonado-inhibitory, in that long photoperiods induce and then maintain photorefractoriness. One possible explanation of this apparent contradiction is that photoperiod has two separate effects on GnRH-I neurons (Dawson *et al.*, 1988). Firstly, an effect on GnRH-I release from the median eminence: the longer the photoperiod, the greater the rate of secretion of GnRH-I. Thus an increase in photoperiod leads to gonadal maturation and, conversely, a decrease in photoperiod (from a photoperiod somewhat shorter than required to induce photorefractoriness) causes gonadal regression (but not photorefractoriness) (Goldsmith and Nicholls, 1984; Dawson, 1993). This effect of photoperiod on GnRH-I secretion is modulated by gonadal steroid feedback. The second effect of photoperiod is on GnRH-I synthesis: chronic exposure to long photoperiods inhibits GnRH-I synthesis (photorefractoriness) and, conversely, exposure to short photoperiods causes it to resume (photosensitivity). This effect is independent of steroid feedback. Which gives rise to the following possible description of the different phases of the annual cycle.

3.2.1. Photorefractory Phase. During this phase, hypothalamic GnRH-I levels are low and little if any GnRH-I is being secreted from the median eminence, since plasma gonadotrophin concentrations are low and the gonads remain fully regressed. The lack of GnRH-I secretion is not due to gonadal steroid feedback since gonadectomy has no effect. We can conclude that little GnRH-I is being synthesised.

3.2.2. Recovery of Photosensitivity. During short photoperiods, stored GnRH-I increases. Since little or no GnRH-I was being secreted during the preceding photorefractory period, this increase cannot be due to a decrease in secretion. Presumably short photoperiods stimulate renewed GnRH-I synthesis. Release of GnRH-I must remain low, since gonadotrophin concentrations show little or no apparent increase. However, unlike the situation during photorefractoriness, the inhibition of GnRH-I release is at least in part due to gonadal steroid feedback. Gonadectomy of photosensitive birds results in an increase in circulating gonadotrophin concentrations, although the magnitude of this increase varies between species. In photosensitive birds therefore, the hypothalamo-pituitary-gonad axis is "switched on" but operating at a low level of activity. It is ready to respond to an increase in photoperiod.

3.2.3. Photostimulation. When photosensitive birds experience an increase in photoperiod, there is an increase in circulating gonadotrophin concentrations, leading to gonadal maturation. Presumably this is due to an increase in GnRH-I secretion. This "photoperiodic drive" on GnRH-I release must be sufficient to overcome the negative feedback of gonadal steroids. The increased secretion of GnRH-I is not associated with a decrease in stored GnRH-I, in fact the reverse is true (Fig. 2). So GnRH-I synthesis must increase at least sufficiently to compensate for the increased release.

3.2.4. Onset of Photorefractoriness. After a period of exposure to photoperiods in excess of a critical photoperiod, circulating gonadotrophin concentrations decrease, presumably reflecting a decrease in GnRH-I secretion. This is not due to an increased sensitivity to steroid feedback (Wilson, 1985). At the same time, stored GnRH-I also decreases, so presumably synthesis of GnRH-I must decrease. Presumably, long photoperiods inhibit GnRH-I synthesis. The other effect of long photoperiods, that on photoperiodic drive specifically affecting GnRH-I secretion rate, may persist. But if no GnRH-I is being synthesised, none can be released.

3.3. GnRH-I Synthesis

Although both short photoperiods, during the recovery of photosensitivity, and long photoperiods, during photostimulation, are associated with an increase in GnRH-I synthesis, the latter is essentially a quantitative difference (gonadal maturation rate increases as photoperiod increases (Lofts *et al.*, 1970)) whereas the difference between the photorefractory and photosensitive phases is of a more qualitative nature. The ideal way to test this hypothesis would be to assess changes in expression of the gene for pre-pro-GnRH-I. This would require the cloning of the gene. This has been achieved for domestic fowl (Dunn *et al.*, 1993) and for another Galliform, the Japanese quail (Boswell, unpublished results). However, despite considerable effort, this gene has so far resisted attempts to clone it from any species which becomes absolutely photorefractory.

In the absence of successful cloning, estimates of changes in the expression of the GnRH-I gene in birds which become photorefractory have been restricted to indirect deductions, based upon changes in the quantity of GnRH-I stored within the hypothalamus. If gonadal regression associated with photorefractoriness is a consequence of inhibition of GnRH-I synthesis, then presumably a decrease in GnRH-I in the hypothalamus should precede gonadal regression. In starlings, this is not the case; gonadal regression occurs at the same time as the decrease in stored GnRH-I (Figure 2). This suggests that rather than being a consequence of inhibition of GnRH-I synthesis, gonadal regression may be caused by an inhibition of GnRH-I secretion, and that only later does this feed back to inhibit synthesis. However, a study by Parry *et al.* (1997) has shown that hypothalamic levels of the GnRH-I precursor, pro-GnRH-GAP, decrease before GnRH-I decreases, implying that inhibition of gene expression is the initial step in the development of photorefractoriness. The delay in the decrease of stored GnRH-I itself may be indicative of an unreleased pool.

If termination of the photorefractory period during short days is a consequence of renewed GnRH-I synthesis, then GnRH-I should appear in the hypothalamus before the return of photosensitivity becomes apparent 'down-stream'. In starlings, the recovery of photosensitivity, as judged by gonadal growth following photostimulation or by the spontaneous increase in plasma gonadotrophins in gonadectomized birds, becomes apparent about 4 weeks after transfer to short days (Dawson, 1991; Boulakoud and Goldsmith, 1994; Boulakoud and Goldsmith, 1995). Moreover, GnRH-I should first appear in the cell bodies in the pre-optic area (POA), rather than in the median eminence. To investigate this, GnRH-I was measured during the recovery of photosensitivity following transfer of photorefractory starlings to short photoperiods. There was indeed an increase in GnRH-I in the POA, and this was significant after only 10 days (Dawson and Goldsmith, 1997). There was no significant increase in the median eminence.

Trying to assess changes in GnRH-I synthesis and secretion during photostimulation is more difficult. Since plasma gonadotrophin concentrations increase markedly following photostimulation, an increase in secretion of GnRH-I can be inferred. As there is also an increase in total GnRH-I within the hypothalamus (Figure 2 and Parry *et al.*, 1997) the increase in secretion must be more than matched by an increase in rate of synthesis. Changes in the quantitative distribution of GnRH-I in this situation may reveal little of the dynamics. However, since the recovery of photosensitivity appears to be gradual (Boulakoud and Goldsmith, 1995), it is possible that synthesis and storage of GnRH-I increases gradually during this recovery. Photostimulation during this period may result in variations in quantitative distribution, from which the changes in GnRH-I synthesis and secretion induced by long days could be inferred. And indeed, photostimulation during the recovery of photosensitivity did result in an increase in GnRH-I in the POA, which preceded an increase in the median eminence, and which in turn preceded an increase in circulating LH (Dawson and Goldsmith, 1997).

One aim of a review by Nicholls *et al.* (1988) was to suggest that the breeding patterns of species of birds showing absolute and relative photorefractoriness, and indeed of mammals also, may share a degree of homology. Studies on changes in GnRH-I neurons provide little evidence for this. In several species of birds which become absolutely photorefractory, hypothalamic GnRH shows a considerable decrease during photorefractoriness (Dawson *et al.*, 1985; Dawson *et al.*, 1986; Foster *et al.*, 1987; Bluhm *et al.*, 1991; Saldanha *et al.*, 1994; Hahn and Ball, 1995). In Japanese quail, gonadal regression caused by relative photorefractoriness, i.e., a decrease in photoperiod following extended exposure to long photoperiods, was not associated with a decrease in hypothalamic GnRH-I (Foster *et al.*, 1988; Follett and Pearce-Kelly, 1990). Indeed, GnRH-I increased, which may suggest that relative photorefractoriness is due to an inhibition of GnRH-I secretion, rather than a decrease in synthesis. Similarly, in opportunistically breeding crossbills, gonadal regression is not associated with decreased hypothalamic GnRH-I (Hahn, personal communication). In House sparrows, which become absolutely photorefractory but in which gonadal regression can be accelerated by a decrease in photoperiod, individuals which had become photorefractory on long photoperiods, showed a decrease in GnRH-I in both the cell bodies and the fibres, whereas in individuals in which gonadal regression was accelerated by transfer to short photoperiods, there was GnRH-I in the fibres but not the cell bodies (Hahn and Ball, 1995). This implies that the decrease in photoperiod had inhibited secretion, whereas spontaneous regression was associated with decreased synthesis. In mammals as in quail, short day-induced gonadal regression is not associated with decreased hypothalamic GnRH (Pickard and Silverman, 1979; Ronchi *et al.*, 1992) but may be a consequence of inhibited secretion (Urbanski *et al.*, 1991).

3.4. Puberty

Young starlings, as other birds, do not breed until spring of the year after that in which they hatch. This is because they are photorefractory and so need to experience short photoperiods before they become photosensitive (Williams *et al.*, 1987). They are photorefractory not because they become photorefractory in response to long photoperiods after hatching, but because they hatch and develop in a photorefractory state (Williams *et al.*, 1989). Juvenile starlings have low hypothalamic GnRH-I and it increases about 3 weeks after exposure to short photoperiods (Goldsmith *et al.*, 1989). In nestlings hand-reared under short photoperiods, GnRH-I increases from about 6 weeks of age (Dawson and Goldsmith, 1989). These pubertal changes are essentially identical to the changes that happen during the recovery of photosensitivity in adults each year. Seasonality is therefore a repeated puberty. Such a marked annual regression to a pre-pubertal condition may be a feature unique to birds, and in particular, to species which show absolute photorefractoriness.

4. NEURAL CONTROL OF GnRH-I NEURONS

The mechanism by which photoperiodic information is relayed to the GnRH neurons in birds differs from that in mammals. In mammals, information on the presence or absence of light is conveyed along a retino-hypothalamic tract to the supra-chiasmatic nuclei (SCN) where it is integrated with information from a circadian clock. This photoperiodic information passes via the superior cervical ganglion to the pineal gland, which produces melatonin during darkness but not during periods of light. The resultant pattern of melatonin is used as the signal to control gonadal maturation.

4.1. Photoreceptors

Little is known about the neural circuitry by which photoperiod controls the activity of the GnRH-I neurons in birds. It is known that light is detected by photoreceptors which are neither retinal nor pineal (Benoit, 1964; Menaker and Keatts, 1968; Wilson, 1991). They are thought to lie within the hypothalamus, possibly in the lateral septum and tuberal hypothalamus (Kuenzel, 1993; Saldanha *et al.*, 1994). Implants of radioluminous paint (Oliver and Baylé, 1982) or light-conducting fibres (Yokoyama *et al.*, 1978) into the hypothalamus of birds on short photoperiods results in testicular growth, and the action spectrum of the photoperiodic response suggests that rhodopsin is the photo-pigment involved (Foster and Follett, 1985).

4.2. The Circadian Clock

There is considerable evidence indicating that circadian rhythms are involved in the measurement of photoperiod in birds, as in mammals (Follett *et al.*, 1998). It is assumed that there is a photoinducible phase which is circadian in nature. This phase begins at about 12h after dawn. If light is perceived during this phase, a photoperiodic response is induced (the external coincidence model). The circadian nature of this photoinducible phase has been demonstrated in White-crowned sparrows and Japanese quail in night interruption experiments (Follett *et al.*, 1974; Follett *et al.*, 1992). In such experiments, birds are moved from short photoperiods into continuous darkness, and exposed to pulses of light a various times thereafter. When the pulse of light coincides with subsequent photoinducible phases, which should begin approximately 12h, 36h, and 60h etc after their last dawn, a photoperiodic response is induced, which is detected as an increase in LH. At other times, there is no increase in LH. Some experiments on quail have produced equivocal results, apparently hinting that quail measure photoperiod using an 'hour-glass' model rather than a circadian clock and external coincidence. However, this may be a consequence of quail having only weakly self-sustaining circadian rhythms which are readily re-entrained by exotic lighting schedules (King *et al.*, 1997).

Most experiments to investigate the involvement of circadian rhythms in photoperiodic time measurement have quantified an end-point related to photostimulation, i.e. an increase in gonadal size or LH secretion, and so demonstrate an involvement of circadian rhythms in photostimulation. King *et al.* (1997) kept starlings on resonance cycles of 6h of light: 30h of darkness. Not only did these birds show testicular maturation, they also then became photorefractory and moulted. Clearly, circadian rhythms are also involved in the development of photorefractoriness.

Although measurement of photoperiod in birds involves a circadian clock, as in mammals, they have no structure analogous to the SCN, which, in mammals, acts as the primary pacemaker. Birds do have pacemakers, located in the eyes, the pineal and in the hypothalamus. Furthermore, although melatonin may have a role in circadian rhythmicity in birds (Juss *et al.*, 1993), and birds also have a well-defined daily pattern of melatonin, unlike mammals, this is not used as a direct signal to regulate gonadal maturation (Chakraborty, 1995). Rather, birds appear to use an entirely neural link between photoreceptors and the GnRH-I neurons.

4.3. Neural Control GnRH-I Secretion

Photostimulation, the increase in GnRH-I secretion induced by long photoperiods, is rapid. Photostimulation of starlings results in an increase in LH concentrations which can be detected 18h after dawn of the first long day (Figure 3). In quail, photo-induction occurs when photoperiod exceeds 14.7h (Follett *et al.*, 1977). An increase in LH can be de-

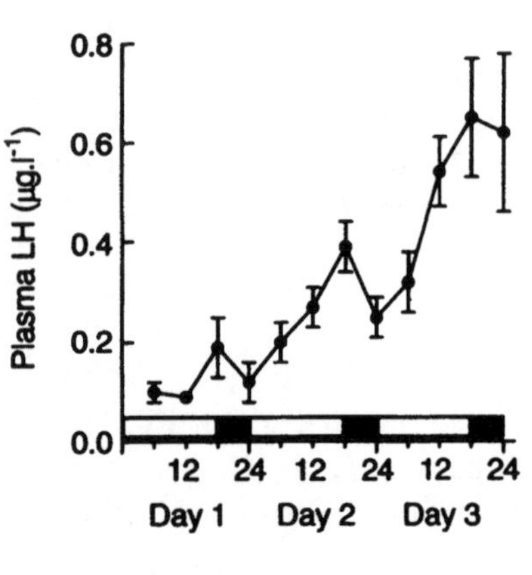

Figure 3. Changes in plasma LH concentration in starlings sampled at 6h intervals during the first three days after being transferred from a short photoperiod of 8h of light per day to a long photoperiod of 18h of light per day. The periods of illumination are shown by the boxes along the horizontal axis.

tected 22h after dawn and an increase in GnRH-I secretion from hypothalami cultured *in vitro* can be detected at the same time (Perera and Follett, 1992). What happens between 14.7h and 22h remains unclear.

Photostimulation appears to act primarily by controlling secretion of GnRH-I from the neuron terminals rather than synthesis in the cell bodies, and this may be controlled from within the infundibular complex. Lesions in this area block the photoperiodic response (Sharp and Follett, 1969; Ohta and Homma, 1987) without affecting the GnRH-I neurons (Juss, 1993). Photostimulation results in the appearance of Fos-like protein within glial cells located in the same region, in advance of the increase in LH secretion (Meddle and Follett, 1997). Since glial cells have a close relationship with GnRH-I neuron terminals, and potential photoreceptor cells directly contact glial cells (Saldanha *et al.*, 1995), it is possible that light could affect GnRH-I secretion via glial cells.

4.4. Neural Control GnRH-I Synthesis

Control of GnRH-I synthesis is less clear. Synaptic input to GnRH-I cell bodies is increased in photorefractory starlings (Parry and Goldsmith, 1993) which may imply that GnRH-I synthesis is actively inhibited. If synaptogenesis is required for photorefractoriness, this may in part account for the delay between exposure to long photoperiods and the manifestation of photorefractoriness. In contrast, the neural circuitry required for photostimulation may be complete in photosensitive birds, permitting a rapid response to a long photoperiod. The recovery of photosensitivity may be due to the withdrawal of neural input to the GnRH-I cell bodies which may passively permit GnRH-I synthesis to resume. If so, this must happen within 10 days of exposure to short photoperiods since an increase in GnRH-I is detectable then (Dawson and Goldsmith, 1997).

High concentrations of circulating prolactin are associated with the onset of photorefractoriness in starlings, although no causal relationship has been demonstrated. In birds, vasoactive intestinal polypeptide (VIP) is the physiological prolactin-releasing hormone. An increase in VIP is associated with a decrease in GnRH-I in photorefractory Juncos (Saldanha *et al.*, 1994). Putative photoreceptive cells, containing proteins characteristic of retinal photoreceptors, also contain VIP (Saldanha *et al.*, 1994). Could VIP-containing cells be linked to the inhibition of GnRH-I synthesis?

5. INVOLVEMENT OF OTHER HORMONES

The evidence provided so far implies that the control of GnRH-I synthesis and secretion may be entirely neural. Nevertheless, there is evidence that two peripheral hormones may have roles in GnRH-I regulation, namely thyroid hormones and prolactin.

5.1. Thyroid Hormones

Thyroid hormones are required at one or more stages in the transmission of photoperiodic information to the GnRH-I neurons. It has been known for many decades that thyroid function can affect reproductive function in birds (e.g., Benoit and Aron, 1934; Woitkewitsch, 1940; Blivaiss, 1947), but the nature of this interaction remains obscure. The reported effects of thyroidectomy on seasonal breeding are often apparently contradictory. In some studies, photo-stimulated gonadal maturation appears to have been largely unaffected by thyroidectomy (Woitkewitsch, 1940; Thapliyal and Pandha, 1965;

Wieselthier and van Tienhoven, 1972; Goldsmith and Nicholls, 1984; Lien and Siopes, 1991), whereas in others, maturation was impaired to varying degrees (Benoit and Aron, 1934; Chaturvedi and Thapliyal, 1983; Follett and Nicholls, 1984; Thapliyal and Lal, 1984; Lien and Siopes, 1989; Devi and Lal, 1992; Wilson and Reinert, 1993). Gonadal regression caused by photorefractoriness, and the associated post-nuptial moult, is normally prevented by thyroidectomy (Woitkewitsch, 1940; Woitkewitsch, 1966; Wieselthier and van Tienhoven, 1972; Follett and Nicholls, 1984; Goldsmith and Nicholls, 1984; Lien and Siopes, 1989; Lien and Siopes, 1991; Pant and Chandola-Saklani, 1993; Wilson and Reinert, 1993) but in some studies, photorefractoriness appears to have been accelerated (Chaturvedi and Thapliyal, 1983; Devi and Lal, 1992).

5.2. Treatment of Birds with Thyroxine

Treatment of birds with thyroxine can mimic the effects of long photoperiods (Follett and Nicholls, 1988; Goldsmith and Nicholls, 1992; Wilson and Reinert, 1995). Reinert and Wilson (1996; Wilson and Reinert, 1996) argue in that tree sparrows, thyroxine plays an active role, i.e. that the increase in plasma thyroxine concentrations following photostimulation (Sharp and Klandorf, 1981; Dawson, 1984; Reinert and Wilson, 1996) serves to programme the subsequent gonadal cycle and moult. The contrary view is taken by Bentley *et al.* (1997). When they maintained thyroxine concentrations at sub-physiological levels by treating thyroidectomized starlings with appropriate amounts of thyroxine, these birds became photorefractory and moulted. This suggests that thyroxine is permissive; thyroxine simply has to be present for the appropriate responses to photoperiod to occur.

The site(s) of action of thyroid hormones must lie within the central nervous system since, in starlings and sparrows, the prevention of photorefractoriness by thyroidectomy is associated with maintenance of the high hypothalamic levels of GnRH-I typical of photosensitive birds (Dawson *et al.*, 1985; Reinert and Wilson, 1996; Dawson, 1998) and thyroidectomy of photorefractory birds results in an increase in GnRH-I (Dawson *et al.*, 1986). Chronic thyroidectomy appears to render starlings 'photoperiodically blind'; the responses to an increase or a decrease in photoperiod are greatly attenuated (Dawson, 1993). Moreover, the gonads remain moderately mature and GnRH-I remains high, as if such birds perceive photoperiod as about 11h of light per day, irrespective of what it actually is. However, in house sparrows, the effects of thyroidectomy are very different. Again thyroidectomized birds appear to be photoperiodically blind; testicular size and GnRH-I are the same whether birds are on long or short photoperiods (Dawson, 1998). However, GnRH-I remains high, typical of photosensitive birds, but testicular size remains minimal, typical of photorefractory birds. The secondary sexual characters also remain indistinguishable from those of photorefractory birds. This may be further evidence of the dual nature of photoperiodic control of GnRH-I neurons, thyroidectomy having different effects on each. Which may account for the apparently contradictory findings on the effects of thyroidectomy amongst different species.

The number of synapses on GnRH-I neuron cell bodies increases in photorefractory starlings (Parry and Goldsmith, 1993) suggesting that neural input may actively inhibit GnRH-I synthesis. There is evidence from mammals that thyroid hormones mediate gene expression for the neurotrophin nerve growth factor (NGF). Could the lack of long day-induced inhibition of GnRH-I synthesis seen in thyroidectomized birds be due to compromised NGF function preventing synaptogenesis? To test this, mouse NGF dissolved in artificial cerebrospinal fluid was infused into the lateral ventricle of thyroidectomized male starlings on long days (Bentley *et al.*, 1997). There was statistically significant testicular regression but this was not as great as normally occurs during the onset of photorefractoriness. Moreover, testicular recrudescence occurred soon after infusions ceased. An alternative strategy was to infuse anti-NGF-antibody into the lateral ventricle of thyroid-intact castrated photorefractory starlings. If the hypothesis was correct, this should result in a termination of photorefractoriness, which would be signalled by an increase in circulating LH. There was indeed a statistically significant and sustained increase in plasma LH, although the magnitude of the increase was somewhat less than normally observed during a return to photosensitivity.

5.3. Prolactin

Free-living starlings have a marked seasonal cycle in prolactin (Dawson and Goldsmith, 1982). In these birds and in starlings in experimental situations (e.g., Ebling *et al.*, 1982; Dawson and Goldsmith, 1983; Dawson and Goldsmith, 1984) prolactin is always high during gonadal regression as birds become photorefractory. Conversely, in thyroidectomized starlings, which do not become photorefractory, there is no increase in prolactin (Goldsmith and Nicholls, 1984). However, a causal relationship between the high levels of prolactin and photorefractoriness could not be demonstrated. This was because there had been no satisfactory procedure to experimentally manipulate prolactin secretion in birds, which is under stimulatory control rather than inhibitory control as in mammals. The identification of VIP as the prolactin-releasing hormone in birds (Mauro *et al.*, 1989; Sharp *et al.*, 1989) presented active immunization against VIP as a strategy to eliminate photo-induced prolactin secretion (El Halawani *et al.*, 1996).

Starlings were immunized against VIP and then photostimulated (Dawson and Sharp, 1998). In half of the VIP-immunized birds the treatment was successful and prolactin secretion was completely suppressed. In the other half it was partially suppressed. In the VIP-immunized birds in which prolactin was completely suppressed, testicular regression was markedly slowed. Regression was slightly slowed in VIP-immunized birds with partial prolactin. Nevertheless, all birds did eventually become photorefractory. This suggests that high prolactin accelerates gonadal regression during the onset of photorefractoriness but that prolactin does not itself cause photorefractoriness. Prolactin may inhibit release of GnRH-I or decrease the pituitary response to GnRH-I, but presumably does not affect GnRH-I synthesis.

6. CONCLUSIONS

Most species of (non-tropical) birds are photoperiodic and use the annual cycle in photoperiod to time both the beginning and end of their breeding seasons. However, breeding seasons are not symmetrical with change in photoperiod, so the photoperiodic control of gonadal regression is not the reverse of that causing gonadal maturation. Gonadal maturation is the result of increasing photoperiods stimulating increased secretion of GnRH-I (Figure 4). This response is rapid and is mediated by photoreceptors within the infundibular complex, which may be closely associated with the GnRH-I neuron terminals. Increased GnRH-I secretion is compensated by increased GnRH-I synthesis. Gonadal regression is due to photorefractoriness, a longer term effect of long photoperiods. This may result from neural input to the GnRH-I cell bodies inhibiting GnRH-I synthesis. The recovery of photosensitivity is signalled by renewed GnRH-I synthesis. This is presumed

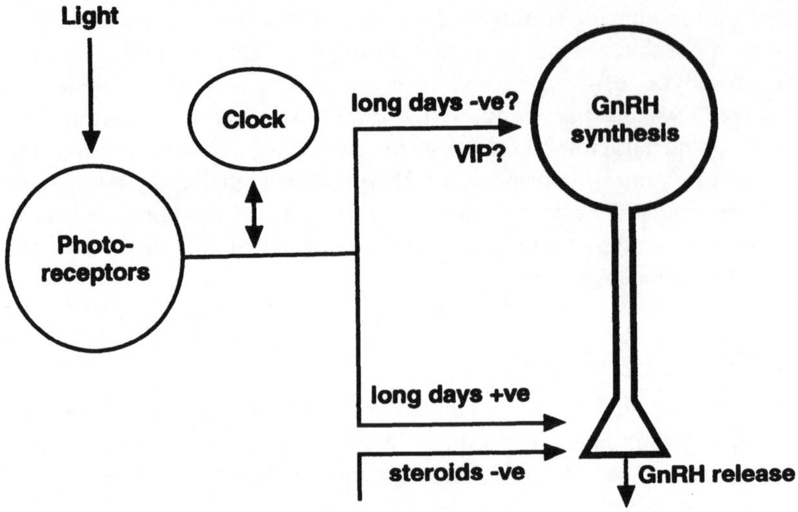

Figure 4. A schematic model of the components of the avian photoperiodic system indicating how they may operate to control GnRH-I secretion during the annual cycle.

to occur passively following the withdrawal of active inhibition of synthesis. Once stored GnRH-I has increased, birds are again able to respond to long photoperiods with an increase in secretion. Thyroid hormones are required for photoperiodic responses, but this may reflect a requirement in neural mechanisms generally rather than a specific active role. Prolactin causes gonadal regression at the onset of photorefractoriness, but does not appear to cause photorefractoriness.

REFERENCES

Baker, J. R. (1938). The evolution of breeding seasons. *In* "Evolution: Essays on Aspects of Evolutionary Biology" (G. R. de Beer, Ed.), pp. 161–177, Oxford University Press, London.

Ball, G. F. (1993). The neural integration of environmental information by seasonally breeding birds. *Am. Zool.* **33,** 185–199.

Benoit, J. (1964). The role of the eye and of the hypothalamus in the photostimulation of gonads in the duck. *Ann. New York Acad. Sci.* **117,** 204–216.

Benoit, J. and Aron, M. (1934). Sur le conditionnement hormonique du développement testiculaire, chez les oiseaux. Résults de la thyroidectomie, chez le coq et le canard. *C.R.Seances Soc. Biol. Ses Fil.* **116,** 221–223.

Bentley, G. E., Goldsmith, A. R., Dawson, A., Glennie, L., Talbot, R. T. and Sharp, P. J. (1997). Photorefractoriness in European starlings (*Sturnus vulgaris*) is not dependent upon the long-day-induced rise in plasma thyroxine. *Gen. Comp. Endocrinol.* **107,** 428–438.

Bentley, G. E., Goldsmith, A. R., Juss, T. S. and Dawson, A. (1997). The effects of nerve growth factor and anti-nerve growth factor antibody on the neuroendocrine reproductive system in the European starling *Sturnus vulgaris*. *Gen. Comp. Endocrinol.* **107,** 428–438.

Blivaiss, B. B. (1947). Development of secondary sexual characters in thyroidectomised brown Leghorn hens. *J. Exp. Zool.* **104,** 267–309.

Bluhm, C. K., Schwabl, H., Schwabl, A., Perera, A., Follett, B. K., Goldsmith, A. R. and Gwinner, E. (1991). Variation in hypothalamic gonadotrophin-releasing hormone content, plasma and pituitary LH, and in-vitro testosterone release in a long-distance migratory bird, the garden warbler (*Sylvia borin*), under constant photoperiods. *J. Endocrinol.* **128,** 339–345.

Boswell, T., Hall, M. R. and Goldsmith, A. R. (1993). Annual cycles of migratory fattening, reproduction and moult in European quail (*Coturnix coturnix*). *J. Zool., London* **231,** 627–644.

Boulakoud, M. S. and Goldsmith, A. R. (1994). Acquisition of photosensitivity in castrated male starlings (*Sturnus vulgaris*) under short day photoperiods. *J. Reprod. Fert.* **100,** 77–79.

Boulakoud, M. S. and Goldsmith, A. R. (1995). The effect of duration of exposure to short days on the gonadal response to long days in male starlings (*Sturnus vulgaris*). *J. Reprod. Fert.* **104,** 215–217.

Chakraborty, S. (1995). Plasma prolactin and luteinizing hormone during termination and onset of photorefractoriness in pinealectomized European starlings (*Sturnus vulgaris*). *Gen. Comp. Endocrinol.* **99,** 185–191.

Chaturvedi, C. M. and Thapliyal, J. P. (1983). Thyroid photoperiod and gonadal regression in the common myna *Acridotheres tristis*. *Gen. Comp. Endocrinol.* **52,** 279–282.

Cho, R. N., Hahn, T. P., MacDougall-Shackleton, S. and Ball, G. F. (1998). Seasonal variation in brain GnRH in free-living breeding and photorefractory House finches (*Carpodacus mexicanus*). *Gen. Comp. Endocrinol.* **109,** 244–250.

Chou, H. F., Johnson, A. L. and Williams, J. B. (1985). Luteinizing hormone releasing activity of (Gln^8)-LHRH and (His^5,Trp^7,Tyr^8) LHRH in the cockerel *in vivo* and *in vitro*. *Life Sci.* **37,** 2459–2465.

Cockrem, J. F. (1995). Timing of seasonal breeding in birds, with particular reference to New Zealand birds. *Reprod. Fert. Developm.* **7,** 1–19.

Dawson, A. (1984). Changes in plasma thyroxine concentrations in male and female starlings (*Sturnus vulgaris*) during a photo-induced gonadal cycle. *Gen. Comp. Endocrinol.* **56,** 193–197.

Dawson, A. (1991). Effect of daylength on the rate of recovery of photosensitivity in male starlings (*Sturnus vulgaris*). *J. Reprod. Fert.* **93,** 521–524.

Dawson, A. (1991). Photoperiodic control of testicular regression and moult in male House Sparrows, *Passer domesticus*. *Ibis* **133,** 312–316.

Dawson, A. (1993). Thyroidectomy progressively renders the reproductive system of starlings, *Sturnus vulgaris*, unresponsive to changes in daylength. *J. Endocrinol.* **139,** 51–55.

Dawson, A. (1998). Photoperiodic control of the termination of breeding and the induction of moult in House sparrows *Passer domesticus*. *Ibis* **140,** 35–40.

Dawson, A. (1998). Thyroidectomy of house sparrows (*Passer domesticus*) prevents photo-induced testicular growth but not the increased hypothalamic gonadotrophin-releasing hormone. *Gen. Comp. Endocrinol.* **110,** 196–200.

Dawson, A., Follett, B. K., Goldsmith, A. R. and Nicholls, T. J. (1985). Hypothalamic gonadotrophin-releasing hormone and pituitary and plasma FSH and prolactin during photostimulation and photorefractoriness in intact and thyroidectomized starlings (*Sturnus vulgaris*). *J. Endocrinol.* **105,** 71–77.

Dawson, A. and Goldsmith, A. R. (1982). Prolactin and gonadotrophin secretion in wild starlings (*Sturnus vulgaris*) during the annual cycle and in relation to nesting, incubation, and rearing young. *Gen. Comp. Endocrinol.* **48,** 213–221.

Dawson, A. and Goldsmith, A. R. (1983). Plasma prolactin and gonadotrophins during gonadal development and the onset of photorefractoriness in male and female starlings (*Sturnus vulgaris*) on artificial photoperiods. *J. Endocrinol.* **97,** 253–260.

Dawson, A. and Goldsmith, A. R. (1984). Effects of gonadectomy on seasonal changes in plasma LH and prolactin concentrations in male and female starlings (*Sturnus vulgaris*). *J. Endocrinol.* **100,** 213–218.

Dawson, A. and Goldsmith, A. R. (1989). Sexual maturation in starlings raised on long or short days: changes in hypothalamic gonadotrophin-releasing hormone and plasma LH concentrations. *J. Endocrinol.* **123,** 189–196.

Dawson, A. and Goldsmith, A. R. (1997). Changes in gonadotrophin-releasing hormone (GnRH-I) in the pre-optic area and median eminence of starlings (*Sturnus vulgaris*) during the recovery of photosensitivity and during photostimulation. *J. Reprod. Fert.* **111,** 1–6.

Dawson, A., Goldsmith, A. R., Nicholls, T. J. and Follett, B. K. (1986). Endocrine changes associated with the termination of photorefractoriness by short daylengths and thyroidectomy in starlings (*Sturnus vulgaris*). *J. Endocrinol.* **110,** 73–79.

Dawson, A., Nicholls, T. J., Goldsmith, A. R. and Follett, B. K. (1988). Comparative endocrinology of photorefractoriness. *In* "Acta XIX Congr. Int. Orn." pp. 634–640.

Dawson, A. and Sharp, P. J. (1998). The role of prolactin in the development of reproductive photorefractoriness and postnuptial molt in the European starling (*Sturnus vulgaris*). *Endocrinology* **139,** 485–490.

Devi, K. P. and Lal, P. (1992). Thyroid involvement in photoperiodically induced testicular and body weight cycles of the black headed bunting, *Emberiza melanocephala*. *J. Exp. Zool.* **264,** 293–297.

Dunn, I. C., Chen, Y., Hook, C., Sharp, P. J. and Sang, H. M. (1993). Characterization of the chicken preprogonadotrophin-releasing hormone-I gene. *J. Mol. Endocrinol.* **11,** 19–29.

Ebling, F. J. P., Goldsmith, A. R. and Follett, B. K. (1982). Plasma prolactin and luteinizing hormone during photoperiodically induced testicular growth and regression in starlings (*Sturnus vulgaris*). *Gen. Comp. Endocrinol.* **48,** 485–490.

El Halawani, M. E., Pitts, G. R., Sun, S., Silsby, J. L. and Sivanandan, V. (1996). Active immunization against vasoactive intestinal peptide prevents photoinduced prolactin secretion in turkeys. *Gen. Comp. Endocrinol.* **104,** 76–83.

Follett, B. K., Davies, D. T. and Gledhill, B. (1977). Photoperiodic control of reproduction in Japanese quail: changes in gonadotrophin secretion on the first day of induction and their pharmacological blockade. *J. Endocrinol.* **74,** 449–460.

Follett, B. K., King, V. M. and Meddle, S. L. (1998). Rhythms and photopereiodism in birds. *In* "Biological Rhythms and Photoperiodism in Plants". (P. J. Lumsden and A. J. Millar Ed.), pp. 231–242, BIOS Scientific Publishers Ltd, Oxford.

Follett, B. K., Kumar, V. and Juss, T. S. (1992). Circadian nature of the photoperiodic clock in Japanese quail. *J. Comp. Physiol. A* **171,** 533–540.

Follett, B. K., Mattocks, P. W. and Farner, D. S. (1974). Circadian function in the induction of gonadotropin secretion in the white-crowned sparrow. *Proc. Natl. Acad. Sci., USA* **71,** 1666–1669.

Follett, B. K. and Nicholls, T. J. (1984). Photorefractoriness in Japanese quail: possible involvement of the thyroid gland. *J. Exp. Zool.* **232,** 573–580.

Follett, B. K. and Nicholls, T. J. (1988). Acute effects of thyroid hormones in mimicking photoperiodically induced release of gonadotropins in Japanese quail. *J. Comp. Physiol.* **B 157,** 837–843.

Follett, B. K. and Pearce-Kelly, A. (1990). Photoperiodic control of the termination of reproduction in Japanese quail (*Coturnix japonica*). *Proc. R. Soc., Lond.* **242,** 225–230.

Foster, R. G. and Follett, B. K. (1985). The involvement of a rhodopsin-like photopigment in the photoperiodic response of the Japanese quail. *J. Comp. Physiol.* **157,** 519–528.

Foster, R. G., Panzica, G. C., Parry, D. M. and Viglietti-Panzica, C. (1988). Immunocytochemical studies on the LHRH system of the Japanese quail: influence by photoperiod and aspects of sexual differentiation. *Cell Tiss. Res.* **253,** 327–335.

Foster, R. G., Plowman, G., Goldsmith, A. R. and Follett, B. K. (1987). Immunohistochemical demonstration of marked changes in the LHRH system of photosensitive and photorefractory European starlings (*Sturnus vulgaris*). *J. Endocrinol.* **115,** 211–220.

Goldsmith, A. R., Ivings, W. E., Pearce-Kelly, A. S., Parry, D. M., Plowman, D. M., Nicholls, T. J. and Follett, B. K. (1989). Photoperiodic control of the LHRH neurosecretory system of European starlings (*Sturnus vulgaris*) during puberty and the onset of photorefractoriness. *J. Endocrinol.* **122,** 255–268.

Goldsmith, A. R. and Nicholls, T. J. (1984). Changes in plasma prolactin in male starlings during testicular regression under short days compared with those during photorefractoriness. *J. Endocrinol.* **102,** 353–356.

Goldsmith, A. R. and Nicholls, T. J. (1984). Recovery of photosensitivity in photorefractory starlings is not prevented by testosterone treatment. *Gen. Comp. Endocrinol.* **56,** 210–217.

Goldsmith, A. R. and Nicholls, T. J. (1984). Thyroidectomy prevents the development of photorefractoriness and the associated rise in plasma prolactin in starlings. *Gen. Comp. Endocrinol.* **54,** 256–263.

Goldsmith, A. R. and Nicholls, T. J. (1992). Thyroxine effects upon reproduction, prolactin secretion and plumage moult in thyroidectomised European starlings *Sturnus vulgaris*. *Ornis Scand.* **23,** 398–404.

Hahn, T. P. (1995). Integration of photoperiodic and food cues to time changes in reproductive physiology by an opportunistic breeder, the Red crossbill, *Loxia curvirostra* (Aves: Carduelinae). *J. Exp. Zool.* **272,** 213–226.

Hahn, T. P. and Ball, G. F. (1995). Changes in brain GnRH associated with photorefractoriness in House sparrows (*Passer domesticus*). *Gen. Comp. Endocrinol.* **99,** 349–363.

Hamner, W. M. (1968). The photorefractory period of the house finch. *Ecology* **49,** 211–227.

Hamner, W. M. (1971). On seeking an alternative to the endogenous reproductive rhythm hypothesis in birds. *In* "Biochronometry". (M. Menaker, Ed.), pp. 448–462, Nat. Acad. Sci. Washington D.C.

Hattori, A., Ishii, S. and Wada, M. (1986). Effects of two kinds of chicken luteinizing hormone releasing hormone (LHRH), mammalian LHRH, and its analogs on the release of LH and FSH in Japanese quail and chickens. *Gen. Comp. Endocrinol.* **64,** 446–455.

Hau, M., Wikelski, M. and Wingfield, J. C. (1998). A neotropical forest bird can measure the slight changes in tropical photoperiod. *Proc. R. Soc., Lond. Series B-Biological Sciences* **1391,** 89–95.

Juss, T. S. (1993). Neuroendocrine and neural changes associated with the photoperiodic control of reproduction. *In* "Avian Endocrinology". (P. J. Sharp Ed.), pp. 47–60, Society for Endocrinology. Bristol.

Juss, T. S., Meddle, S. L., Servant, R. S. and King, V. M. (1993). Melatonin and photoperiodic time measurement in Japanese quail (*Coturnix coturnix japonica*). *Proc. R. Soc., Lond. Series B-Biological Sciences* **254,** 21–28.

King, J. D. and Millar, R. P. (1982). Structure of chicken hypothalamic luteinizing hormone-releasing hormone. II. Isolation and characterization. *J. Biol. Chem.* **257,** 10729–10732.

King,V. M., Bentley,G.E. and Follett,B.K. (1997) A direct comparison of photoperiodic time measurement and the circadian system in European starlings and Japanese quail. *J. Biol. Rhythms* **12,** 431–442.

Kuenzel, W. J. (1993). The search for deep encephalic photoreceptors within the avian brain, using gonadal development as a primary indicator. *Poult. Sci.* **72,** 959–967.

Lien, R. J. and Siopes, T. D. (1989). Effects of thyroidectomy on egg production, molt, and plasma thyroid hormone concentrations of turkey hens. *Poult. Sci.* **68,** 1126–1132.

Lien, R. J. and Siopes, T. D. (1991). Influence of thyroidectomy on reproductive responses of male domestic turkeys (*Meleagris gallopavo*). *Br. Poult. Sci.* **32,** 405–414.

Lofts, B., Follett, B. K. and Murton, R. K. (1970). Temporal changes in the pituitary-gonadal axis. *Mem. Soc. Endocrinol.* **18,** 545–575.

Mauro, L., Elde, R. P., Youngren, O. M., Phillips, R. E. and El Halawani, M. E. (1989). Alterations in hypothalamic vasoactive intestinal peptide-like immunoreactivity are associated with reproduction and prolactin release in the female turkey. *Endocrinology* **125,** 1795–1804.

Meddle, S. L. and Follett, B. K. (1997). Photoperiodically driven changes in Fos expression within the basal tuberal hypothalamus and median eminence of Japanese quail. *J. Neurosci.* **17,** 8909–8918.

Menaker, A. H. and Keatts, H. (1968). Extraretinal light perception in the sparrow II. Photoperiodic stimulation of testis growth. *Proc. Natl. Acad. Sci.USA.* **60,** 146–151.

Mikami, S., Yamada, S., Hasegawa, Y. and Miyamoto, K. (1988). Localisation of avian LHRH-immunoreactive neurons in the hypothalamus of the domestic fowl, *Gallus domesticus*, and the Japanese quail, *Coturnix coturnix*. *Cell Tiss. Res.* **251,** 51–58.

Millar, R. P., Milton, R. C. d. L., Follett, B. K. and King, J. A. (1986). Receptor binding and gonadotropin-releasing activity of a novel chicken gonadotropin-releasing hormone ([His^5,Trp^7,Tyr^8] GnRH) and a D-Arg^6 analog. *Endocrinology* **119,** 224–231.

Millar, R. P., Troskie, B., Sun, Y.-M., Ott, T., Wakefield, I., Myburgh, D., Pawson, A., Davidson, J. S., Flanagan, C., Katz, A., Hapgood, J., Illing, N., Weinstein, H., Seaflon, S. C., Peter, R. E., Terasawa, E. and King, J. A. (1997). Plasticity in the structural and functional evolution of GnRH: a peptide for all seasons. *In* "Advances in Comparative Endocrinology" (S. Kawashima and S. Kikuyama, Eds.), pp. 15–27, Monduzzi Editore. Bologna (Italy).

Miyamoto, K., Hasegawa, Y., Minegishi, T., Nomura, M., Takahashi, Y., Igarashi, M., Kanagawa, K. and Matsuo, H. (1982). Isolation and characterization of chicken hypothalamic luteinizing hormone-releasing hormone. *Biochem. Biophys. Res. Comm.* **107,** 820–827.

Miyamoto, K., Hasegawa, Y., Nomura, M., Igarashi, M., Kanagawa, K. and Matsuo, H. (1984). Identification of the second gonadotropin-releasing hormone in the chicken hypothalamus: evidence that gonadotropin secretion is probably controlled by two distinct gonadotropin-releasing hormones in avian species. *Proc. Natl. Acad. Sci.,U.S.A.* **81,** 3874–3878.

Murton, R. K. and Westwood, N. J. (1977). "Avian Breeding Cycles" Clarendon, Oxford.

Nicholls, T. J., Goldsmith, A. R. and Dawson, A. (1988). Photorefractoriness in birds and comparison with mammals. *Physiol. Rev.* **68,** 133–176.

Ohta, M. and Homma, K. (1987). Detection of neuronal connections to the infundibular complex by partial or complete hypothalamic deafferentation in male quail. *Gen. Comp. Endocrinol.* **68,** 286–292.

Oliver, J. and Baylé, J. D. (1982). Brain photoreceptors for the photoinduced testicular responses in birds. *Experimentia* **28,** 1021–1029.

Pant, K. and Chandola-Saklani, A. (1993). A role for thyroid hormones in the development of premigratory disposition in redheaded bunting, *Emberiza bruniceps*. *J. Comp. Physiol. B* **163,** 389–394.

Parry, D. M. and Goldsmith, A. R. (1993). Ultrastructural evidence for changes in synaptic input to the hypothalamic luteinizing hormone-releasing hormone neurons in photosensitive and photorefractory starlings. *J. Neuroendocrinol.* **5,** 387–395.

Parry, D. M., Goldsmith, A. R., Millar, R. P. and Glennie, L. M. (1997). Immunocytochemical localisation of GnRH precursor in the hypothalamus of European starlings during sexual maturation and photorefractoriness. *J. Neuroendocrinol.* **9,** 235–243.

Perera, A. D. and Follett, B. K. (1992). Photoperiodic induction in vitro: the dynamics of GnRH release from hypothalamic explants of the Japanese quail. *Endocrinology* **131,** 2898–2908.

Perrins, C. M. and Birkhead, T. (1983). "Avian Ecology" Blackie, London.

Pickard, G. E. and Silverman, A. J. (1979). Effects of photoperiod on hypothalamic luteinizing hormone-releasing hormone in the male hamster. *J. Endocrinol.* **83,** 421–428.

Reinert, B. D. and Wilson, F. E. (1996). The thyroid and the hypothalamus-pituitary- ovarian-axis in American tree sparrows (*Spizella arborea*). *Gen. Comp. Endocrinol.* **103,** 60–70.

Reinert, B. D. and Wilson, F. E. (1996). Thyroid dysfunction and thyroxine-dependent programming of photoinduced ovarian growth in American tree sparrows (*Spizella arborea*). *Gen. Comp. Endocrinol.* **103,** 71–81.

Robinson, J. E. and Follett, B. K. (1982). Photoperiodism in Japanese quail: the termination of seasonal breeding by photorefractoriness. *Proc. R. Soc., Lond.* **215,** 95–116.

Ronchi, E., Aoki, C., Krey, L. C. and Pfaff, D. W. (1992). Immunocytochemical study of GnRH and GnRH-associated peptide in male Syrian hamsters as a function of photoperiod and gonadal alterations. *Neuroendocrinology* **55,** 134–145.

Rowan, W. (1926). On photoperiodism, reproductive periodicity, and the annual migrations of certain birds and fishes. *Proc. Boston Soc. Nat. Hist.* **38,** 147–189.

Saldanha, C. J., Deviche, P. J. and Silver, R. (1994). Increased VIP and decreased GnRH expression in photorefractory dark-eyed Juncos *Junco hyemalis*. *Gen. Comp. Endocrinol.* **93,** 128–136.

Saldanha, C. J., Leak, R. K. and Silver, R. (1994). Detection and transduction of daylength in birds. *Psychoneuroendocrinology* **19,** 641–656.

Saldanha, C. J., Silverman, A. J. and Silver, R. (1995). Deep brain photoreception; relation between opsin and GnRH expressing neurons. *Soc. Neurosci. Abstr. 21: 46.3* **19,** 641–656.

Seeburg, P. H. and Adelman, J. P. (1984). Characterization of cDNA for precursor of human luteinising hormone releasing hormone. *Nature* **311,** 666–668.

Sharp, P. J. (1996). Strategies in avian breeding cycles. *Anim. Reprod. Sci.* **42,** 505–513.

Sharp, P. J., Dunn, I. C. and Talbot, R. T. (1987). Sex differences in the LH responses to chicken LHRH-I and -II in the domestic fowl. *J. Endocrinol.* **115,** 323–331.

Sharp, P. J. and Follett, B. K. (1969). The effect of hypothalamic lesions on gonadotrophin release in Japanes quail. *Neuroendocrinology* **5,** 205–218.

Sharp, P. J. and Klandorf, H. (1981). The interaction between day length and the gonads in the regulation of levels of plasma thyroxine and triiodothyronine in the Japanese quail. *Gen. Comp. Endocrinol.* **45,** 504–512.

Sharp, P. J., Sterling, R. J., Milton, R. C. d. L. and Millar, R. P. (1986). Effect of luteinizing hormone releasing hormone and its analogues on plasma luteinizing hormone concentrations in incubating bantam hens. *Br. Poult. Sci.* **27,** 129–135.

Sharp, P. J., Sterling, R. J., Talbot, R. T. and Huskisson, N. S. (1989). The role of hypothalamic vasoactive intestinal polypeptide in the maintenance of prolactin secretion in incubating bantam hens: observations using passive immunization, radioimmunoassay and immunohistochemistry. *J. Endocrinol.* **122,** 5–13.

Sharp, P. J., Talbot, R. T., Main, G. M., Dunn, I. C., Fraser, H. M. and Huskisson, N. S. (1990). Physiological roles of chicken LHRH-I and -II in the control of gonadotrophin release in the domestic chicken. *J. Endocrinol.* **124,** 291–299.

Storey, C. R. and Nicholls, T. J. (1978). Observations on the regulation of sexual quiescence in juvenile and adult male mallard (*Anas platyrhynchos*). *J. Zool., Lond.* **184,** 181–186.

Thapliyal, J. P. and Lal, P. (1984). Role of thyroid and male hormones in photorefractory state of the migratory redheaded bunting *Emberiza bruniceps*. *Indian J. Exp. Biol.* **22,** 295–297.

Thapliyal, J. P. and Pandha, S. K. (1965). Thyroid-gonad relationship in spotted munia, *Uroloncha punctulata*. *J. Exp. Zool.* **158,** 253–262.

Urbanski, H. F., Doan, A. and Pierce, M. (1991). Immunocytochemical investigation of luteinizing hormone-releasing hormone neurons in Syrian hamsters maintained under long and short days. *Biol. Reprod.* **44,** 687–692.

Wada, M. (1993). Low temperature and short days together induce thyroid activation and suppression of LH release in Japanese quail. *Gen. Comp. Endocrinol.* **90,** 355–363.

Wieselthier, A. S. and van Tienhoven, A. (1972). The effect of thyroidectomy on testicular size and on the photorefractory period in the starling (*Sturnus vulgaris* L.). *J. Exp. Zool.* **179,** 331–338.

Williams, T. D., Dawson, A. and Nicholls, T. J. (1989). Sexual maturation and moult in juvenile starlings in response to different daylengths. *Ibis* **131,** 135–140.

Williams, T. D., Dawson, A., Nicholls, T. J. and Goldsmith, A. R. (1987). Short days induce premature reproductive maturation in juvenile starlings, *Sturnus vulgaris*. *J. Reprod. Fert.* **80,** 327–333.

Wilson, F. E. (1985). Androgen feedback-dependent and -independent control of photoinduced LH secretion in male tree sparrows (*Spizella arborea*). *J. Endocrinol.* **105,** 141–152.

Wilson, F. E. (1991). Neither retinal nor pineal photoreceptors mediate photoperiodic control of seasonal reproduction in American tree sparrows. *J. Exp. Zool.* **259,** 117–127.

Wilson, F. E. and Donham, R. S. (1988). Daylength and control of seasonal reproduction in male birds. *In* "Processing of Environmental Information in Vertebrates". (M. H. Stetson, Ed.), pp. 101–119, Springer-Verlag. New York.

Wilson, F. E. and Reinert, B. D. (1993). The thyroid and photoperiodic control of seasonal reproduction in American tree sparrows (*Spizella arborea*). *J. Comp. Physiol. B* **163,** 563–573.

Wilson, F. E. and Reinert, B. D. (1995). A one-time injection of thyroxine programmed seasonal reproduction and post-nuptial molt in chronically thyroidectomized male American tree sparrows *Spizella arborea* exposed to long days. *J. Avian Biol.* **26,** 225–233.

Wilson, F. E. and Reinert, B. D. (1996). The timing of thyroid dependent programming in seasonally breeding male American tree sparrows (*Spizella arborea*). *Gen. Comp. Endocrinol.* **103,** 82–92.

Wingfield, J. C. (1980). Fine temporal adjustment of reproductive functions. *In* "Avian Endocrinology". (A. Epple and M. H. Stetson, Eds.), pp. 367–389, Academic Press. New York.

Wingfield, J. C., Hahn, T. P., Wada, M. and Schoech, S. J. (1997). Effects of day length and temperature on gonadal development, body mass, and fat depots in white-crowned sparrows, *Zonotrichia leucophrys pugetensis*. *Gen. Comp. Endocrinol.* **107,** 44–62.

Woitkewitsch, A. A. (1940). Dependence of seasonal periodicity in gonadal changes on the thyroid gland in *Sturnus vulgaris* L. *Compt. Rend. (Doklad.) l'Acad. Sci. de l'URSS* **27,** 741–745.

Woitkewitsch, A. A. (1966). "The feathers and plumage of birds" Sidgewick and Jackson, London.

Yokoyama, K., Oksche, A., Darden, T. R. and Farner, D. S. (1978). The sites of encephalic photoreception in photoperiodic induction of growth of the testes in the white-crowned sparrow. *Cell Tiss. Res.* **189,** 441–467.

11

CHANGES IN THE EXPRESSION OF AVIAN CENTRAL STEROID RECEPTORS WITH RESPECT TO THE BREEDING CYCLE

J. A. Clark, M. D. C. Belle, G. C. Georgiou, and R. W. Lea

University of Central Lancashire
Preston, PR1 8EJ, England

1. GENERAL INTRODUCTION

In many species, a dramatic change in behaviour occurs as the animal progresses through different stages of the reproductive cycle. These changes are often associated with equally significant alterations in the animal's endocrine profile, with much evidence to indicate a close causal relationship.

Over the last thirty years or so, a considerable body of information has been accumulated on the subject of behaviour-endocrine interactions. Recently these studies have extended to the identification of the neural substrate sensitive to endocrine influence and to its role in the expression of the various components of reproductive behaviour.

This review reports upon some recent findings of the expression and possible function of central steroid receptors which occur during the breeding cycle of a species which has proved a very popular animal model for this area of research, the ring dove.

1.1. The Ring Dove as an Animal Model

A monomorphic member of the order Columbiformes, the ring dove (*Streptopelia risoria*) is a domesticated form of the African Collared dove (*S. roseogrisa*). It is well suited for laboratory study and breeds successfully for most of the year in an indoor colony where day length and environmental temperature can be controlled. Study of the ring dove's highly specialised biparental reproductive cycle by a number of prominent research groups has made important contributions to the elucidation of the neural and hormonal regulation of avian parental behaviour, and of endocrine-behaviour interactions (Lehrman, 1959; Hutchison, 1970; Silver *et al.*, 1973; Balthazart *et al.*,1980; Lea *et al.*, 1986; Buntin, 1993; Sharp *et al.*, 1996).

Neural Regulation in the Vertebrate Endocrine System, edited by Prasada Rao and Peter, Kluwer Academic / Plenum Publishers, New York, 1999.

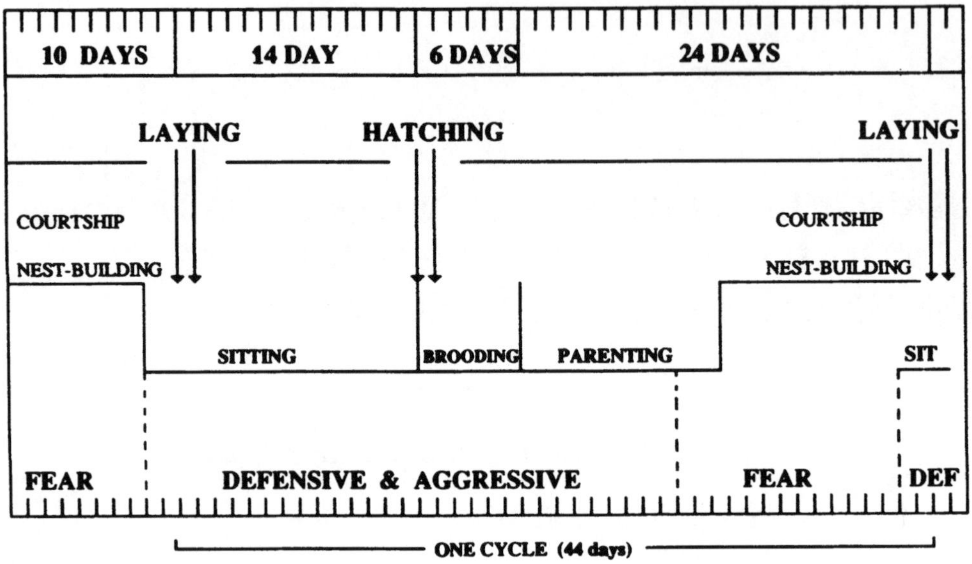

Figure 1. Schematic diagram to illustrate the ring dove breeding cycle.

1.2. Breeding Cycle of the Ring Dove

The complete breeding cycle of the ring dove progresses from courtship, through nest building, egg-laying, incubation and hatching, to feeding, rearing and protecting the young (squabs) (Figure 1). These events are synchronously timed between male and female in response to visual, auditory and tactile stimuli, and are accompanied by stereotyped behaviour patterns (Cheng, 1979). Under controlled conditions the entire cycle may be repeated at intervals of from six to seven weeks throughout the year (Silver *et al.*, 1974).

In the male and female dove testosterone and estrogen respectively are involved in the initiation of courtship, and testosterone with intraspecific aggression in the male (Erickson, 1970, Cheng, 1973, Bennet, 1940). In both sexes progesterone and estrogen are involved in nest-displays, nestbuilding and the initiation of incubation (Lehrman *et al.*, 1961. A sexually dimorphic temporal pattern of incubation is expressed, the male incubating the eggs for a period during the daylight hours and the female incubating for the rest of the time (Wallman *et al.*, 1979). In contrast to gallinaceous species, in which an increase in prolactin secretion coincides with the initiation of incubation (Riddle *et al.*, 1935), this occurs later in the ring dove and is thought to be responsible for the maintenance of incubation, initiate the care of the squabs after hatching and the manufacture of food by both sexes in the form of crop "milk" (Lea *et al.*, 1981).

Ring doves reach sexual maturity from the age of five to nine months. When a pair of male and female doves in breeding condition are placed together with nest materials and nest bowl, they enter into a sequence of predictable behavioural changes which culminates in the arrival of a clutch of two eggs, laid at an interval of approximately forty hours apart. The duration of the courtship phase of the cycle may vary but is normally between seven and eleven days (Lehrman, 1964).

The preliminary observable event of courtship is androgen-induced bow-cooing and hop-charging behaviour of the male which stimulates an increase in circulating estrogen in the female leading to ovarian development (Lehrman *et al.*, 1961). It has been shown that audiofeedback of male-induced cooing by the female promotes the ovarian response (Cheng, 1986). Aggressive behaviour by the male subsequently declines and is replaced by nest-soliciting which peaks around the third day. It has been proposed that the promotion of nesting behaviour over aggressive behaviour at this time is induced by increased aromatase activity in the ventromedial preoptic area, which may be initiated by the interaction of pairing (Dudley *et al.*, 1984; Lea and Armstrong, 1986). Earlier studies suggest that the inhibition of bow-cooing behaviour may be brought about as a result of progesterone secretion from a non gonadal source (Erickson *et al.*, 1967).

Copulation takes place throughout the courtship phase, and the two to four days before egg laying are occupied by nest-building by both partners. Increasing nest-building and associated behaviours, exhibited by the female as courtship proceeds, occur in response to the high levels of circulating estrogen combined with an increase in the secretion of progesterone (Cheng, 1979). Just before laying there is a sharp decline in the overall activity in both male and female doves which is closely associated with the occurrence of ovulation and signals the onset of readiness to incubate (Lehrman et al., 1961).

In avian species successful reproduction is dependent upon effective incubation of the eggs. In the ring dove, incubation starts some time between the laying of the first and of the second egg. After the two eggs have been laid, the parents share the incubation period of 14 days to hatching. The behaviour of both parents is characteristically quiet during incubation and both become difficult to dislodge from the nest if disturbed (Lehrman, 1964).

Environmental and social factors, such as previous breeding experience, act in an additive fashion with each other and with hormones to facilitate the transition from courtship to incubation in both male and female ring doves (Michel and Moore, 1985; Silver *et al.*, 1973). Participation in courtship and contact with nesting material together bring about the physiological changes underlying the onset of readiness to incubate. These events are under hormonal control (Lehrman, 1958a).

Incubation behaviour occurs after several days of gonadal hormone-stimulated activity (Lehrman, 1958b). It is well established that progesterone induces incubatory behaviour in both male and female doves and that progestational compounds are involved in the transition from courtship to incubation, via stimulation of the hypothalamo-pituitary-ovarian axis (Lehrman and Brody, 1957; Komisaruk, 1967; Stern, 1974). In the female dove plasma progesterone levels increase at the time of egg laying and initiation of incubation. Male ring doves show a basal level of plasma progesterone throughout courtship and incubation, although this level is higher than during the non-breeding state (Silver *et al.*, 1974; Lea *et al.*, 1986).

1.3. Endocrine Involvement in the Dove Breeding Cycle

Steroid hormones play an important role in the expression of reproductive behaviour and evidence has accumulated that gonadal steroid hormones are primary components of the basis of change through reproductive stages in avian species, including budgerigar, turkey, hen, quail and dove. The specific contribution of steroid hormones in this regard may vary with sex, species and breeding stage (Cheng and Lehrman, 1975; Hutchison, 1975; Wilson and Sharp, 1976; Peczely and Pethes, 1979; Bottoni *et al.*, 1985; Lea *et al.*, 1986; Ramos and Silver, 1992; El-Halawani and Rozenboim, 1993).

The effects of steroid hormones on reproductive physiology and behaviours are widely attributed to interaction with intracellular hormone receptors within specific brain regions. Binding studies using autoradiography have found that the distribution of androgens, estrogens and progestagens are similar in all vertebrates and the cells that retain sexual steroids are localised in particular areas in the preoptic region, hypothalamus and midbrain (Signoret and Balthazart, 1993).

The intracellular distribution of steroid hormone receptors has been the subject of intense investigation to elucidate the cellular mechanisms of action of steroid hormones (Kawashima *et al.*, 1979; Blaustein and Feder, 1979; Blaustein, 1994). Immunocytochemical and autoradiographic techniques, and *in vitro* binding studies have demonstrated steroid hormone receptors in the nucleus and in the cytoplasm of cells in many vertebrate species and there have been observations of immunostaining of primary neuronal processes of steroid-receptor-containing cells (Blaustein and Turcotte, 1989).

The intensity of immunostaining observed in cytoplasm is low in contrast with that seen in cell nuclei, suggesting that steroid hormone receptors are predominantly nuclear proteins which act as genomic regulators and mediate the intracellular responses to steroid hormones in target tissues (Schlinger and Callard, 1989). Other steroid hormone effects have been reported that are not compatible with a slower genomic mechanism and alternative cellular functions for cytoplasmic steroid hormone receptors present in soma and cytoplasmic processes have been postulated in the regulation of reproductive behaviours. It has been suggested that some of the immunoreactivity observed may be associated with a membrane receptor (Blaustein and Turcotte,1989)

1.4. Objectives Covered in This Review

This review will report on studies carried out to localise receptors for both androgen and progesterone in the brain of the ring dove. Changes in the expression of these receptors with respect to the reproductive cycle are demonstrated together with evidence for colocalisation of these two steroid receptors in discrete nuclei of the brain.

Secondly, the relationship of these receptors with the central dopaminergic neurotransmitter system, often claimed to be associated with the expression of parental behaviour, and the neuropeptide vasoactive intestinal polypeptide (VIP), known to be the physiological prolactin-releasing factor during incubation and brooding in the dove, has been explored.

2. CHANGES IN CENTRAL PROGESTERONE AND ANDROGEN RECEPTORS

For these studies, an immunofluorescent method was developed which utilised different fluorophores for the antigens and a computerised image analysis system, comprising of a low light digital camera and an image archiving system.

2.1. Immunocytochemical Methods Used

Mature male and female ring doves (*Streptopelia risoria*) with good breeding experience were used. Samples were selected and labelled according to their breeding stage which were, non-breeding (isolated birds), courting (3 days), day 1 of incubation, and brooding (2 days post-hatch).

Birds were anaesthetised and perfused with heparinised saline, followed by a Zamboni's fixative (4% paraformaldehyde and 2% picric acid), and finally with 10% sucrose.

The brain, pituitary and oviduct/testis were removed and immersed in 10% sucrose for a further 4–6 hours. The tissues were then frozen and stored at -80°C.

Following sectioning, cut to the orientation of the pigeon atlas of Karten and Hodos, 1967, the slide-mounted brain sections (16 µm thick) were re-hydrated and subjected to a strict protocol of antigen presentation and labelling. For the androgen receptor, labelling resulted from the incubation of the sections with an anti-androgen receptor antibody (rabbit polyclonal from Novocastra Ltd, Newcastle-upon-Tyne, England) overnight at 4°C. Visualisation was achieved by a further incubation of the sections with a secondary antibody conjugated with a CY3 fluorochrome. To investigate the co-expression with progesterone receptor, the sections were incubated further with an anti-progesterone receptor antibody (mouse monoclonal) and then visualised with a secondary antibody conjugated with an FITC fluorochrome. During microscopic analysis, a filter system was used to discriminate between the two wavelengths, and images were captured and stored.

Co-expression of these two receptors was investigated by tagging the images electronically, followed by a blending process. Co-expression was then quantified.

2.2. Distribution of Androgen Receptor Immunoreactivity (AR-ir) in CNS; Expression with Respect to Breeding Condition

In total, 38 birds were used. Three sections per every brain nuclei quantified were analysed and the average taken. No differences between brain hemispheres was observed.

Brains of courting birds exhibited a widespread distribution of androgen receptor. Over 40 brain nuclei were identified as possessing AR-ir. Those areas expressing intense or medium staining are depicted in Tables 1 and 2.

Generally, AR-ir cells were more numerous in the frontal and midbrain, compared to the hindbrain, with no discernible differences between males and females.

Particularly large nuclear immunofluorescent labelling was observed in the archistriatum, cortex piriformis, the area corticoidea dorsolateralis, area parahippocampalis, hippocampus, neostriatum, hyperstriatum ventral, hyperstriatum dorsal, archistriatum, pars dorsalis archistriatum, pars ventralis, nucleus taeniae and the neostriatum caudal. In the olfactory bulb some small cells showing immunofluorescent labelling were found along the internal and mitral layer.

It was noted that the fluorescent intensity of the AR-containing neurons was higher in the courting bird when compared to the non-breeding, incubating and brooding birds. Generally, the density of AR-ir was either significantly decreased or absent in brooding birds for all brain regions.

2.3. Changes in Central Progesterone and Androgen Receptor Expression with Respect to Reproductive Cycle

The influence of the breeding cycle upon the androgen receptor expression was more prominent in four hypothalamic areas of the brain, namely the preoptic area, nucleus paraventricularis magnocellularis, nucleus periventricularis magnocellularis and the tuberal hypothalamus. An analysis of variance (ANOVA) performed on all data showed no significant differences between the sexes in these four hypothalamic regions. However, the amount of AR-ir, in all four hypothalamic regions (including the ventral, lateral and dorsal region of the nucleus tuberalis (Tu), was significantly higher in courting birds (male: n=6; female: n=5) ($p < 0.05$), when compared to the same regions of non-breeding

Table 1. Brain areas showing intense AR-ir in ring doves at two stages of the breeding cycle

Abbreviations	Courting	Brooding
Archistriatum	++++	+
Archistriatum, pars dorsalis	++++	+
Archistriatum, pars ventralis	++++	+
Area parahippocampalis	++++	+
Area temporo-parieto-occipitalis (Edinger, Wallenberg, and Holmes)	++++	+
Cortex piriformis	++++	+
Hippocampus	++++	+
Hyperstriatum dorsale	++++	+
Hyperstriatum ventrale	++++	+
Infundibulum	++++	
Neostriatum	++++	+
Nucleus accumbens	+++++	
Nucleus anterior medialis hypothalami	++++	
Nucleus basalis	++++	
Nucleus dorsointermedius posterior thalami	++++	
Nucleus dorsolateralis posterior thalami	++++	
Nucleus dorsomedialis anterior thalami	++++	
Nucleus dorsomedialis posterior thalami	++++	
Nucleus ectomamillaris	++++ (cytoplasmic)	
Nucleus habenularis medialis	++++	+
Nucleus lateralis hypothalami	++++	
Nucleus lateralis hypothalami posterioris	+++++	
Nucleus mamillaris lateralis	++++	
Nucleus ovoidalis	++++	
Nucleus periventricularis magnocellularis	++++	
Nucleus preopticus anterior	+++++	
Nucleus preopticus medialis (van Tienhoven)	+++++	
Nucleus preopticus paraventricularis magnocellularis (van Tienhoven)	+++++	
Nucleus septalis lateralis	+++++	+
Nucleus subhabenularis lateralis	++++	
Nucleus superficialis parvocellularis (Nucleus tractus septomesencephalici)	+++++	
Nucleus supraopticus (Ralph)	+++++	
Nucleus taeniae	+++++	+
Nucleus tuberis	+++++	
Nucleus ventrolateralis thalami	++++	
Paleostriatum augmentatum	++++	+
Polus caudalis telencephali	++++	
Stratum cellulare internum	++++	+
Stria medullaris	++++	
Tractus infundibularis	++++	

(male: n=4; female: n=5), day 1 of incubation (male: n=4; female n=4) and brooding (male: n=5; female n=5). Although at day 1 of incubation, the level of androgen receptor was still moderately high, it was not significantly higher than in non-breeding and brooding birds.

In an earlier study (Askew *et al.*, 1997) an immunohistochemical method was used to localise progesterone receptor immunoreactivity (PR-ir) in the brains of male and female ring doves over a breeding cycle. Distribution of PR-ir was similar for both sexes but, in contrast to AR-ir, localisation was restricted to just four hypothalamic regions: the preoptic area (POA), nucleus preopticus paraventricularis magnocellularis (PPM), nucleus hypothalami lateralis (PLH) and the tuberal region (Tu).

Table 2. Brain areas showing medium AR-ir in ring doves at two stages of the breeding cycle

Abbreviations	Courting	Brooding
Area corticoidea dorsolateralis	+++	+
Area ventralis (Tsai)	+++	
Brachium colliculi superioris	+	
Bulbus olfactorius	+	
Capsula interna occipitalis	+	
Cerebellum	+	+
Hyperstriatum accessorium	+++	+
Hyperstriatum intercalatus superior	+++	+
Hyperstriatum ventrale dorso-ventrale	+++	+
Lobus parolfactorius	+	
Neostriatum caudale	+++	+
Nucleus habenularis lateralis	+++	
Nucleus intercollicularis	+	
Nucleus intrapeduncularis	+++	
Nucleus isthmi, pars magnocellularis	+	
Nucleus nervi oculomotorii, pars dorsalis	+++	
Nucleus nervi oculomotorii, pars ventralis	+++	
Nucleus nervi oculomotorii, pars dorsolateralis	+++	
Nucleus olfactorius anterior	+	
Nucleus papillioformis	+	
Nucleus pontis lateralis	+	
Nucleus pontis medialis	+	
Nucleus septalis medialis	+	
Paleostriatum primitivum	+	+
Stratum griseum centrale	+	
Stratum opticum	+	
Substantia grisea et fibrosa periventricularis	+	
Torus semicircularis	+	
Tractus fronto-archistriatalis	+	+
Tractus vestibulo-mesencephalicus (Papez)	+	

Quantitative analysis demonstrated that the density of PR-containing cell nuclei in the TR of both sexes had significantly decreased in brooding birds compared to non-breeding and birds at the onset of incubation. Interestingly, in context, the density of cell nuclei containing PR-ir in the POA, PPM, and PLH in brooding birds was significantly higher than in birds at the onset of the reproductive cycle.

A double immunofluorescent technique revealed a high degree of co-expression of AR-ir and PR-ir within the POA, PPM, nucleus periventricularis magnocellularis (PVM) and the tuberal hypothalamus. Furthermore, co-localisation was affected by the breeding state of the birds, with the greatest amount of co-localisation observed in courting birds of both sexes (80%). A lower degree of co-localisation was seen in the brooding birds, and an intermediate level in the non-breeding and incubating birds of both sexes.

Measurement of fluorescent intensity (corresponding to receptor density) was also performed for both AR-ir and PR-ir. Image analysis revealed that the highest fluorescent intensity occurred in courting birds for both progesterone and androgen receptors. Moreover, between 60 and 70 percent of those cells demonstrating high fluorescent intensity for progesterone receptor were found to co-express androgen receptor, whereas only 30 percent of the low intensity progesterone receptor-containing neurons also expressed androgen receptor.

3. THE ROLE OF DOPAMINE IN REPRODUCTIVE FUNCTION

The neurotransmitter dopamine has often been implicated in reproductive function. Evidence for a role of dopamine in the reproductive function of avian and mammalian species has been accumulating over many years (Nistico and Stephenson, 1979; Labrie *et al.*, 1980; MacNamee and Sharp, 1989; Barclay and Cheng, 1992; Petitclerc, Bedard et al., 1995). Dopamine may act indirectly through its effects on hormone secretion at the hypothalamo-pituitary axis (Peczely and Pethes, 1979; Piroli *et al.*, 1996; Youngren *et al.*, 1995), and there is evidence for a direct effect of dopamine on sexual behaviour via stimulation and interaction of dopamine receptor subtypes (Melis and Argiolas,1995). It has been proposed that a common feature of dopaminergic action is enhancement of sensorimotor integration by the removal of tonic inhibition (Wynne and Gunturkun, 1995).

In mammals dopamine in the medial preoptic area has a facilitative role in masculine sexual performance and in the stimulation of sexual motivation (van Furth *et al.*, 1995; Hull *et al.*, 1995). In estrous female rats, mesolimbic dopamine depletion in the ventral striatum, has been found to impair sexually appetitive behaviour, (Robbins and Everitt, 1992). Functional aspects of dopamine on avian behaviour include the inducement of altered states of attention and aggression (de Lanerolle and Millam, 1980), the production of stereotyped behaviour (Goodman, 1981), and the control of posture and motor activity (Nistico and Stephensen, 1979).

Increasing acknowledgement of the role of dopamine in reproductive function gives rise to the need to progress understanding of the mechanisms by which dopamine exerts its behavioural and physiological effects. Despite the biochemical documentation more than three decades ago of significant amounts of the primary catecholamines, dopamine and noradrenaline, in the avian brain (Bertler *et al.*, 1964; Fuxe and Ljunggren, 1965; Jorio and Vogt, 1967), relatively little is known of the organisation and distribution of central dopamine in avian species. Interest in the elucidation of dopaminergic mechanisms subserving aspects of reproductive behaviour, using the ring dove as a model, prompted a detailed investigation of the organisation of dopamine-containing neuronal systems in the ring dove brain by means of immunocytochemical methods to demonstrate tyrosine hydroxylase.

Tyrosine hydroxylase (TH) is the rate-limiting enzyme in catecholamine biosynthesis. Dopaminergic cells contain TH for the conversion of tyrosine into L-dihydroxyphenylalanine (L-Dopa) and thus dopaminergic neurons are also TH positive (Wynne and Gunturkun, 1995). Specificity for the localisation of dopamine with the use of antibodies against TH is evidenced by studies demonstrating that TH is present in immunocytochemically detectable quantities only in dopamine neurons, and that dopamine-β-hydroxylase-containing neurons (the enzymatic marker for noradrenaline) are not detected (Lewis et al., 1986). This is supported by mammalian and avian studies demonstrating correspondence in the immunohistochemistry for TH with that for anti-dopamine antibodies (Divac *et al.*, 1994; Zoli *et al.*, 1993).

As well as a regulatory role of dopamine in reproductive function in birds (Martin *et al.*, 1982), there is a considerable body of research to determine how structural sex differences in the brain contribute to functional differences in avian behaviour. Studies have revealed that in certain areas dopamine synthesis, content and metabolism is sexually differentiated and under the influence of sex steroid levels in development and adulthood (Balthazart *et al.*, 1992). Gender specific structural differences in catecholaminergic systems have been demonstrated in telencephalic nuclei in song birds, and there are sex differences in the medial preoptic area in the quail and zebra finch, consisting of size of nuclei, cell density, arborization and connectivity (Bailhache and Balthazart, 1993; Bottjer, 1993; De Vries, 1990).

The integrative role of central monoamines in the regulation of both broodiness and aggression in avian reproduction has received considerable attention. Distribution and concentration of monoamines have been examined in the pigeon, chicken and quail (Wynne and Gunturkun, 1995; MacNamee and Sharp, 1989; Balthazart et al., 1992). A study of monoamine activity in the ring dove demonstrated that temporal changes in release and/or degradation of dopamine contribute to the diurnal rhythm of this amine in the hypothalamus, paleostriatum and hyperstriatum (Lea *et al.*, 1987). However, the neural substrates that subserve dopaminergic function in the ring dove breeding cycle remain to be determined. Their elucidation will rely on a knowledge of the neuroanatomy and organisation of dopaminergic neuronal systems in the brain of male and female doves during stages of the breeding cycle.

The aims of this study were to use immunocytochemical techniques to map and report the major areas of dopaminergic cell body and terminal labelling in the ring dove brain, using antibodies raised against tyrosine hydroxylase (TH-ir); and subsequently investigate the co-localisation of TH-ir with PR-ir.

3.1. Distribution of Tyrosine Hydroxylase Immunoreactivity (TH-ir) in Ring Dove CNS

A comprehensive mapping of TH-ir in both male and female doves was achieved. No obvious differences were observed in the pattern of staining between male and female birds, nor in the distribution of labelling between non-breeding and brooding birds. Therefore, the following description applies to all birds unless exceptions are specifically noted. Evaluation of morphological and quantitative differences in neuronal cell populations in individual brain nuclei between brooding and non-breeding birds, and between males and females, is on-going.

From the rostral aspect, the first TH-ir noted was in the parolfactory lobe (LPO), which is a major component of avian striatum. The restriction of immunoreactivity to the LPO was clearly defined at its dorsal margin by the lamina medullaris dorsalis (LMD). Beyond this little staining was evident, except for a fine tract ascending from the ventricle, between the lamina frontalis suprema (LFM) and the medial portion of the hyperstriatum accessorium (HA). Throughout the rostral/caudal extent of the LPO, the character of the staining was represented by fine grained TH-ir, which was generally uniform in nature and appeared to represent a terminal field. No cell bodies or neuronal processes were distinguished, although staining intensity tended to be greater along the ventral borders.

Proceeding caudally through the LPO, a distinct area of adjacent staining became apparent in the paleostriatum augmentatum (PA). This staining was finer than in the LPO but defined the boundaries of the PA throughout the whole of its extent. Caudally the staining in the PA became continuous with the adjoining LPO. Sparse neuronal processes were present in the tractus quintofrontalis (QF) and the paleostriatum primitivum (PP), with little TH-ir observed dorsally in the ectostriatum (E).

A region devoid of labelling appeared along the midline at the level of the rostral nucleus septalis lateralis (SL) and, caudally, fine terminal labelling appeared restricted to the LPO and PA, forming a characteristic pattern of staining surrounding the PP in all birds. The entire SL, and an area ventral to the nucleus accumbens (Ac), showed dense, coarse-grained TH-positive fibres, with lighter areas of stained fibres observed lateral to the Ac and ventral to the LPO, continuing into the PP and Nucleus intrapeduncularis

(INP). At this level, groups of fibres appeared to extend from the LPO into the SL (Figure 2a and b).

TH immunoreactive cell bodies were observed in the rostral nucleus preopticus anterior (POA), and were evident throughout the preoptic area. Labelled cells and neuronal processes were seen in all birds between the tractus septomesencephalicus (TSM) and the optic chiasm (CO), surrounding the third ventricle. Rostrally, cell bodies in the POA were scattered throughout the ventral portion of the defined region and were few in number. As the POA extended caudally cell bodies became more numerous and assumed a characteristic pattern of distribution, with a distinct group of cells in the lateral POA adjacent to the TSM, and a separate dorso-medial grouping, surrounding the midline, and encompassing the nucleus preopticus medialis (POM) (Fig. 2a and b).

Caudal to the POM, the medial group of immunoreactive perikarya extended into the nucleus periventricularis magnocellularis (PVM), and assumed a progressively more dorsal position to follow the rostro/caudal extent of this nucleus. Cells in the lateral pre-optic area became continuous with the nucleus lateralis hypothalami (LHy) and spread dorso-laterally towards the midline to assume an arc-like distribution. A separate and distinct group of cell bodies became apparent in the nucleus ventrolateralis thalami (VLT) and was found in an apparently uninterrupted manner until the mesencephalic level (Fig 2a-c).

At the level of the commissurra anterior (CA), an area of densely-stained, striated fibres became apparent, confined to the QF and ansa lenticularis (AL) and appearing to represent a fibre tract which continued to the most caudal areas of the brain.

3.2. Co-Localization of Progesterone Receptors in TH-ir Neurons

Neurotransmitter and progesterone receptor interactions are involved in the modulation of sexual behaviour in mammalian species and dopamine is known to play a major role in mediating steroid hormone-induced sexual and aggressive behaviours (Mani *et al.*, 1997; Lee *et al.*, 1995).

Progesterone treatment has been shown to significantly increase the concentration of dopamine in the anterior hypothalamus-preoptic region in avian species (Knight *et al.*, 1982). Turnover of catecholamines differs in sexually dimorphic brain structures implicated in the control of reproductive function in quail and in Zebra Finch, where steroid hormones cause striking changes in steady state catecholamine levels and catecholaminergic turnover rates in the hypothalamus and vocal control nuclei, with differential effects in different brain regions (Balthazart *et al.*, 1992; Barclay and Harding, 1990).

Dopamine neurons in the hypothalamus have been found to show selective patterns of gonadal steroid hormone accumulation, and there are region specific differences in the extent of co-localisation of steroid hormone receptors with dopaminergic neurons in different species (Asmus and Newman, 1993; Herbison *et al.*, 1995; Kritzer, 1997). Progesterone receptors have been co-localised with tyrosine hydroxylase in hypothalamic regions in several mammalian species, suggesting a direct action of progesterone on dopaminergic neurons (Blaustein and Turcotte, 1989; Sar, 1988). No studies have been reported in avian species.

Figure 2a-d. Diagrams of coronal sections through the ring dove brain, illustrating TH-ir and PR-ir. TH-ir perikarya are represented by solid circles, terminal fields by shaded areas and fibers as thicker lines. Immunoreactive PR are represented by small shaded dots. Sections are presented in a rostral to caudal order from a to d. Co-ordinates are given as in the stereotaxic atlas of the pigeon brain (Karten and Hodos, 1967). Insets show detail of hypothalamic regions where co-localization was investigated.

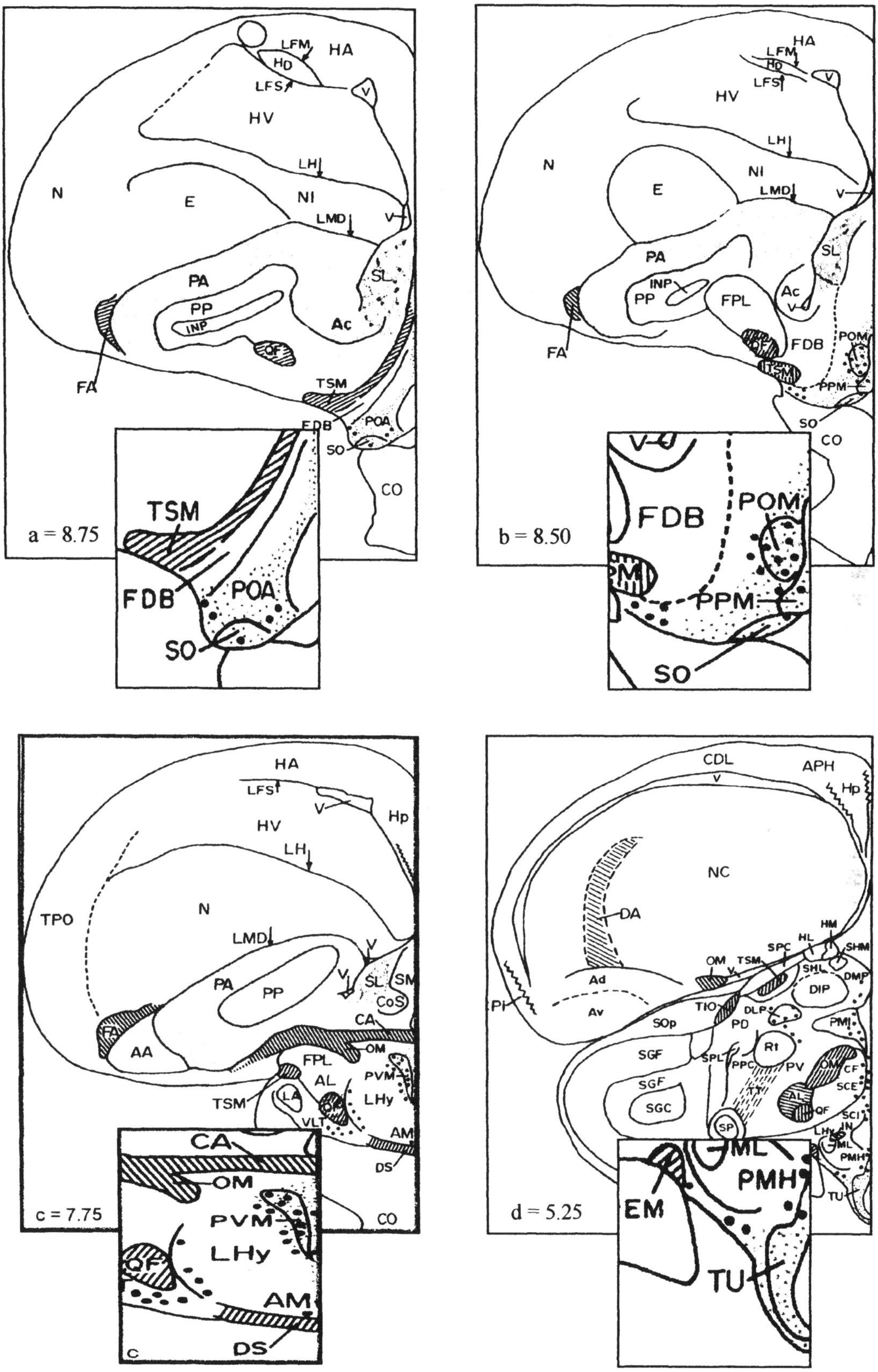
a = 8.75
HA
LFM
HD
LFS
V
HV
N
LH
E
NI
LMD
PA
SL
PP
INP
Ac
QF
FA
TSM
FDB
POA
SO
CO
b = 8.50
FPL
POM
PPM
c = 7.75
Hp
TPO
SM
CoS
CA
AA
OM
PVM
AL
LA
LHy
VLT
AM
DS
d = 5.25
CDL
APH
NC
DA
HM
HL
SPC
SHM
SHL
DMP
Ad
DIP
Av
TIO
DLP
PD
SOp
PMI
SGF
SPL
PPC
PV
CF
SCE
SGC
SCI
SP
IN
ML
PMH
TU
EM

The aims of the study were to localise and compare the distribution of progesterone receptors (PR) and (TH-ir) in the ring dove hypothalamus, and to employ a double immunofluorescence protocol for the investigation of cells expressing TH and co-localisation of PR. It was envisaged that a comparison of PR and TH immunoreactivity and co-localisation in incubating, brooding and non-breeding doves of both sexes might further elucidate mechanisms of dopaminergic and progesterone interactions underlying the onset of nest defence behaviour during the ring dove breeding cycle.

Tissues for the study were taken from adult male and female ring doves (*Streptopelia risoria*). Breeding birds were allowed to commence a normal breeding cycle. For incubating birds, tissues were taken on the second day of incubation and for brooding birds, on the second day of brooding of the squabs.

TH neurons were present in all regions where PR were visualised, with the exception of the Tu. TH-fluorescent staining appeared to be cytoplasmic, with the cell nucleus remaining unlabelled. TH-ir cell bodies and neuronal processes were observed within dense populations of PR, particularly in the POA, POM, PPM, PVM and at the dorso-lateral margins of the Tu. PR were visualised and recorded in close association to TH-ir perikarya and neuronal processes. No co-localisation of PR within TH cell bodies was observed (Figure 2).

Co-localisation of PR- and TH-ir neurons has been described in several mammalian species and has been reported in ventral and medial periventricular nuclei and arcuate nucleus (Kohama and Bethea, 1995). There is region-specific species difference in the extent of co-localisation and the number of co-labelled neurons has been shown to vary depending on endocrine state. This study revealed no co-localisation of PR in TH-ir perikarya in the POA, POM, PVM or Tu of the ring dove during progressive stages of the reproductive cycle, or in non-breeding birds, and suggests the possibility of non-receptor mediated or interneuronal mechanisms of progesterone regulation of dopamine neurons.

The presence of TH neuronal processes in close association with labelled PR may further suggest an anatomical relationship between dopaminergic terminals and the cell bodies of neurons exhibiting PR within their cell nuclei. These results are in agreement with previous findings proposing dopaminergic modulation of progesterone sensitive neurons in the hypothalamus (Blaustein and Turcotte, 1989). However, this study suggests that any possible progesterone modulation of dopamine activity in the ring dove hypothalamus is not mediated by a direct genomic action in dopaminergic neurons.

4. THE RELATIONSHIP BETWEEN PR AND VIP IN THE RING DOVE BRAIN

In the dove (Lea *et al.*, 1991) as with many other species, it is clearly established that VIP has an important physiological prolactin-releasing function at the level of the pituitary. Furthermore, it is detectable in regions of the avian hypothalamus. In the ring dove brain there is a systematic change in the size of immunoreactive VIP cell bodies in the ventrolateral tuberal hypothalamus as a function of the stage of the reproductive cycle (Silver and Ramos, 1991). These cells increase in both size (diameter) and staining intensity. The authors report that this increase is slight during courtship, anticipating the onset of incubation, and rises substantially from the second to the fifth day of incubation, anticipating the onset of prolactin secretion. VIP cell size remains enlarged beyond the period of prolactin secretion, lasting to the fifteenth day of incubation. However, the current data indicates that this increase in cell size during courtship does not provide a detected function in the expression of incubation in the ring dove. There is evidence, however, that the VIP-ergic cell bodies in the tuberal region of the hypothalamus are those responsible for stimulating the release of prolactin.

We report here a study to determine the relationship between the immunolocalisation of progesterone receptor and VIP in the brain of the ring dove at various stages during the breeding cycle. Brains were taken from male and female doves at two days following the onset of incubation (D2), seven days (D7) and two days following the hatching of the eggs (D2 brooding).

4.1. Changes in the Distribution of VIP-ir and PR-ir in the Tuberal Hypothalamus (Tu) during the Breeding Cycle

The distribution of VIP in the Tu observed in this study was similar to that described in previous studies on the ring dove (Cloues *et al.*, 1990; Silver and Ramos, 1991). Likewise, PR-ir was also detected in the Tu at different stages of the breeding cycle in a similar pattern to that demonstrated in a previous study (Askew *et al.*, 1997).

Two days following the onset of incubation, neurons containing VIP were observed in the dorsal region of the medial segment of the Tu in both male and female doves. These neurons were small in diameter and possessed short processes. VIP was not observed in any other region of the Tu at this time of the breeding cycle. Numerous cells containing PR were observed in the dorsolateral region of the medial Tu and the ventrolateral region of the posterior Tu at this stage of the breeding cycle in both sexes.

Seven days following the onset of incubation, numerous neurons containing VIP were seen in the dorsolateral and ventrolateral regions of the medial Tu in both male and female doves. The size of these neurons was larger than that seen at D2, with larger processes extending towards the ventricle. PR were detected in cells in the dorsolateral and ventrolateral regions of the medial Tu. Cells expressing PR were in higher numbers in the ventrolateral region of the Tu in both sexes. The number of cells expressing PR at this time of the breeding cycle was less than that observed at D2 of incubation.

Two days following the hatch of the eggs, VIP was observed in the dorsolateral region of the anterior Tu, and in both the dorsolateral and ventrolateral portions of the medial and posterior Tu in both male and female doves. These cells were the largest in size, number and fluorescent intensity of all the three stages of the breeding cycle compared in this study. In contrast, only a few cells expressing PR were detected in the dorsolateral region of the Tu at this time of the breeding cycle in both sexes.

4.2. Co-Localization of PR in VIP Neurons

No co-localisation of VIP with PR was detected at D2 of incubation in either sex (Table 3). The majority of cells expressing PR at this time of the breeding cycle were present in the dorsolateral and ventrolateral portions of the Tu, whereas VIP neurons were limited to the dorsal region of this region of the hypothalamus.

During mid-incubation (D7), fifty per cent of VIP neurons also expressed PR in both male and female doves (Table 3). This was seen to occur in both the dorsolateral and ventrolateral regions of the Tu. Cell nuclei expressing PR were observed which fluoresced green, with a surrounding orange-red fluorescence of the cytoplasm of the axons which contained VIP.

Sections of brain examined from birds during the brooding period displayed ten per cent co-localisation in the ventrolateral region of the medial Tu in the female ring dove.

Table 3. Percentage co-localization of VIP with PR in the Tu of male and female ring doves during various stages of the breeding cycle

Reproductive state	Sex	(*n*)	No. of sections examined	No. of sections showing VIP+PR	% Co-localization of VIP+PR
D2 incubation	M	1	9	3	0
D2 incubation	F	1	8	2	0
D7 incubation	M	2	16	11	50
D7 incubation	F	2	17	8	50
D2 brooding	M	3	22	9	20
D2 brooding	F	3	20	7	10

Sections examined from the male ring dove displayed twenty per cent co-localisation in the same region of the Tu (Table 3).

The progesterone receptor co-localised with VIP in the tuberal hypothalamus occurs in an area known to be involved in the stimulatory control of prolactin release in the ring dove (Cloues *et al.*, 1990; Silver and Ramos, 1991). In this species, prolactin secretion begins to increase after 5 days of incubation, reaching a peak when the squabs have hatched after 14–15 days of incubation (Goldsmith *et al.*, 1981; Cheng and Burke, 1983; Lea and Armstrong, 1986). Circulating progesterone may function through progesterone receptor in the Tu to inhibit VIP secretion and prolactin release, until incubation behaviour has become firmly established. This may be under the influence of plasma estrogen, which decreases once incubation behaviour is established and the gonads have regressed. This may be the mechanism which prevents premature development of the crop sac at the onset of incubation. Consistent with this view, is the observation that administration of progesterone to incubating ring doves prevents crop sac development but does not disrupt incubation behaviour itself (Silver and Feder, 1973). Once incubation has been established and the inhibitory effect of estrogen and progesterone on VIP secretion is removed, the release of prolactin to stimulate crop sac growth may proceed. This co-localisation study has identified cells which appear to be involved in this neuroendocrine regulation are present in the dorsolateral and ventrolateral regions of the Tu.

SUMMARY

AR-ir expression changes dramatically in both sexes over the ring dove breeding cycle, with maximal density observed in courting birds. PR-ir shows a less dynamic fluctuation but exhibits a high degree of co-localisation with AR-ir.

In the preoptic area, an area thought to be involved in the expression of parental behaviour, no co-localisation of PR was found with tyrosine hydroxylase. In contrast, in the tuberal hypothalamus, an area of the brain concerned with neuroendocrine regulation, a pattern of co-localisation of PR together with the prolactin-releasing factor, VIP, was discerned which indicates a possible inhibitory action of progesterone upon VIP release and subsequent prolactin secretion.

ACKNOWLEDGMENTS

This work has been supported by the BBSRC, The British Council, and the University of Central Lancashire.

REFERENCES

Askew, J.A., Georgiou, G.C., Sharp, P.J. and Lea. R.W. (1997). Localization of progesterone receptor in brain and pituitary of the ring dove: Influence of breeding cycle and estrogen. *Hormones and Behavior* **32**, 105–113

Asmus, S.E. and Newman, S.W. (1993). Tyrosine hydroxylase neurons in the male hamster chemosensory pathway contain androgen receptors and are influenced by gonadal hormones. *J. Comparative Neurology* **331**, 445–457

Bailhache, T. and Balthazart, J. (1993). The catecholaminergic system of the Quail brain: Immunoctyochemical studies of dopamine β-hydroxylase and tyrosine hydroxylase. *J. Comparative Neurology* **329**, 230–256

Balthazart, T., Blaustein, J.D., Cheng, M.F. and Feder, H.H. (1980). Hormones modulate the concentration of cytoplasmic progestin receptors in the brain of male Ring Doves (*Streptopelia Risoria*). *J. Endocrinology* **86**, 251–261

Balthazart, J., Foidart, A., Sante, P.H. and Hendrick, J.C. (1992). Effects of α-methyl-para-tyrosine on monoamine levels in the Japanese Quail: Sex differences and testosterone effects. *Brain Research Bulletin* **28**, 275–288.

Barclay, S. R. and Cheng, M-F. (1992). Role of catecholamines in the courtship behavior of male ring doves. *Pharmacol. Biochem., and Behavior* **41**, 739–747

Barclay, S.R. and Harding, C.F. (1990). Differential modulation of monoamine levels and turnover rates by estrogen and/or androgen in hypothalamic and vocal control nuclei of male Zebra Finches. *Brain Research* **523**, 251–262

Bennett, M.A. (1940). The social hierarchy in ring doves. II. The effect of treatment with testosterone propionate. *Ecology* **21**, 148–165

Bertler, A.B., Falck,C.G, Gottfries, L., Llunggren, L. and Rosengren, E. (1964). Some observations on adrenergic connections between mesencephalon and cerebral hemispheres. *Acta Pharmacol. Toxicol.* **21**, 283–289

Blaustein, J.D. (1994). Estrogen receptors in neurons: new subcellular locations and functional implications. *Endocrine Journal* **2**, 249–258

Blaustein, J.D. and Feder, H.H. (1979). Cytoplasmic progestin receptors in female guinea pig brain and their relationship to refractoriness in expression of female sexual behaviour. *Brain Research* **177**, 489–498

Blaustein, J.D. and Turcotte, J.C. (1989). Estrogen receptor-immunostaining of neuronal cytoplasmic processes as well as cell nuclei in guinea pig brain. *Brain Research* **495**, 75–82

Bottjer, S.W. (1993). The distribution of tyrosine hydroxylase immunoreactivity in the brains of male and female Zebra Finches. *J. Neurobiology* **24**, 51–69

Bottoni, L., Lucini, V. and Massa, R. (1985). Effect of progesterone on the sexual behavior of the male Japanese Quail. *General and Comparative Endocrinology* **57**, 345–351

Buntin, J.D. (1993). Prolactin-brain interactions and reproductive function. *Amer. Zool.* **33**, 229–243

Cheng, M. F. (1973). Effect of estrogen on behavior of ovariectomized ring doves (*Streptopelia risoria*). *J. of Comp. and Physiological Psychology* **83**, 234–239

Cheng, M. F. (1979). Progress and prospects in ring dove research: A personal view. *Advances in the Study of Behavior* **9**, 97–129

Cheng, M. F. (1986). Female cooing promotes ovarian development in ring doves. *Physiology and Behavior* **37**, 371–374

Cheng, M. F. and Burke, W. H. (1983). Serum prolactin levels and crop-sac development in ring doves during a breeding cycle. *Hormones and Behaviour* **17**, 54–65

Cheng, M. F. and Lehrman, D. (1975). Gonadal hormone specificity in the sexual behavior of ring doves. *Psychoneuroendocrinology* **1**, 95–102

Cloues, R., Ramos, C. and Silver, R. (1990). Vasoactive intestinal polypeptide-like immunoreactivity during reproduction in doves: Influence of experience and number of offspring. *Hormones and Behaviour* **24**, 215–231

De Vries, G. (1990). Sex differences in neurotransmitter systems. *Journal of Neurendocrinology* **2**, 2–13

Divac, I., Thibault, J. Skageberg, G., Palacios, J.M. and Dietl, M.M. (1994). Dopaminergic innervation of the brain in pigeons. The presumed 'Prefrontal Cortex'. *Acta Neurobiol. Exp.* **54**; 227–234

Dudley, S. P., Salisbury, R. S., Adkins-Reagan, E. R., Weisz, J., (1984). Courtship stimulates aromatase activity in preoptic area of brain in male ring doves. *Endocrinology*, 115, 3, 1224–1226

El-Halawani, M.E., Rozenboim, I., (1993) The ontogeny and control of incubation behavior in Turkeys. *Poultry Science*, **72**, 906–911

Erickson, C. J. (1970) Induction of ovarian activity in female ring doves by androgen treatment of castrated males. *J. of Comparative and Physiological Psychology*, **71**, 210–215

Erickson, C.J., Bruder, R.H., Komisaruk, B.R., Lehrman, D.S. (1967) Selective inhibition by progesterone of androgen-induced behavior in male ring doves (*Streptopelia risoria*) *Endocrinology*, **81**, 39–44

Fuxe, K., Ljunggren, L. (1965) Cellular localisation of monoamines in the upper brain stem of the pigeon. *J. Comp. Neuro.* **125**, 355–382

Goldsmith, A.R., Edwards, C., Koprucu, M., Silver, R.. (1981) Concentrations of prolactin and luteinizing hormone in plasma of doves in relation to incubation and development of the crop gland. *J. of Endocrinology*, **90**, 437–443

Goodman, I .J., (1981) Amphetamine and apomorphine induced stereotyped behavior in adult pigeons. *Pharmacology, Biochemistry and Behavior*, **15**, 701–704

Herbison, A.E., Horvath, T.L., Naftolin, F., Leranth, C. (1995) Distribution of estrogen receptor-immunoreactive cells in monkey hypothalamus: Relationship to neurons containing luteinizing hormone-releasing hormone and tyrosine hydroxylase. *Neuroendocrinology*, **61**, 1–10

Hull, E.M., Du., J., Lorrain, D.S., Matuszewich, L., (1995) Extracellular dopamine in the medial preoptic area: implications for sexual motivation and hormonal control of copulation. *J. of Neuroscience*, **15**, 7465–7471

Hutchison, J.B.(1970). Influence of gonadal hormones on the hypothalamic integration of courtship behaviour in the Barbary Dove, *J. Reprod. Fert. Suppl.* 15–41

Hutchison, R.E. (1975). Influence of estrogen on the initiation of nesting behaviour in female budgerigars. *J. Endocrinology* **64**, 417–428

Jurio, A.V. and Vogt, M. (1967). Monoamines and their metabolites in the avian brain. *J. Physiol.* (London) **189**, 489–518

Karten, H. J. and Hodos, W. (1967). "Stereotaxic atlas of the brain of the pigeon (*Columba Livia*)." Baltimore, John Hopkins University Press.

Kawashima, M., Kamiyoshi, M. and Tanaka, K. (1979). Effects of progesterone, estradiol and testosterone on cytoplasmic progesterone receptor concentrations in the hen hypothalamus and pituitary. *Biology of Reproduction* **21**, 639–646

Knight, P.J., Francies, P.T., Holman, R.B. and Gladwell, R.T. (1982). Changes in hypothalamic monoamine concentrations accompany the progesterone-induced release of luteinizing hormone in the domestic hen. *Neuroendocrinology* **35**, 359–362

Kohama. S.G. and Bethea, C.L. (1995). Steroid regulation of tyrosine hydroxylase messenger ribonucleic acid in dopaminergic subpopulations of Monkey hypothalamus. *Endocrinology* **136**, 1790–1800

Komisaruk, B.R. (1967). Effects of local brain implants of progesterone on reproductive behavior in ring doves. *J. of Comparative and Physiological Psychology* **64,** 219–224

Kritzer, M.F. (1997). Selective colocalization of immunoreactivity for intracellular gonadal hormone receptors and tyrosine hydroxylase in the ventral tegmental area, substantia nigra, and retrorubral fields in the Rat. *J. Comparative Neurology* **379**, 247–260

Labrie, F., Ferland, L., Denizeau, F. and Beaulieu, M. (1980). Sex steroids interact with dopamine at the hypothalamic and pituitary levels to modulate prolactin secretion. *J of Steroid Biochemistry* **12**, 323–330

Lea, R.W. and Armstrong, D.G. (1986). Changes in aromatase activity in the brain of the male ring dove (*Streptopelia risoria*) during the breeding cycle, *Comp. Biochem. Physiol.* **831**, 693–696

Lea, R.W., Dods, A.S.M., Sharp, P.J. and Chadwick, A. (1981). The possible role of prolactin in the regulation of nesting behaviour and the secretion of luteinizing hormone in broody bantams. *J. Endocrinology* **91**, 89–97.

Lea, R.W., Hall, T.R. and Vowles, D.M. (1987). Daily variations in monoamine turnover in the brain of the ring dove. *Neuroscience* **21**, 639–645

Lea, R.W., Hall, T.R. and Sharp, P.J. (1992). Hypothalamic monoamine turnover in ring doves (*Streptopelia Risoria*) courting, incubating and brooding young. *Comp. Biochem. Physiol.* **103C**, 345–350.

Lea. R.W., Talbot, R.T. and Sharp, P.J. (1991). Passive immunisation against chicken Vasoactive Intestinal Polypeptide suppresses plasma prolactin and crop sac development in incubating ring doves. *Hormones and Behaviour* **25**, 283–294

Lea, R.W., Vowles, D.M. and Dick, H.R. (1986). Factors affecting prolactin secretion during the breeding cycle of the ring dove (*Streptopelia risoria*) and its possible role in incubation. *J. Endocrinology* **110**, 447–458

Lee, C.Y., Double, K.L. and Crocker, A.D. (1995). Expression of stereotyped behaviour requires stimulation of nigral D1 dopamine receptors. *Brain Research* **681**, 205–208

Lehrman, D. S. (1958a). Induction of broodiness by participation in courtship and nest-building in the ring dove (*Streptopelia risoria*) *J. of Comparative and Physiological Psychology* **51**, 1, 32–36

Lehrman, D.S. (1958b). Effect of female sex hormones on incubation behavior in the ring dove *(Streptopelia risoria*) *J. of Comparative and Physiological Psychology* **51**, 142–145)

Lehrman, D.S.(1959). On the origin of the reproductive behavior cycle in doves. *Transactions of the New York Academy of Sciences* Ser II, **21**, 682–688

Lehrman, D.S. (1964). The reproductive behavior of ring doves. *In* "Animal Behaviour". (T. Eisner and E. O. Wilson, Eds.), pp. 210–216

Lehrman, D.S. and Brody, P.N. (1957). Oviduct response to estrogen and progesterone in the ring dove (*Streptopelia risoria*) *Proceedings of the Society for Experimental Biology and Medicine* **95**, 373–375

Lehrman, D.S., Brody, P.N. and Wortis, R.P. (1961). The presence of the mate and of nesting material as stimuli for the development of incubation behavior and for gonadotropin secretion in the ring dove (*Streptopelia Risoria*). *Endocrinology* **68**, 507–516

Lewis, D.A., Campbell, M.J., Foote, S.L. and Morrison, J.H. (1986). The monoaminergic innervation of primate neocortex. *Human Neurobiol.* **5**, 181–188

Macnamee, M. C. and Sharp, P.J. (1989). The functional activity of hypothalamic dopamine in broody bantam hens. *J. Endocrinology* **121**, 67–74

Mani, S.K., Blaustein, J.D. and O'Malley, B.W. (1997). Progesterone receptor function from a behavioral perspective. *Hormones and Behavior* **31**, 244–255

Martin, J.T., El Halawani, M. and Phillips, R.E. (1982). Diurnal variation in hypothalamic monoamines and plasma corticosterone in the turkey after inhibition of tyrosine hydroxylase or tryptophan hydroxylase. *Neuroendocrinology* **34**, 191–196

Melis, M.R. and Argiolas, A. (1995). Dopamine and sexual behavior. *Neuroscience and Biobehavioral Reviews* **19**, 19–38

Michel, G.F. and Moore, C.L. (1985). Contribution of nesting experience to progesterone-induced incubation in ring doves (*Streptopelia risoria*). *J. Comparative Psychology* **99**, 259–265

Miller, W.L. (1988). Molecular Biology of Steroid Hormone Synthesis. *Endocrine Reviews* **9**, 295–317

Nistico, G. and Stephenson, J.D. (1979). Dopaminergic mechanisms and stereotyped behaviour in birds. *Pharmacological Research Communications* **11**, 555–569.

Peczely, P. and Pethes, G. (1979). Alterations in plasma sexual steroid concentrations in the Collared dove (*Streptopelia Decaocto*) during the sexual maturation and reproduction cycle. *Acta Physiologica Academiae Scientarum Hungaricae* **54**, 161–170

Petitclerc, M., Bedard, P.J. and Di Paolo, T. (1995). Progesterone releases dopamine in male and female rat striatum: A behavioural and microdialysis study. *Prog. Neuro-Psychophamacol. & Biol. Psychiat.* **19**, 491–497.

Piroli, G.G., Grillo, C.A. Ferrini, M.G., Lux-Lantos, V. and De Nicola, A.F. (1996). Antagonism by progesterone of diethylstilbestrol-induced pituitary tumorigenesis in Fischer 344 rats: Effects on sex steroid receptors and tyrosine hydroxylase mRNA. *Neuroendocrinology* **63**, 530–539.

Ramos, C. and Silver, R. (1992). Gonadal hormones determine sex differences in timing of incubation by doves. *Hormones and Behavior* **26**, 586–601

Riddle, O., Bates, R.W. and Lahr, E.L. (1935). Prolactin induces broodiness in fowl. *American J. of Physiology* **111**, 352–360

Robbins, T.W. and Everitt, B.J. (1992). Functions of dopamine in the dorsal and ventral striatum. *The Neurosciences* **4**, 119–127

Sar, M. (1988). Distribution of progestin-concentrating cells in rat brain: colocalisation of [3H]ORG.2058, a synthetic progestin, and antibodies to tyrosine hydroxylase in hypothalamus by combined autoradiography and immunocytochemistry. *Endocrinology* **123**,1110–1118

Schlinger, B.A. and Callard, G.V. (1989). Estrogen receptors in quail brain: A functional relationship to aromatase and aggressiveness. *Biology of Reproduction* **40**, 268–275

Sharp, P.J., LI, Q., Georgiou, G. and Lea, R.W. (1996). Expression of fos-like immunoreactivity in the hypothalamus of the ring dove (*Streptopelia risoria*) at the onset of incubation. *J. of Neuroendocrinology* **8**, 291–298

Signoret, J-P. and Balthazart, J. (1993). Sexual behaviour. *In* "Reproduction in Mammals and Man". (C. Thibault, M. C. Lavasseur, R.H.F. Hunter, Eds.), pp. 531–551. Ellipses, Paris

Silver, R., Feder, H.H. and Lehrman, D.S. (1973). Situational and hormonal determinants of courtship, aggressive and incubation behavior in male ring doves (*Streptopelia risoria*). *Hormones and Behaviour* **4**, 163–172.

Silver, R., Reboulleau, C., Lehrman, D.S. and Feder, H.H. (1974). Radioimmunoassay of plasma progesterone during the reproductive cycle of male and female ring doves (*Streptopelia risoria*). *Endocrinology* **94**, 437–444

Stern, J. M. (1974). Estrogen facilitation of progesterone-induced incubation behavior in castrated male ring doves. *J of Comparative and Physiological Psychology* **37**, 332–337

Van Furth, W.R., Wolterink, G. and Van Ree, J.M. (1995). Regulation of masculine sexual behavior: involvement of brain opioids and dopamine. *Brain Research Reviews* **21**, 162–184

Vowles, D.M. and Harwood, D. H. (1966). The effect of exogenous hormones on aggressive and defensive behaviour in the ring-dove. *J. Endocrinology* **36**, 35–51.

Wallman, J., Grabon, M. and Silver, R. (1979). What determines the pattern of sharing of incubation and brooding in ring doves?. *J. of Comparative and Physiological Psychology* **93**, 481–492

Wilson, S.C. and Sharp, P.J. (1976). The effects of progesterone on oviposition and ovulation in the domestic fowl (*Gallus Domesticus*). *Br. Poult. Sci.* **17** 163–173

Wynne, B. and Gunturkun, O. (1995). Dopaminergic innervation of the telencephalon of the pigeon (*Columba livia*): A study with antibodies against tyrosine hydroxylase and dopamine. *J. of Comp. Neurol.* **357**: 446–464

Youngren, O.M., Pitts, G.R., Phillips, R.E. and El Halawani, M.E. (1995). The stimulatory and inhibitory effects of dopamine on prolactin secretion in the turkey. *General and Comparative Endocrinology* **98**, 111–117

Zoli, M., Agnata, L.F., Tinner, B., Steinbusch, H.W.M. and Fuxe, K. (1993). Distribution of dopamine-immunoreactive neurons and their relationship to transmitter and hypothalamic hormone-immunoreactive neuronal systems in the rat mediobasal hypothalamus. A morphometric and microdensitometric analysis. *J. of Chem. Neuroanatomy* **6**, 293–310

12

NEURAL ELEMENTS IN PINEAL ORGANS OF VERTEBRATES

Tetsuji Sato, Hiroki Fujieda, and Kenjiro Wake

Department of Anatomy
School of Medicine
Tokyo Medical and Dental University
1-5-45, Yushima, Bunkyo-ku
Tokyo 113-0032, Japan

1. INTRODUCTION

Phylogenetically, the pineal organ changes from a photoreceptor organ endowed with true photoreceptor cells to an endocrine gland influenced by light stimuli perceived in the retina. In poikilothermic vertebrates, the pineal complex contains photoreceptor cells which are synaptically connected to intrapineal second-order neurons. The axons of these second-order neurons constitute the central pinealofugal innervation which connects the pineal complex with various nuclei located in the diencephalon and mesencephalon. Changes in the structure and function of pineal photoreceptors are accompanied by the reduction of intrapineal neurons in the course of evolution. In mammals, pinealocytes receive information from a biological clock system—present inside the suprachiasmatic nucleus of the hypothalamus—entrained by the photoperiod and have lost structural and functional characteristics of photoreceptor cells. Nevertheless, mammalian pinealocytes are considered as derivatives of the pineal photoreceptor cells of anamniotic vertebrates (Oksche, 1971; Collin, 1971). This functional shift of the pinealocyte has been confirmed by immunohistochemical and biochemical studies showing that neuroendocrine pinealocytes of mammals express photoreceptor-specific molecules which otherwise are synthesized only by true (retinal and pineal) photoreceptor cells (Korf *et al.*, 1986; Kramm *et al.*, 1993). Concomitantly with the loss of the direct photosensitivity of pinealocytes, the central pinealofugal innervation is reduced in mammals. Alternatively, the pineal gland of mammals receives a conspicuous autonomic innervation (Cajal, 1904; Kappers, 1960), which is essential for the regulation of melatonin metabolism in these species. The autonomic nervous system also innervates the pineal organ of reptiles and birds, but in these species it appears to be less important than in mammals. Autonomic nerve fibers are

Neural Regulation in the Vertebrate Endocrine System, edited by Prasada Rao and Peter, Kluwer Academic / Plenum Publishers, New York, 1999.

scarce or even absent in the pineal organ of poikilothermic vertebrates such as fishes and amphibians (see Vollrath, 1981 for details).

The change of the structures and components of the pineal organ in the course of the evolution is reflected to the innervation pattern of the organ. This paper is intended to review some morphological, physiological and biochemical data related to the interactions between the reduction of the photoreceptor-specific characteristics and the transformation from the central pinealofugal innervation to the pinealopetal innervation.

2. CENTRAL PINEALOFUGAL INNERVATION

The distributional patterns of the intrapineal neurons and central pinealofugal pathways in the pineals of submammalian animals have been well demonstrated with the acetylcholinesterase (AChE) method (teleosts, Figure 1a; Wake, 1973; Korf, 1974; Falcon, 1979; Ohba *et al.*, 1979; Omura, 1980; Matsuura and Herwig, 1981: amphibians, Fig. 1b; Wake *et al.*, 1974; Hetherington, 1981; Korf *et al.*, 1981: birds, Fig. 1c; Ueck and Kobayashi, 1972; Sato and Wake, 1983). The general trend corresponds to a decrease in number of intrapineal (intrinsic) neurons during vertebrate evolution (Table 1). Especially in the avian pineal organ, the progressive reduction of the pinealofugal component and the spectacular increase in pinealopetal sympathetic innervation occur in parallel (Fig. 1d; Sato and Wake 1983). Both sensory and sympathetic neural representations appear to be complementary in the avian pineal organ, as could be shown convincingly by reconstructions of serial sections (Figure 2). In avian pineal organs, the number of intrapineal neurons varies extremely from species to species; however, their topographic distribution in the different morphologic types of pineal organs basically represents a similar pattern in spite of the interspecific differences in the numerical presentation of these neurons, which show a clear-cut preference for the proximal portion of the organ (Figure 3). In the domestic fowl, the number of intrapineal AChE-positive neurons decreases rapidly during ontogenetic development, whereas the sympathetic innervation becomes more prominent (Sato and Wake, 1984). The peculiar comparative position of the avian pineal organ has been discussed in detail by Oksche and Hartwig (1979), Collin and Oksche (1981), Korf *et al.* (1982), and Sato and Wake (1983, 1984, and 1986). The outer segments of avian pinealocytes exhibit distinct signs of structural regression; this regression coincides with a spectacular increase in the secretory capacity of the pinealocytes. On the other hand, in the avian pineal, melatonin synthesis is controlled by light (Axelrod *et al.*, 1964), but sympathetic nerves are not involved in this control in contrast to the situation in mammals (Gaston and Menaker, 1968; Zimmerman and Menaker, 1975). The isolated chicken pineal organ in vitro displays rhodopsin-like photosensitivity (Deguchi, 1981), and an oscillatory activity of enzymes is involved in melatonin synthesis (Binkley *et al.*, 1978; Deguchi, 1979a,b).

Figure 1. **a** Horizontal section through the pineal organ of the goldfish, *Carassius auratus*, showing a number of AChE-positive nerve cells. (Wake, 1973) **b** A number of AChE-positive pseudounipolar cells distributed in the pineal organ of *Rana esculenta*. **c** AChE-positive ganglion cells and their processes form a network at the posterior aspect of the pineal stalk of *Padda oryzivora*. **d** Catecholamine-containing elements in the pineal organ of *Padda oryzivora*. *L* lumen. **e** NSE-immunoreaction in the pineal organ of a 1-week-old chicken. Pineal follicle with positive bipolar-shaped cells (*arrowheads*) encircling the pineal lumen. **f** NSE-immunoreactive pinealocytes in the pineal organ of a 6-month-old (adult) domestic fowl. **g** Sagittal section of an E20 pineal organ (*PO*) immunolabeled for SY38. **h** Nerve fibers immunoreactive for tyrosine hydroxylase (TH), the rate limiting enzyme in catecholamine synthesis which converts tyrosine to dihydroxy-phenylalanine and dopamine-hydroxylase, in the pineal gland of 9-day-old rat. *Bars*: 50 μm.

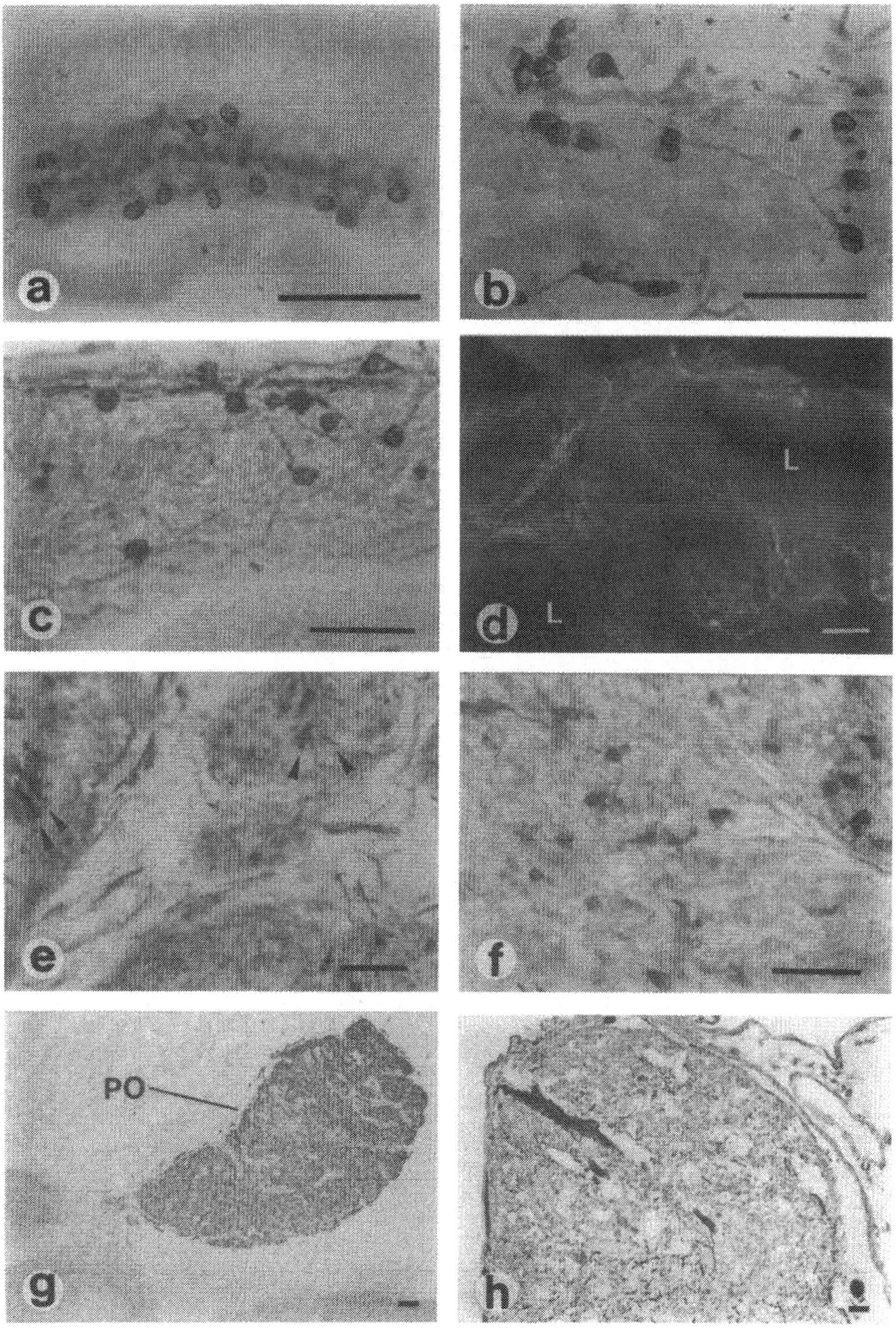
a
b
c
L
d
L
e
f
PO
g
h

Table 1. The number of AChE-positive nerve cells in the pineal organ of anamniotes

I. Teleosts	
Salmo gairdneri	1325 ~ 1725 (Korf, 1974)
Carassius gibelio langsdorfi	532 (Ohba et al., 1979)
S. fontinalis	230 ~ 780 (Falcon, 1979)
Anguilla anguilla	370 (Matsuura and Herwig, 1981)
Plecoglossus altiveis	313 (Omura, 1988)
II. Amphibians	
R. ridibunda	310 (Wake et al., 1974)
r. esculenta	215 ~ 226 (Wake et al., 1974)
Xenopus laevis	100 (Korf et al., 1981)
Enasatina eschscholtzi	12 ~ 86 (Hetherington, 1981)
III. Birds (Sato and Wake, 1983)	
Chrysolophus pictus	362
Taeniopygia castanotis (albino)	249
Emberiza aureora or. Shulpin	208
Padda oryzivora	181
Melopsittacus undulatus	160
Padda oryzivora (albino)	142
Lonchura striata var. domestica	96 ~ 128
Estrilda cinerea	109
Lonchura maja	94
Steganura paradisea	92
Serinus canaria (red)	92
Taeniopygia castanotis	91
Larus crassirostris	72
Estrilda amandava	71
Oena capensis	15
Zosterops palpebroas	9
Coturnix coturnix japonica	2
Gallus gallus	0

In the pineal organ of the domestic fowl, neuron-specific enolase (NSE) is strictly localized both in a limited number of pinealocytes and intrinsic afferent neurons (Sato *et al.*, 1995); the former rapidly increase in number during post-hatching development, whereas the latter are only rarely found in the adult animal. The distributional pattern of the NSE-immunoreactive nerve cells is very similar to that of AChE-positive elements in the post-hatching chicken pineal (Sato and Wake, 1984; Sato *et al.*, 1995). Both types of reactive neurons form the pineal tract and the pineal ganglion associated with it (Sato and Ebisawa, 1988). Moreover, both the NSE-immunoreactive and the AChE-positive elements display an ontogenetic numerical decrease (Sato and Wake, 1984; Sato *et al.*, 1988, 1995). Thus, it may be that they represent one and the same population of neurons. Twenty to 40 NSE-immunoreactive pinealocytes are found exclusively in the distal portion of the pineal organ of the 2-day-old animal. In the 1-week- (Fig. 1e), 2-week-, and 1-month-old chicken, immunohistochemistry for NSE revealed a distal to proximal decreasing gradient in the number of stained pinealocytes. In the adult animal, these immunoreactive cells are more diffusely distributed throughout the organ (Fig. 1f). In addition to this ontogenetic change in the distributional pattern, the number of reactive pinealocytes rapidly increases with age (Table 2). The distributional pattern of NSE-stained pinealocytes appears to be

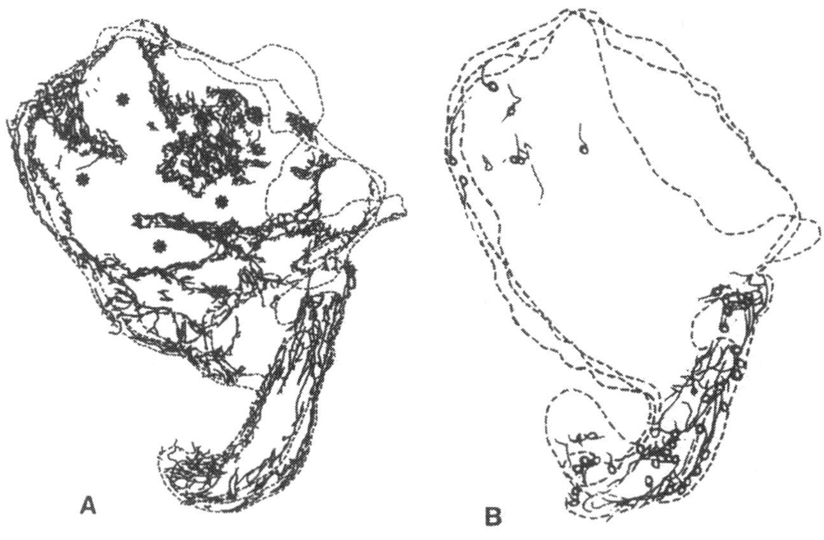

Figure 2. Combination of glyoxylic acid (GA)- and AChE-methods. Sympathetic (**a**) and sensory (**b**) neural elements are distributed in a complementary fashion. *Black dotted lines* contours of the pineal organ; *asterisks* lumen (Sato and Wake, 1983).

similar to the representation of the sympathetic nervous system as previously demonstrated (Sato and Wake, 1981, 1983, 1984; Sato *et al.*, 1990, 1995).

It is suggested that the NSE-stained pinealocytes, being concentrated distally, are under sympathetic control. These results are supported by a distal to proximal decrease in staining intensity for NSE in the rat pineal gland (McClure *et al.*, 1986). Furthermore, it has been suggested that NSE is useful as a cytoplasmic marker of neurons and neuroendocrine cells (Marangos *et al.*, 1978, 1982; Schmechel *et al.*, 1978). The intense NSE reaction in the pineal organ of the domestic fowl may, therefore, reflect the neurosecretory nature (cf. Oksche *et al.*, 1987) of the pinealocytes composing the organ. The NSE is the most acidic isoenzyme of the glycolytic enzyme enolase and is found in neurons, neuroen-

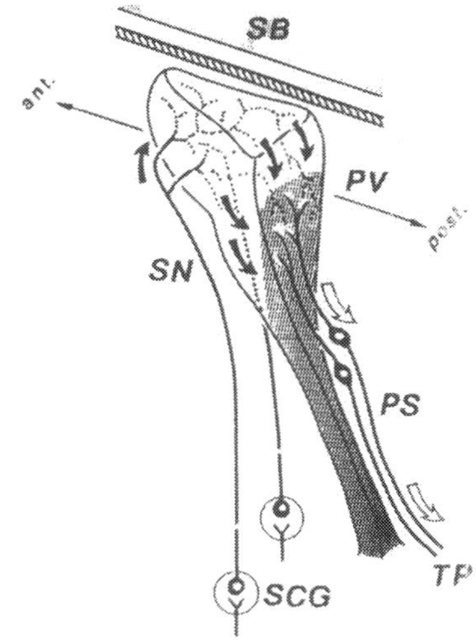

Figure 3. Schematic diagram illustrating a possible phylogenetic trend in the innervation pattern of the avian pineal organ. *White bent arrows* indicate the reduction of the sensory apparatus and the central pinealofugal projections (*dark area*) toward the proximal portion of the pineal stalk (*PS*), whereas *black bent arrows* show the invasion of the postganglionic sympathetic nerves (*SN*) originating from superior cervical ganglion (*SCG*) at the anterolateral aspects of the pineal vesicle (*PV*). *SB* skull bone; *TP* tractus pinealis (Sato and Wake, 1983).

Table 2. An increase in the number of NSE-positive pinealocytes during the course of post-hatching development

Days after hatching	Animal no.	Number of NSE-positive cell/mm^3
2	1	116
	2	46
	3	39
	4	87
14	1	6922
	2	4101
Adult	1	11067
	2	11966

docrine, and APUD (amine precursor uptake and decarboxylation) tissues (Marangos *et al.*, 1978; Schmechel *et al.*, 1978; Trapp *et al.*, 1981). As described by Erickson and Moore (1980), axonal transport of NSE takes place with the proteins of the "slow component b" (SCb). Characteristic immunohistological localization of the NSE antigen is seen in the rat cerebellum and cerebrum in the form of a cytoplasmic staining of Purkinje, granule, stellate, basket, and pyramidal cells (Ghandour *et al.*, 1981; Schmechel *et al.*, 1980). General NSE-positive staining of thyroid C cells, pituitary, adrenal medullary chromaffin cells, and pinealocytes has also been shown (Marangos *et al.*, 1982). According to Cimino *et al.* (1977), not all neurons stain immunohistologically for NSE.

We have succeeded in the identification of NSE-immunoreactive pinealocytes and afferent neurons in the same sections by use of immunohistochemistry for NSE (Sato *et al.*, 1995). Furthermore, an analysis of the chicken pineal organ by the use of a computer-assisted three-dimensional imaging technique revealed the reverse distributional pattern of both elements. These results are in accord with the innervation pattern of the chicken pineal as revealed by means of the AChE- and glyoxylic acid- methods (Sato and Wake, 1984; Sato *et al.*, 1990).

3. EFFERENT (PINEALOPETAL) INNERVATION

3.1. Sympathetic Innervation

The majority of the autonomic nerve fibers innervating the pineal organ are of sympathetic origin. In the mouse pineal gland, Cajal (1904) reported that the autonomic fibers entered the surface of the organ together with blood vessels and invaded the parenchyma from a perivascular location. A similar innervation was found in the pineal organ of other mammalian species (rabbit, Pines, 1927; cat, Antonow, 1925; dog, Hartmann, 1957; cattle, Antonow, 1925). In 1960 Kappers demonstrated the conarian nerve ("nervus conarius") in the rat and showed it to extend into the pineal parenchyma from a bilateral aspect and to degenerate after bilateral removal of the superior cervical ganglia, thus verifying its sympathetic nature and origin from the superior cervical ganglion. This study provided a most valuable and stimulating morphological basis for a series of elegant biochemical and physiological experiments which elucidated the interactions between the sympathetic innervation and the melatonin synthesis of the mammalian pineal gland.

The best investigated neuroendocrine principle of the pineal is melatonin, an indoleamine, and the targets of melatonin have been identified in various mammalian and non-mammalian species with the use of 2-[^{125}]iodomelatonin as a ligand (Laudon and Zisapel, 1986; Reppert *et al.*, 1988; Vanecek, 1988; Weaver *et al.*, 1989; Laitinen and Saavedra, 1990). Although these targets vary considerably among the species, the suprachiasmatic nucleus of the hypothalamus and the pars tuberalis of the hypophysis are consistent sites of melatonin binding in nearly all mammals investigated. In non-mammals, melatonin binding sites are obviously more widespread than in mammals (Ekström and Vanecek, 1992). In mammals, the direct photosensitivity of the pineal is lost, and the retina of the lateral eyes is required for the light-dependent control of the melatonin biosynthesis. Light stimuli perceived by retinal photoreceptors are transmitted via the retinohypothalamic tract to the suprachiasmatic nucleus (Moore, 1978). Signals from the suprachiasmatic nucleus are conveyed to the pineal organ via a complex neuronal chain which involves the paraventricular nucleus and the intermediolateral nucleus of the spinal cord. The last part of this chain is the postganglionic sympathetic nerve originating from the superior cervical ganglion (Figure 4; Kappers, 1960, 1965). Noradrenaline (NA), the neurotransmitter of the pinealopetal sympathetic nerve fibers, stimulates the melatonin biosynthesis in the mammalian pineal organ primarily through serotonin-N-acetyltransferase (NAT) (Axelrod, 1974; Klein, 1985). In poikilothermic vertebrates the autonomic (sympathetic) innervation is either missing or rather inconspicuous and does not seem to play an important functional role for the control of pineal functions. Instead, pineal direct photoreception and phototransduction appear as the prevailing regulatory mechanism in these species.

3.2. Parasympathetic Innervation

Several authors have reported on the existence of ganglia-like clusters of neurons in the pineal organ of different mammalian species (Marburg, 1909; Pastori, 1928; Møllgard and Møller, 1973; Møller, 1978). Nothing is known about the neurotransmitters employed by such neurons; also their function remains enigmatic. Although these ganglia may be considered rudiments of the central pinealofugal apparatus, fully established in the pineal sense organs of poikilothermic vertebrates, it has also been suggested that they might belong to the parasympathetic system. Nerve cells found in the pineal organ of the rabbit receive preganglionic input from the superior salivary nucleus via the conarian nerve

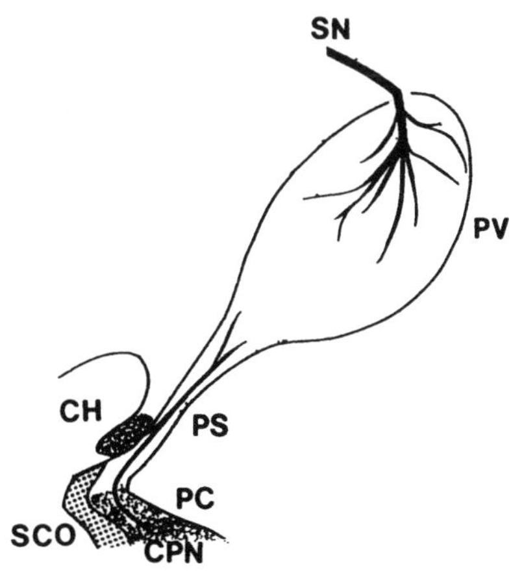

Figure 4. Neural elements involved in the innervation of the rat pineal gland. *CH* habenular commissure; *CPN* central pinealopetal nerves originating from the habenular nucleus and the paraventricular nucleus; *PC* posterior commissure; *PS* pineal stalk; *PV* pineal vesicle; *SCO* subcommissural organ; *SN* postganglionic sympathetic nerves originating from superior cervical ganglion.

(Romijn, 1973, 1975). Also, lesions of the greater petrosal nerve induce degeneration of nerve fibers in the conarian nerve and intrapineal nerve cells (Kenny, 1961). According to Kenny (1961) and Romijn (1975) the conarian nerve would comprise not only postganglionic sympathetic nerve fibers, but also pre- and postganglionic parasympathetic nerve fibers. Since labeled perikarya were observed in the pterygoparatine ganglion after injection of fluorescent tracers into the gerbil pineal organ, some of peptidergic nerve fibers innervating the pineal organ represent postganglionic parasympathetic nerve fibers (Shiotani *et al.*, 1986). Shiotani *et al.* (1986) also observed labeled perikarya in the trigeminal ganglia. The precise course of these fibers, however, remains to be elucidated.

3.3. Peptidergic Innervation

Evidences for the existence of peptidergic nerve fibers and terminals in the mammalian pineal organ have been obtained with the use of immunohistochemical techniques. These include neuropeptide Y (NPY), substance P, somatostatin, vasoactive intestinal peptide (VIP), LHRH, calcitonin gene-related peptide (CGRP), vasopressin (AVP) and oxytocin (OXT). Peptidergic fibers and terminals can be most frequently found in juxtaposition to intrapineal capillaries of different calibers; only a few nerve fibers are observed within the pineal parenchyma. Their number and density varies with the neuropeptide and also the species of mammals. The amount of immunoreactive material in the AVP-and OXT-containing nerve fibers of the hedgehog pineal organ exhibits a characteristic seasonal variation (Nürnberger and Korf, 1981). The peptidergic nervous system in the pineal organ is thought to originate from three different sources: superior cervical ganglion, parasympathetic and sensory ganglia, and diencephalic and mesencephalic nuclei. NPY is generally considered as a neuromodulator which is often colocalized with NA in sympathetic nerve fibers (Lundberg *et al.*, 1982; Ekblad *et al.*, 1984). In the pineal gland, the majority of the NPY-positive nerve fibers obviously belong to the sympathetic innervation, since they degenerate after bilateral superior cervical ganglionectomy (Zhang *et al.*, 1991; Cozzi *et al.*, 1992; Møller *et al.*, 1994). However, some NPY-immunoreactive nerve fibers persist in the pineal gland after removal of bilateral superior cervical ganglia. These elements probably originate from diencephalic areas and represent a part of the central pinealopetal innervation (Figure 4). The origin of the remaining peptidergic nerve fibers innervating the pineal gland is less clear. From experiments combining tracing techniques and immunohistochemistry Shiotani *et al.* (1986) concluded that, in the gerbil, VIP-immunoreactive nerve fibers originated from the pterygopalatine ganglion and the substance P- and CGRP-immunoreactive nerve fibers originated from the trigeminal ganglion. However, numerous VIP-immunoreactive neurons are also present in the superior cervical ganglia (Sasek and Zigmond, 1989) and the hypothalamic paraventricular nucleus (Mikkelsen and Møller, 1988) both of which project to the pineal gland. The perikarya of vasopressin- and oxytocin-immunoreactive nerve fibers may be located in the paraventricular nucleus of the hypothalamus. To date, the functional significance of the peptidergic nerves in the pineal gland is open to discussion. No electrophysiological or biochemical data are available concerning the influence of the peptidergic input on the electrical or secretory activity of the pinealocytes.

3.4. Other Pinealopetal Innervation

A direct connection between the pineal and the brain has been demonstrated in hamster and mouse. This is established by long axonlike processes of pinealocytes, located in the deep pineal, which extend into the habenular nucleus and the pretectal area (Korf *et al.*, 1986; Korf *et al.*, 1990; Sato *et al.*, 1991, 1994). Processes of pinealocytes were immunolabeled using antisera against the S-antigen (arrestin) (Figure 5a). S-antigen belongs to a family of proteins which are selectively expressed by retinal photoreceptors and pinealocytes. These morphological data indicate that signals from the mammalian pineal organ may be conveyed not only by means of the release of melatonin into the general circulation, but also via neuronlike pathways. Long processes of pinealocytes are in intimate connection with the central nervous system within the habenular nucleus, because they form the postsynaptic element in conventional synapses involving presynaptic terminal provided with clear synaptic vesicles (Figure 5b; Korf *et al.*, 1990).

4. EPITHELIAL REMODELING OF MAMMALIAN PINEAL GLAND

The pineal derives from an evagination of the primitive neuroepithelium which has apical-basal polarity, maintains contiguity via lateral cell junctions and sits on a well organized basal lamina. The process of epithelial remodeling in the embryonic pineal gland has been characterized by local dissolution of the basal lamina or of the lateral cell junctions, which is light-microscopically demonstrable by laminin and cadherin immunostaining (Fujieda *et al.*, 1997). Two modes of cell emigration, one into the stroma and the other into the pineal lumen, are characteristic features of the pineal epithelial remodeling. The emigrating cells form cell clusters maintaining epithelial integrity, but they do not show a basal lamina stained for laminin nor cadherin-labeled adherent junctions. The epithelial remodeling of the pineal is similar to mesenchyme formation in that the epithelium gives rise to a new cell population free of epithelial polarity. However, the pineal emigrating cells maintain continuity with the remaining epithelium and do not disperse through the stroma like mesenchyme. This tissue integrity as an organ is considered to be preserved by cell-cell adhesion in the pineal parenchyma.

It remains to be studied whether the epithelial remodeling characteristic of the rat pineal occurs also in other mammalian species. Since the pineal gland of most mammalian species possesses a solid structure like that of rat, it is likely to undergo similar epithelial remodeling during its embryonic development. From a phylogenetic point of view, the pineal organ of chicks gives another interesting example of epithelial remodeling (Calvo and Boya, 1978, 1979). The chicken pineal organ develops by continual formation of follicles. The follicular epithelium gives rise to parafollicular zone composed of parafollicular cells detached from the follicular lumen. The parafollicular zone is a transient structure and is destined to be rearranged into follicles, forming a lumen in the center. The cell emigration and the resulting disintegration of epithelial organization during rat pineal development might also be a transient process in the pineal tissue remodeling. Since the emigrating parenchymal cells have neither basal lamina nor cell junctions, they might create those structures de novo to gain morphological and functional stability during their subsequent differentiation. Previous reports that junctional structures connecting pinealocytes with each other increase with age postnatally (Zimmerman and Tso, 1975; Calvo and Boya, 1983) are consistent with this possibility.

Synaptophysin (SY38) is a major calcium-binding glycoprotein of synaptic vesicles (Jahn *et al.*, 1985; Wiedenmann and Franke, 1985; Rehm *et al.*, 1986). It is present in almost all nerve terminals, where it is specifically associated with small synaptic vesicles. It has also been identified in association with secretory vesicles of neuroendocrine cells, but is not present in other cell types (Navone *et al.*, 1986). Pinealocytes immunoreactive for SY38

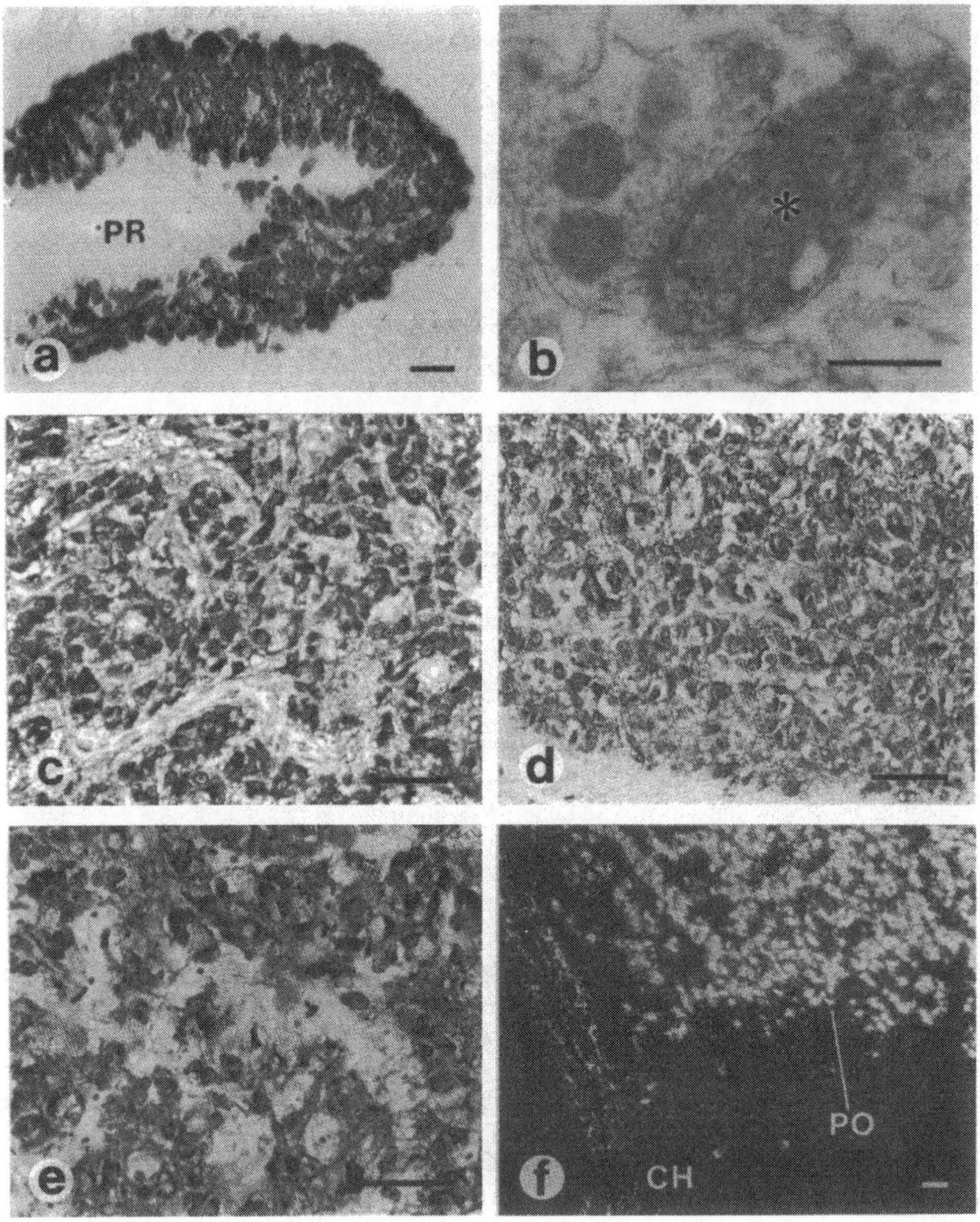

Figure 5. **a** Immunoreaction for S-antigen in the deep pineal of the pigmented mouse. *PR* pineal recess **b** An S-antigen-immunoreactive process (*asterisk*) of the pinealocyte receives a synaptic contact in the medial habenular nucleus of the pigmented mouse. (Korf *et al.*, 1990) **c** NSE-positive pinealocytes in the bovine pineal gland (cortex). **d** SY38-immunoreactive pinealocytes in the bovine pineal gland (cortex). **e** HIOMT-immunoreactivity in the bovine pineal gland (cortex). **f** *In situ* hybridization of HIOMT-B18 to the section of the bovine epithalamic region. Dark-field photomicrograph of the transitional portion from the deep pineal (*PO*) to the habenular commissure (*CH*). *Bars*: 50 μm for **a**, **c-f**; 0.5 μm for **b**.

first occur at E18 and for NSE at E20; both cells mainly scatter in the distal periphery of the organ (Figure 1g). SY38- or NSE-labeled cells are rapidly increased around large blood vessels. At adult, the immunoreactivity for SY38 or NSE is stronger in pinealocytes of the peripheral zone (cortex) of the organ than in those of the central zone (medulla). Tyrosine hydroxylase (TH)-immunoreactive (sympathetic) nerve fibers (Figure 1h) are more densely distributed in the periphery of the organ than in the central portion. The majority of emigrating cells are mitotically active until E18, judging from their BrdU immunoreactivity, and they form new proliferative cell populations detached from the luminal cell junctions. However, as the epithelial remodeling progresses, SY38-stained cells conspicuously increase in number within the emigrating cell population compared to within the rest of the epithelium, which results in the uneven distribution of SY38 immunoreactivity in the pineal parenchyma. The possibility that the infolding is produced by regional difference of mitotic activity is unlikely, because mitotic activity assessed by BrdU incorporation was homogeneously detected along the overall length of the pineal epithelium. The parallelism between the direction of epithelial bending and the direction of cell emigration strongly suggests that tensile forces generated in the folded epithelium play a role in the cell extrusion from the pineal epithelium. The remodeling pattern was closely correlated with the three-dimensional configuration of the pineal epithelium and its topographical relationship with the surrounding structures (Figure 6). Despite this regional difference the essential characteristics of this tissue remodeling, loss of epithelial polarity and maintenance of epithelial integrity, were common, which lead to an apparent homogeneity of tissue architecture in the postnatal pineal. The epithelial remodeling may play significant roles not only in the pineal morphogenesis but also in the phenotype of pinealocytes, implying possible relationship between cell shape and differentiation. These morphological data suggest that epithelial remodeling in the fetal pineal is induced by epithelial infolding and that this remodeling apparently promotes the differentiation of secretory pinealocytes independent of sympathetic innervation.

5. HETEROGENEITY WITHIN THE MAMMALIAN PINEAL GLAND

Although there can be no doubt that the indoleamine melatonin is formed within the main cellular constituent of the pineal gland, viz., the pinealocyte, it remains an open

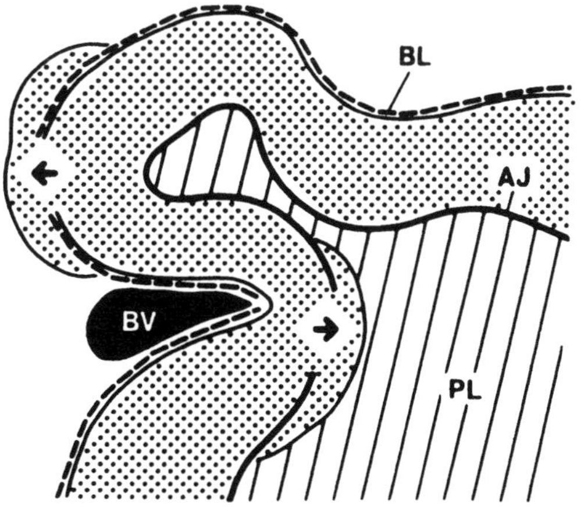

Figure 6. Schematic diagram showing the pattern of epithelial remodeling in the rat pineal gland during embryonic development. *Arrows* directions of cell emigration from the pineal epithelium; *AJ* adherent junctions; *BL* basal lamina; *BV* blood vessel; *PL* pineal lumen (Fujieda *et al.*, 1997).

question whether all the pinealocytes are involved in melatonin formation or whether some are engaged in different functions. Bubenik *et al.* (1974) have noted fluorescence in almost all parenchymal cells of the rat pineal, albeit with different intensities, whereas Freund *et al.* (1977) have revealed fluorescent parenchymal cells in the outer regions of the organ only. The latter observation is not the first indication of structural differences between the center and the periphery of the rat pineal. Quay (1957, 1958) has demonstrated a higher lipid and a greater blood content in the periphery than in the center (Quay and Renzoni, 1966; Blumfield and Tapp, 1970). Dafny *et al.* (1975) have reported that electrical recordings from the pineal are similar between the surface and 0.8 mm in depth, but change 1.2 mm below the surface. Spontaneous spike activity is detectable only within the first 200 μm below the surface (McClung and Dafny, 1975). The functional significance of these differences is obscure.

In addition to the morphological heterogeneity seen in the staining properties of pineal parenchymal cells (Tapp and Blumfield, 1970; McClure *et al.*, 1986) and in the varying electron densities of light and dark pinealocytes (Arstila, 1967; Benson and Satterfield, 1975), electrophysiological data from the pineal gland of the guinea pig reveal that several cell types are synchronized to different phases of the circadian rhythm of the gland (Semm and Vollrath, 1979, 1980). Biochemical support for cellular differences is demonstrated by the tentative immunohistochemical localization of melatonin to the marginal zone of the rat pineal (Freund *et al.*, 1977) and by the inhibition of 40% of guinea pig pinealocytes by microelectrophoretically supplied serotonin (Semm and Vollrath, 1982). Although a majority of bovine pinealocytes stains for both hydroxyindole O-methyltransferase (HIOMT), a specific marker for melatonin synthesis, and NSE, a few stain for neither (Kuwano *et al.*, 1983). In the rat pineal gland, neighboring pinealocytes differ from one another in their staining for the NSE antigen (McClure *et al.*, 1986). They also demonstrate a proximal to distal gradient of staining intensity. From these findings, it has been suggested that the more strongly stained cells, being concentrated distally, are under sympathetic control. In the bovine pineal, however, pinealocytes do not demonstrate a proximal to distal gradient in their staining intensity for NSE (Sato *et al.*, 1994). This result may be explained by the central pinealopetal innervation via the habenular or posterior commissure as revealed by immunohistochemistry for NFP. In those studies describing morphological heterogeneity (Arstila, 1967; Pévet, 1977), no specific distribution of cell groups has been mentioned with the exception of variations among neighboring pinealocytes. Although various authors have described a cortical and medulla arrangement of cells and lipid droplets in the rat pineal gland (Freund *et al.*, 1977; Quay, 1957; Tapp and Blumfield, 1970), this arrangement has not been observed in the study on the NSE immunoreactivity of the rat pineals by McClure *et al.* (1986).

Intense immunoreaction for SY38 is found in the cytoplasm of pinealocytes in addition to the labeling of the nerve terminals around the blood vessels. In some areas of the central nervous system (Dennis *et al.*, 1981; Knaus *et al.*, 1986), the formation of synapses and the expression of SY38 occur simultaneously, suggesting that SY38 is a useful marker for synaptogenesis. With this in view, the expression of SY38 in the pinealocytes might be used as a differentiation marker for neurosecretory pinealocytes. The intercellular differences in the degree of SY38 immunostaining may reflect specific states of cellular activity. It has been claimed that the immunohistochemically detectable level of NSE depends on the functional state of a cell (Bishop *et al.*, 1985). However, there seems to be no simple linear correlation between NSE (Figure 5c), SY38 (Figure 5d), and HIOMT (Figure 5e) immunoreactivities in pinealocytes.

The expression of HIOMT mRNA (Fig. 5f) is more intense in the cortex than in the medulla (Sato *et al.*, 1994). The level of mRNA transcripts for HIOMT displays an overall decline from the distal to proximal portion within the cortex of the organ. The differences between the cortex and the medulla in the degree of expression of HIOMT mRNA may reflect different states of melatonin synthesis. In the medial habenular nucleus, the hybridization-positive cells do not exhibit HIOMT immunolabeling (Sato *et al.*, 1991). This result may be explained by the lower expression of HIOMT-mRNA transcripts in this region than in the pineal proper.

GFAP is the main component of glial filaments; it is considered to be an antigenic marker of astrocytes (Bignami and Dahl, 1974; Eng *et al.*, 1971; Gown and Gabbiani, 1984). The finding that pineal supportive cells contain GFAP suggests that they are closely related to the astrocytic macroglia cell line. The lack of pinealocytes staining for this marker protein implies that these cells are related to other cell types, presumably neurons. In the bovine pineal, the GFAP-immunoreactive cells in the cortex are characterized by long, thin, smooth, infrequently branched expansions, whereas those in the medulla have smaller soma and numerous thin cytoplasmic processes (Sato *et al.*, 1994). Although the former is similar to the fibrous astrocyte, the latter resembles the protoplasmic astrocyte in its morphological characteristics. Furthermore, the morphological differences of both have been confirmed by using the Golgi method. The glial nature of the pineal supportive cells may have important functional implications. Thus, this cell type in the pineal gland may have functions similar to those of the glial cells in nervous tissues. Pineal glial cells, in addition to a supporting function, may also play a role in the exchange of substances between the pineal parenchyma and blood. Morphological evidence supporting this function is provided by the close relationship between the pineal glial cells and the connective tissue spaces of the gland (Wolfe, 1965; Calvo and Boya, 1984). These results provide further insights into the multiple functional aspects of the mammalian pineal gland.

6. SUMMARY AND CONCLUDING REMARKS

Innervation as well as structure and function of the pineal organ have undergone conspicuous transformations in the course of vertebrate evolution. In anamniotes, the pineal complex comprises true photoreceptor cells synaptically connected to second-order neurons which give rise to central pinealofugal innervation. The central pinealofugal system transmits signals from the intrapineal photoreceptors to various nuclei in the diencephalon and mesencephalon. Pineal photoreceptor cells, intrapineal neurons and central pinealofugal pathways are gradually reduced during phylogeny. The reduction of central pinealofugal elements is accompanied by the progressive development of sympathetic innervation; this may be related to the changes of the pineal phenotype. Irrespective of these transformations, the pineal organ functions as a neuroendocrine gland in all vertebrates: it represents rhythmical production and release of melatonin, a neurochemical analog of darkness. Melatonin synthesis and secretion are influenced by the environmental lighting conditions and endogenous oscillators. In nonmammalian species, photoreceptors and oscillators regulating the melatonin synthesis are located in the pineal itself, while in mammals, the photoreceptors are located in the retina and the (or at least a major) oscillator is present inside the suprachiasmatic nucleus of the hypothalamus. Signals from both the retina and the suprachiasmatic nucleus are conveyed to the mammalian pineal through a complex neuronal chain involving central structures and the postganglionic sympathetic innervation originating from the superior cervical ganglion. In mammals the sympathetic innervation of the pineal organ is required to maintain the photoneural control of the melatonin biosynthesis, but not in any other vertebrates. Sympathetic noradrenergic nerve fibers are scarce or even lacking in the pineal organ of poikilothermic animals such as lampreys, teleosts and amphibians. In avian pineals they become more prominent and innervate the parenchyma to a varying degree. The pineal of mammals receive the most prominent sympathetic (noradrenergic) innervation, as demonstrated by fluorescence- and electron-microscopy as well as immunohistochemistry. The sympathetic nervous system enter the pineal organ either over the capsule in juxtaposition to pial blood vessels or as a specialized nerve, the "conarian nerve" running in the tentorium cerebelli. Also, the mammalian pineal gland appears to be the target of parasympathetic nerve fibers. However, the course and the origin of these fibers remain to be clarified. It is not yet determined whether such fibers represent pre- or post ganglionic elements and whether they occur also in non-mammals.

The diversity of the pinealopetal innervation in mammals is also evident from immunohistochemical investigations showing the occurrence of various neuropeptides (e.g., NPY, VIP, substance P, AVP or OXT) in pineal nerve fibers and terminals. The peptidergic innervation of the pineal is thought to originate from different sources. NPY appears to be colocalized with NA in sympathetic nerve fibers originating from the superior cervical ganglion. VIP may be localized in nerve fibers originating from parasympathetic ganglia or from nuclei of the central nervous system. The perikarya giving rise to AVP- and OXT-immunoreactive pinealopetal nerve fibers are apparently located in the paraventricular nucleus of the hypothalamus. The functional role of these substances in the pineal gland has been elucidated only to a limited extent.

Pinealocytes immunoreactive for SY38 first occur at E18 and for NSE at E20; both cells mainly scatter in the distal periphery of the organ. SY38- or NSE-labeled cells are rapidly increased around large blood vessels. At adult, the immunoreactivity for SY38 or NSE is stronger in pinealocytes of the peripheral zone (cortex) of the organ than in those of the central zone (medulla). Tyrosine hydroxylase-immunoreactive (sympathetic) nerve fibers are more densely distributed in the periphery of the organ than in the central portion. Furthermore, epithelial remodeling in the rat pineal during fetal development was immunohistochemically analyzed by using antibodies for laminin and cadherin as molecular markers of basal lamina and intercellular junctions, respectively. These results suggest that epithelial remodeling in the fetal pineal is induced by epithelial infolding and that this remodeling promotes the differentiation of secretory pinealocytes independent of sympathetic innervation.

The intercellular differences in the degree of SY38 immunostaining may reflect specific states of cellular activity. It has been claimed that the immunohistochemically detectable level of NSE depends on the functional state of a cell. However, there seems to be no simple linear correlation between SY38, NSE, and HIOMT immunoreactivities in pinealocytes. The expression of HIOMT mRNA is more intense in the cortex than in the medulla. The level of mRNA transcripts for HIOMT display an overall decline from the distal to proximal portion within the cortex of the organ. The differences between the cortex and the medulla in the degree of expression of HIOMT mRNA may reflect different states of melatonin synthesis. GFAP-immunoreactive pineal supportive cells are closely related to the astrocytic macroglia cell line. The lack of pinealocyte staining for this marker protein implies that these cells are related to other cell types, presumably neurons. In the bovine pineal, the GFAP-immunoreactive cells in the cortex are characterized by long, thin, smooth, infrequently branched expansions, whereas those in the medulla have smaller soma and numerous thin cytoplasmic processes. The glial nature of the pineal supportive cells may have important functional implications. Thus, this cell type in the pineal gland may have functions similar to those of the glial cells in nervous tissues. Pineal glial cells, in addition to a supporting function, may also play a role in the exchange of substances between the pineal parenchyma and blood. Morphological evidence supporting this function is provided by the close relationship between the pineal glial cells and the connective tissue spaces of the gland. These results provide further insights into the multiple functional aspects of the mammalian pineal gland.

REFERENCES

Antonow, A. (1925). Zur Frage von dem Bau der Glandula pinealis. *Anat. Anz.* **60,** 21–31.

Arstila, V. A. (1967). Electron microscopic studies on the structure and histochemistry of the pineal gland of the rat. *Neuroendocrinol. Suppl.* **2,** 1–101.

Axelrod, J. (1974). The pineal gland: a neurochemical transducer. *Science* **184,** 1341–1348.

Axelrod, J., Wurtman, R. J., and Winget, C. M. (1964). Melatonin synthesis in the hen pineal gland and its control by light. *Nature* **201,** 1134.

Benson, R. and Satterfield, V. (1975). Ultrastructural characteristics of mouse pinealocytes following optic enucleation or continuous illumination. *Anat. Rec.* **181,** 312.

Bignami, A. and Dahl, D. (1974). Astrocyte-specific protein and neuroglial differentiation. An immunofluorescence study with antibodies to glial fibrillary acidic protein. *J. Comp. Neurol.* **153,** 27–37.

Binkley, S. A., Riebman, J. B., and Reilly, K. B. (1978). The pineal gland: A biological clock in vitro. *Science* **202,** 1198- 1201.

Bishop, A. E., Carlei, F., Lee, V., Trojanowski, J., Marangos, P. J., Dahl, D., and Polak, J. M. (1985). Combined immunostaining of neurofilaments, neuron specific enolase, GFAP and S-100. A possible means for assessing the morphological and functional status of the enteric nervous system. *Histochemistry* **82,** 93–97.

Blumfield, M. and Tapp, E. (1970). Measurements of pineal parenchymal cells and their nuclei in the albino rat at different ages. *Acta Morphol. Neerl-Scand.* **8,** 1–8.

Bubenik, G. A., Brown, G. M., Uhlir, I., and Grota, L. J. (1974). Immunohistological localization of N-acetylindolealkylamines in pineal gland, retina and cerebellum. *Brain Res.* **81,** 233–242.

Cajal, R. S. (1904). Textura del Sistema Nervioso del Hobmre y de los Vertebrados. II, 2nd part, Moya, Madrid.

Calvo, J. and Boya, J. (1978). Embryonic development of the pineal gland of the chicken (*Gallus gallus*). *Acta Anat.* **101,** 289–303.

Calvo, J. and Boya, J. (1979). Ultrastructural study of the embryonic development of the pineal gland of the chicken (*Gallus gallus*). *Acta Anat.* **103,** 39–73.

Calvo, J. and Boya, J. (1983). Postnatal development of cell types in the rat pineal gland. *J. Anat.* **137,** 185–195.

Calvo, J. and Boya, J. (1984). Ultrastructure of the pineal gland in the adult rat. *J. Anat.* **138,** 405–409.

Cimino, M., Hartman, B., and Moore, B. (1977). Immunohistochemical localization of 14–3–2 protein during development in chick and rat. *Proc. Int. Soc. Neurochem.* **6,** 304.

Collin, J. P. (1971). Differentiation and regression of the cells of the sensory line in the epiphysis cerebri. *In* "The Pineal Gland" (G. E. M. Wolstenholme and J. Knight, Eds.), pp. 79–125. Churchill-Livingstone, Edinburgh.

Collin, J. P. and Oksche, A. (1981). Structural and functional relationships in the nonmammalian pineal gland. *In* "The Pineal Gland" (R. J. Reiter, Ed.), Vol. 1, Anatomy and Biochemistry, pp. 27- 67. CRC Press, Boca Raton, Florida.

Cozzi, B., Mikkelsen, J. D., Ravault, J. P., and Møller, M. (1992). Neuropeptide Y (NPY) and C-flanking peptide of NPY in the pineal gland of normal and ganglionectomized sheep. *J. Com. Neurol.* **316,** 238–250.

Dafny, N., McClung, R., and Strada, S. J. (1975). Neurophysiological properties of the pineal body. 1. Field potentials. *Life Sci.* **16,** 611–620.

Deguchi, T. (1979a). Circadian rhythm of serotonin N- acetyltransferase activity in organ culture of chicken pineal gland. *Science* **203,** 1245–1247.

Deguchi, T. (1979b). A circadian oscillator in cultured cells of chicken pineal gland. *Nature* **282,** 94–96.

Deguchi, T. (1981). Rhodopsin-like photosensitivity of isolated chicken pineal gland. *Nature* **290,** 702–704.

Dennis, M. J., Ziskind-Conhaim, L., and Harris, A. J. (1981). Development of neuromuscular junctions in rat embryos. *Devl. Biol.* **81,** 266–279.

Ekblad, E., Edvinsson, L., Wahlestedt, C., Uddman, R., Hakanson, R., and Sundler, F. (1984). Neuropeptide Y coexists and co-operates with noradrenaline in perivascular nerve fibers. *Regul. Pept.* **8,** 225–235.

Ekström, P. and Vanecek, J. (1992). Localization of 2-[^{125}]iodomelatonin binding sites in the brain of the atlantic salmon, *Salmo salar L. Neuroendocrinology* **55,** 529–537.

Eng, L. F., Vanderhaeghen, J. J., Bignami, A., and Gerstl, B. (1971). An acidic protein isolated from fibrous astrocytes. *Brain Res.* **28,** 351–354.

Erickson, P. and Moore, B. (1980). Investigation of the axonal transport of three acidic soluble proteins (14–3–2, 14–3–3, and S-100) in the rabbit visual system. *J. Neurochem.* 35, 232–241.

Falcon, J. (1979). Unusual distribution of neurons in the pike pineal organ. *In* "The Pineal Gland of Vertebrates including Man" (J. A. Kappers and P. Pévet, Eds.), Progr. in Brain Research, Vol. 52, pp. 89–91. Elsevier, Amsterdam.

Freund, D., Arendt, J., and Vollrath, L. (1977). Tentative immunohistochemical demonstration of melatonin in the rat pineal gland. *Cell Tissue Res.* **181,** 239–244.

Fujieda, H., Sato, T., Shi, J., and Wake, K. (1997). Remodeling of pineal epithelium in the fetal rat as delineated by immunohistochemistry of laminin and cadherin. *Cell Tissue Res.* **287,** 263–274

Gaston, S. and Menaker, M. (1968). Pineal function: The biological clock in the sparrow? *Science* **160,** 1125- 1127.

Ghandour, M., Langley, O., and Keller, A. (1981). A comparative immunohistological study of cerebellar enolase. *Exp. Brain Res.* **41,** 271–279.

Gown, A. M. and Gabbiani, G. (1984). Intermediate-sized (10-nm) filaments in human tumours. *In* "Advances in Immunohistochemistry" (R. A. DeLellis, Ed.), pp. 89–109. Masson, New York.

Hartmann, F. (1957). Uber die Innervation der Epiphysis cerebri einiger Saugetiere. *Z. Zellforsch.* **45,** 416–429.

Hetherington, T. (1981). Morphology of the pineal organ in the salamander, *Ensatina eschscholtzi*. *J. Morphol.* **169,** 191–206.

Jahn, R., Schiebler, W., Ouimet, C., and Greengard, P. (1985). A 38000 dalton membrane protein (p38) present in synaptic vesicles. *Proc. Natl. Acad. Sci. USA* **82,** 4137–4141.

Kappers, J. A. (1960). The development, topographical relations and innervation of the epiphysis cerebri in the albino rat. *Z. Zellforsch.* **52,** 163–215.

Kappers, J. A. (1965). Survey of the innervation of the epiphysis cerebri and the accessory pineal organs of vertebrates. *In* "Structure and Function of the Epiphysis Cerebri" (J. A. Kappers and J. P. Schadé, Eds.), Progr. in Brain Research, vol. 10, pp. 87–153. Elsevier, Amsterdam.

Kenny, G. C. T. (1961). The "nervus conarii" of the monkey. An experimental study. *J. Neuropath. Exp. Neurol.* **20,** 563–570.

Klein, D. C. (1985). Photoneural regulation of the mammalian pineal gland. *In* "Photoperiodism, Melatonin and the Pineal Gland" (D. Evered and S. Clark, Eds.), pp. 38–56. Pitman, London.

Knaus, P., Betz, H., and Rehm, H. (1986). Expression of synaptophysin during postnatal development of the mouse brain. *J. Neurochem.* **47,** 1302–1304.

Korf, H. W. (1974). Acetylcholinesterase-positive neurons in the pineal and parapineal organs of the rainbow trout, *Salmo gairdneri*, with special reference to the pineal tract. *Cell Tissue Res.* **155,** 475–489.

Korf, H. W., Liesner, R., Meissl, H., and Kirk, A. (1981). Pineal complex of the clawed toad, *Xenopus laevis* Daud: Structure and function. *Cell Tissue Res.* **216,** 113–130.

Korf, H. W., Oksche, A., Ekström, P., van Veen, T., Zigler, J. S., Gery, I., Stein, P., and Klein, D. C. (1986). S-antigen immunocytochemistry. *In* "Pineal and Retinal Relationships" (P. O'Brien and D. C. Klein, Eds.), pp. 343–355. Academic Press, New York.

Korf, H. W., Sato, T., and Oksche, A. (1990). Complex relationships between the pineal organ and the medial habenular nucleus-pretectal region of the mouse as revealed by S-antigen immunocytochemistry. *Cell Tissue Res.* **261,** 493–500.

Korf, H. W., Zimmerman, N. H., and Oksche, A. (1982). Intrinsic neurons and neural connections of the pineal organ of the house sparrow, *Passer domesticus*, as revealed by anterograde and retrograde transport of horseradish peroxidase. *Cell Tissue Res.* **222,** 243–260.

Kramm, C. M., De Grip, W. J., Korf, H. W. (1993). Rod-opsin immunoreaction in the pineal organ of the pigmented mouse does not indicate the presence of a functional photopigment. *Cell Tissue Res.* **274,** 71–78.

Kuwano, R., Iwanaga, T., Nakajima, T., Masuda, T., and Takahashi, Y. (1983). Immunocytochemical demonstration of hydroxyindole O-methyltransferase (HIOMT), neuron- specific enolase (NSE) and S-100 protein in the bovine pineal. *Brain Res.* **274,** 171–175.

Laitinen, J. T. and Saavedra, J. M. (1990). Characterization of melatonin receptors in the rat suprachiasmatic nuclei: Modulation of affinity with cations and guanine nucleotides. *Endocrinology* **126,** 2110–2115.

Laudon, M. and Zisapel, N. (1986). Characterization of central melatonin receptors using ^{125}I-melatonin. *FEBS Lett.* **197,** 9–12.

Lundberg, J. M., Terenius, L., Hökfelt, T., Martling, C. R., Tatemoto, K., Mutt, V., Polak, J., Bloom, S., and Goldstein, M. (1982). Neuropeptide Y-like immunoreactivity in peripheral noradrenergic neurons and effects of NPY on sympathetic function. *Acta Physiol. Scand.* **116,** 477–480.

Marangos, P., Zis, A., Clark, R., and Goodwin, F. (1978). Neuronal, non-neuronal and hybrid forms of enolase in brain: Structural properties of neuronal and glial isoenzymes of brain enolase. *Brain Res.* **150,** 117–133.

Marangos, P., Polak, J., and Pearse, A. (1982). Neuron-specific enolase: A probe for nervous and neuroendocrine cells. *TINS* **June,** 193–196.

Marburg, O. (1909). Zur Kenntnis der normalen und pathologischen Histologie der Zirbeldruse. Die Adipositas cerebralis. *Arb. Neurol. Inst. Univ. Wien* **17,** 217–279.

Matsuura, T. and Herwig, H. J. (1981). Histochemical and ultrastructural study of the nervous elements in the pineal organ of the eel, *Anguilla anguilla. Cell Tissue Res.* **216,** 545–555.

McClung, R. and Dafny, N. (1975). Neurophysiological properties of the pineal body. II. Single unit recording. *Life Sci.* **16,** 621–628.

McClure, C. D., McMillan, P. J., and Miranda, A. C. (1986). Demonstration of differential immunohistochemical localization of the neuron-specific enolase antigen in rat pinealocytes. *Amer. J. Anat.* **176,** 461–467.

Mikkelsen, J. D. and Møller, M. (1988). Vasoactive intestinal peptide in the hypothalamohypophysial system of the mongolian gerbil. *J. Com. Neurol.* **273,** 87–98.

Møller, M. (1978). Presence of a pineal nerve (Nervus pinealis) in the human fetus; a light and electron microscopical study of the innervation of the pineal gland. *Brain Res.* **154,** 1–12.

Møller, M., Phansuwan-Pujito, P., Pramanlkijja, S., Kotchabhakdi, N., and Govitrapong, P. (1994). Innervation of the cat pineal gland by neuropeptide Y-immunoreactive nerve fibers: an experimental immunohistochemical study. *Cell Tissue Res.* **276,** 545–550.

Møllgard, K. and Møller, M. (1973) On the innervation of the human fetal pineal gland. *Brain Res.* **52,** 428–432.

Moore, R. Y. (1978). Neuronal control of pineal function in mammals and birds. *J. Neural Trans.* **13,** 47–58.

Navone, F., Jahn, R., DiGioia, G., Stukenbrok, H., Greengard, P., and De Camilli, P. (1986). Protein p38: an integral membrane protein specific for small vesicles of neurons and neuroendocrine cells. *J. Cell Biol.* **103,** 2511–2527.

Nürnberger, F. and Korf, H. W. (1981). Oxytocin- and vasopressin-immunoreactive nerve fibers in the pineal gland of the hedgehog, *Erinaceus europaeus* L. *Cell Tissue Res.* **220,** 87–97.

Ohba, S., Wake, K., Ohnishi, R., and Ueck, M. (1979). Neue Befunde am pinealen Sinnesapparat von funa, *Carassius gibelio langsdorfi* (Teleostei). *Verh. Anat. Ges.* **73,** 953–959.

Oksche, A. (1971). Sensory and glandular elements of the pineal organ. *In* "The Pineal Gland" (G. E. W. Wolstenholme and J. Knight, Eds.), pp. 127–146. Churchill-Livingstone, Edinburgh.

Oksche, A. and Hartwig, H. G. (1979). Pineal sense organs - components of photoneuroendocrine systems. *In* "The Pineal Gland of Vertebrates including Man" (J. A. Kappers and P. Pévet, Eds.), Progr. in Brain Research, Vol. 52, pp. 113–130. Elsevier, Amsterdam.

Oksche, A., Korf, H. W., and Rodriguez, E. M. (1987). Pinealocytes as photoneuroendocrine units of neuronal origin: concepts and evidence. *In* "Advances in Pineal Research" (R. J. Reiter and F. Fraschini, Eds.), Vol. 2, pp. 1–18. John Libbey, London, Paris.

Omura, Y. (1980). Histochemical and ultrastructural studies on the nervous organization of the pineal organ of the ayu, *Plecoglossus altivelis*. Bull. *Jap. Soc. Sci. Fish* **46,** 1483–1488.

Pastori, G. (1928). Ein bis jetzt noch nicht beschriebenes sympathisches Ganglion zum Nervus conarii sowie zur Vena magna Galenii. *Z. Ges. Neurol. Psychial.* **123,** 81–90.

Pévet, P. (1977). On the presence of different populations of pinealocytes in the mammalian pineal gland. *J. Neural. Transm.* **40,** 289–304.

Pines, G. (1927). Uber die Innervation der Epiphyse. *Z. gesamt. Neurol.* **11,** 365–369.

Quay, W. B. (1957). Cytochemistry of pineal lipids in rat and man. *J. Histochem. Cytochem.* **5,** 145–153.

Quay, W. B. (1958). Pineal blood content and its experimental modification. *Am. J. Physiol.* **195,** 391–395.

Quay, W. B. and Renzoni, A. (1966). Twenty-four hour rhythms in pineal mitotic activity and nuclear and nucleolar dimensions. *Growth* **30,** 315–324.

Rehm, H., Wiedenmann, B., and Betz, H. (1986). Molecular characterization of synaptophysin, a major calcium-binding protein of the synaptic vesicle membrane. *EMBO J.* **5,** 535–541.

Reppert, S. M., Weaver, D. R., Rivkees, S. A., and Stopa, E. G. (1988). Putative melatonin receptors in a human biological clock. *Science* **242,** 78–81.

Romijn, H. J. (1973). Structure and innervation of the pineal gland of the rabbit, *Oryctolagus cuniculus* (L). I. A light microscopic investigation. *Z. Zellforsch.* **139,** 473–485.

Romijn, H. J. (1975) Structure and innervation of the pineal gland of the rabbit, *Oryctolagus cuniculus* (L). III. An electron microscopic investigation of the innervation. *Cell Tissue Res.* **157,** 25–51.

Sasek, C. A. and Zigmond, R. E. (1989). Localization of vasoactive intestinal peptide- and peptide histidine isoleucine amide-like immunoreactivities in the rat superior cervical ganglion and its nerve trunks. *J. Com. Neurol.* **280,** 522–532.

Sato, T., Deguchi, T., Ichikawa, T., Fujieda, H., and Wake, K. (1991). Localization of hydroxyindole O-methyltransferase-synthesizing cells in bovine epithalamus: immunocytochemistry and in-situ hybridization. *Cell Tissue Res.* **263,** 413–418.

Sato, T. and Ebisawa, S. (1988). A pineal ganglion associated with the pineal tract in the domestic fowl. *Cell Tissue Res.* **252,** 287–292.

Sato, T., Ebisawa, S., and Wake, K. (1988). Neuronal degeneration in the pineal ganglion during the post-hatching development. *Cell Tissue Res.* **254,** 25–30.

Sato, T., Kaneko, M., Fujieda, H., Deguchi, T., and Wake, K. (1994). Analysis of the heterogeneity within bovine pineal gland by immunohistochemistry and in situ hybridization. *Cell Tissue Res.* **277,** 201–209.

Sato, T., Kaneko, M., Ekataksin, W., and Wake, K. (1995). Expression of neuron-specific enolase in the pineal organ of the domestic fowl during post-hatching development. *Cell Tissue Res.* **279,** 25–36.

Sato, T. and Wake, K. (1981). Organizations of the sensory and sympathetic nerves in the avian pineal organs. *Jikeikai Med. J.* **28(Suppl),** 7–12.

Sato, T. and Wake, K. (1983). Innervation of the avian pineal organ - A comparative study. *Cell Tissue Res.* **233,** 237- 264.

Sato, T. and Wake, K. (1984). Regressive post-hatching development of acetylcholinesterase-positive neurons in the pineal organs of *Coturnix coturnix japonica* and *Gallus gallus*. *Cell Tissue Res.* **237,** 269–275.

Sato, T. and Wake, K. (1986). Separation of the pineal vesicle from the wall of the third ventricle during the post-hatching development of the chicken. *Cell Tissue Res.* **244,** 321–326.

Sato, T., Wake, K., Kramm, C., and Korf, H. W. (1990). Chicken pineal organs during post-hatching development: photoreceptor-specific characteristics and innervation. *In* "Neuroendocrinology: Quid Novi" (D. Gupta, H. A. Wollmann, and M. B. Ranke, Eds.), pp. 191–200. Brain Research Promotion, Tübingen, London.

Schmechel, D., Brightman, M., and Goodwin, F. (1978). Brain enolase as specific markers of neuronal and glial cells. *Science* **199,** 313–315.

Schmechel, D., Brightman, M., and Marangos, P. (1980). Neurons switch from non-neuronal enolase to neuron-specific enolase during differentiation. *Brain Res.* **190,** 195–214.

Semm, P. and Vollrath, L. (1979). Electrophysiology of the guinea-pig pineal organ: sympathetic influenced cells responding differently to light and darkness. *Neurosci. Lett.* **12,** 93–96.

Semm, P. and Vollrath, L. (1980). Electrophysiological evidence for circadian rhythmicity in a mammalian pineal organ. *J. Neural. Transm.* **47,** 181–190.

Semm, P. and Vollrath, L. (1982). Alterations in the spontaneous activity of cells in the guinea pig pineal gland and visual system produced by pineal indoles. *J. Neural. Transm.* **53,** 265–275.

Shiotani, Y., Yamano, M., Shiosaka, S., Emson, P. C., Hillyard, C. J., Girgis, S., and MacIntyre I. (1986). Distribution and origins of substance P (SP)-, calcitonin gene-related peptide (CGRP)-, vasoactive intestinal polypeptide (VIP)- and neuropeptide Y (NPY)-containing nerve fibers in the pineal gland of gerbils. *Neurosci. Lett.* **70,** 187–192.

Tapp, E. and Blumfield, M. (1970). The parenchymal cells of the rat pineal gland. *Acta Morphol. Neerl-Scand.* **8,** 119–131.

Trapp, B., Marangos, P., and Webster, H. (1981). Immunocytochemical localization and developmental profile of neuron-specific enolase during differentiation. *Brain Res.* **220,** 121–130.

Ueck, M. and Kobayashi, H. (1972). Vergleichende Untersuchungen über acetylcholinesterasehaltige Neurone im Pinealorgan der Vögel. *Z. Zellforsch.* **129,** 140–160.

Vanecek, J. (1988). Melatonin binding sites. *J. Neurochem.* **51,** 1436–1440.

Vollrath, L. (1981). The pineal organ. *In* "Handb. Mikrosk. Anat. Mensch." (A. Oksche and L. Vollrath, Eds.), VI/7, Springer, Berlin Heidelberg New York.

Wake, K. (1973). Acetylcholinesterase-containing nerve cells and their distribution in the pineal organ of the goldfish, *Carassius auratus*. *Z. Zellforsch.* **145,** 287–298.

Wake, K., Ueck, M., and Oksche, A. (1974). Acetylcholinesterase-containing nerve cells in the pineal complex and subcommissural area of the frogs, *Rana ridibunda* and *Rana esculenta*. *Cell Tissue Res.* **154,** 423–442.

Weaver, D. R., Rivkees. S. A., and Reppert, S. M. (1988). Localization and characterization of melatonin receptors in rodent brain by in vitro autoradiography. *J. Neurosci.* **9,** 2581–2590.

Wiedenmann, B. and Franke, W. W. (1985). Identification and localization of synaptophysin, an integral membrane glycoprotein of M 38000 characteristic of presynaptic vesicles. *Cell* **41,** 1017–1028.

Wolfe, D. E. (1965). The epiphyseal cell: an electron-microscopic study of its intercellular relationships and intracellular morphology in the pineal body of the albino rat. *In* "Structure and Function of the Epiphysis Cerebri" (J. A. Kappers and J. P. Schadé, Eds.), Progr. in Brain Research, Vol. 10, pp. 332–386. Elsevier, Amsterdam.

Zhang, E. T., Mikkelsen, J. D., and Møller, M. (1991). Tyrosine hydroxylase- and neuropeptide Y-immunoreactive nerve fibers in the pineal complex of untreated rats and rats following removal of the superior cervical ganglia. *Cell Tissue Res.* **265,** 63–71.

Zimmerman, N. H. and Menaker, M. (1975). Neural connections of sparrow pineal: Role in circadian control of activity. *Science* **190,** 477–479.

Zimmerman, B. L. and Tso, M. O. M. (1975). Morphologic evidence of photoreceptor differentiation of pinealocytes in the neonatal rat. *J. Cell Biol.* **66,** 60–75.

13

INTERACTIONS BETWEEN THE ENDOCRINE AND EXOCRINE PANCREAS

Effects of Islet Hormones, Secretagogues, and Nerve Stimulation

Maria Dolores Yago,[1] Ernest Adeghate,[2] and Jaipaul Singh[1]

[1]Department of Applied Biology
University of Central Lancashire
Preston, PR1 2HE, England, United Kingdom
[2]Department of Human Anatomy
Faculty of Medicine and Health Sciences
United Arab Emirates, PO Box 17666
Al-Ain, United Arab Emirates

1. INTRODUCTION

The pancreas is a soft elongated organ lying posterior to the greater curvature of the stomach. It contains both exocrine and endocrine portions and, thus, it is regarded as two separate organ systems. The interrelationship and interaction between the endocrine and exocrine parts of the pancreas have not been properly investigated. However, it is worthwile to note that the pancreas is an integrated organ which is involved in secretion of important hormones and digestive enzymes and in the digestion and assimilation of food stuffs. Both the exocrine and endocrine portions of this organ evolved from a primordial outgrowth of the primitive gut and the pancreas is normally referred to as an integral part of the gastrointestinal system. The literature on the interaction between the endocrine and exocrine pancreas (insulin-acinar relationship) is meagre (see review by Owyang, 1993; Williams and Goldfine, 1993). However, the available studies have investigated mainly the effects of exogenous and endogenous insulin on secretagogue-evoked enzyme secretion. Furthermore, a few other studies have also investigated the effects of glucagon and somatostatin on exocrine pancreatic function. The cellular mechanism of action and interactions of the islet hormones with the classical secretagogues acetylcholine (ACh) and cholecystokinin (CCK) have not been investigated and, moreover, there is little or no evidence on the interaction of the endocrine and exocrine pancreas during diabetes. This is a very important area since diabetes is a global problem and it affects more than 100 million

Neural Regulation in the Vertebrate Endocrine System, edited by Prasada Rao and Peter, Kluwer Academic / Plenum Publishers, New York, 1999.

people and its prevalence is increasing worldwide. The present review is different from previous ones in that it also focuses on the morphology and on the cellular mechanisms of the interactions of the islet hormones with nerve-mediated and secretagogue-evoked secretory responses in the pancreas of both normal and diabetic animals. In addition, this review is also concerned with the role of the brain and nervous system in the control of pancreatic exocrine function since pancreatic innervation plays a dominant role in the minute-to-minute control of pancreatic secretion not only for the cephalic phase but also for the intestinal phase. The literature on a detailed analysis of the brain and nervous system in the control of pancreatic secretion has been reviewed elsewhere (Richens, 1945; Holst, 1993).

2. THE PANCREAS

The mammalian pancreas is situated in the duodenal loop with a head, body and tail. Its tail extends as far as the spleen onto which it has morphological relations. The pancreas has two ducts, the main and accesory ducts, which open into the duodenum with the common bile duct from the liver and gallbladder. The pancreas is a retroperitoneal organ with a mass of 85 ± 15 g in human female and 90 ± 16 g in the male. The pancreas is a mixed gland, with a larger exocrine and a much smaller endocrine glands. The exocrine pancreas forms the greatest volume of 84%. Ductal cells and blood vessels form around 4%, while the endocrine part comprises of 2%. The other part (10%) is occupied by extracellular matrix (Gorelick and Jamieson, 1987). The endocrine cells are arranged into small islands of cells called the islets of Langerhans. Acinar tissue in the pancreas is in close contact with the islets since no significant capsule surrounds the islets (Williams and Goldfine, 1993). Thus, marked interaction must occur between the exocrine and endocrine pancreas. The function of both the exocrine and the endocrine parts are particularly important for the normal functioning of the body. The exocrine gland harbours important digestive enzymes (pancreatic amylase, lipase, trypsin) which are produced by acinar cells and transported into the duodenum by the pancreatic ducts. The endocrine cells produce indispensable hormones (insulin, glucagon, somatostatin) which are crucial to the normal functioning of body metabolism. The pancreas is also one of the most investigated organs in human scientific history. The pancreas is well innervated by autonomic nerves rich in different types of peptides including vasoactive intestinal peptide, neuropeptide Y, galanin, leucine-enkephalin, calcitonin-gene-related peptide, serotonin and CCK (Adeghate and Donáth, 1990; Adeghate *et al.*, 1995; Adeghate *et al.*, 1996; Adeghate, 1997; Adeghate *et al.*, 1997).

3. CONTROL OF EXOCRINE PANCREATIC SECRETION

3.1. Hormonal Control

The control of exocrine pancreatic secretion involves the action of neural, paracrine and endocrine factors. Among the latter, secretin and CCK are considered to be the major hormonal stimulators of the response to meals. Following the arrival of acidic chyme from the stomach, both are released from specialised cells in the upper small intestinal mucosa into the circulation. The hormones are then transported to the pancreas, where they regulate the secretion of pancreatic juice (Singh *et al.*, 1992). It is generally accepted that the action of secretin involves the secretion of bicarbonate from ductal cells and, in a lesser extent, enzymes from acinar cells (Jensen and Gardner, 1981). Certainly, secretin is the most potent and efficacious stimulant of pancreatic flow and bicarbonate secretion in all species in which it has been tested. In doses that mimic the postprandial plasma levels, exogenous secretin has been shown to stimulate pancreatic secretion of water and bicarbonate (Gullo *et al.*, 1987; Gyr *et al.*, 1987). Administration of secretin antiserum to conscious dogs greatly reduces, though not abolishes, the pancreatic response to a meal (Chey *et al.*, 1975). CCK, in turn, stimulates mainly pancreatic acinar cells to release a juice which is rich in digestive enzymes. Earlier studies examined the effect of different doses of CCK that reproduced postprandial plasma levels, but, now, the availability of potent and specific CCK receptor antagonists has enabled investigators to clearly determine the physiological role of circulating CCK in pancreatic exocrine secretion. In healthy volunteers, the CCK receptor antagonist MK329 (Cantor *et al.*, 1991) inhibited by 74 and 89% trypsin and amylase secretion respectively stimulated by physiological doses of CCK-8 and secretin. Fried *et al.* (1991) found a suppression by loxiglumide of CCK-8-stimulated enzyme secretion, as well as a reduction equal to 50 and 75% in trypsin and lipase output after a mixed meal. In this study, an attenuation in these parameters during fasting was also recorded. CCK receptor blockade reduced trypsin and chymotrypsin output (by 70%) and amylase output (by 55%) stimulated by intraduodenal infusion of a mixed meal (Schwarzendrube *et al.*, 1991). Thus, circulating CCK is a major stimulant of the intestinal phase of pancreatic enzyme secretion.

3.2. Neural Control

The importance of the neural control of exocrine pancreatic secretion has been implicated in physiological studies (Pearson *et al.*, 1984; Holst, 1993). The main sources of extrinsic nerves are the vagus nerves and the splanchnic nerves. Preganglionic parasympathetic fibers reach the pancreas directly or pass the celiac ganglion uninterrupted to terminate in intrapancreatic ganglia. Postganglionic parasympathetic fibers are ultimately distributed to acini, ducts and islets. Presynaptic sympathetic fibers are interrrupted in the celiac plexus, and postganglionic fibers terminate in nonadrenergic intrapancreatic ganglia, ducts and vessels. Noradrenergic cell bodies have been also identified within the pancreas in rats (Holst, 1993). However, nerve endings on or between acinar cells and around ducts are believed to represent mainly terminations of cholinergic neurons. Recently, the concept of an important peptidergic component is well established (Solomon, 1987; Holst, 1993; Adeghate *et al.*, 1996; Adeghate *et al.*, 1997; Juma *et al.*, 1997a).

3.2.1. Vagal Cholinergic Control. Activation of efferent vagal nerve impulses by electrical stimulation of the vagal trunks causes strong stimulation of pancreatic enzyme secretion (Kaminski *et al.*, 1975; Holst *et al.*, 1979). Volume and bicarbonate responses are less pronounced (Holst *et al.*, 1979). Similar stimulation of enzyme secretion occurs after vagal activation induced by insulin hypoglycemia or administration of the glucose antagonist 2-deoxy-D-glucose (Eisenberg and Orahood, 1971; Rosenberg *et al.*, 1976). All three methods cause enzyme secretion due to release of ACh, because administration of atropine inhibits the response in most studies (Holst *et al.*, 1979). Consistent with these findings are the observations that administration of ACh or muscarinic cholinergic agonists to intact animals or humans also stimulates pancreatic enzyme secretion (Holst *et al.*, 1979). Although cholinergic mechanisms certainly have other indirect effects on pancreatic secretion in the intact organism, several lines of evidence indicate that a major site of action of vagally-induced ACh release in on the pancreas itself: intrapancreatic ganglia and cholinergic postganglionic neurons have been demonstrated near acini and ducts (Coupland, 1958); the exocrine pancreas contains enzymes involved in the synthesis and inactivation of ACh (Godfrey and Matschinski, 1975); muscarinic cholinergic drugs and ACh are full agonists for enzyme secretion by pancreatic acinar cells *in vitro* (Williams *et al.*, 1978; Bommelaer *et al.*, 1981); and specific muscarinic receptors on pancreatic acinar cells have been demonstrated (Larose *et al.*, 1983). There is strong evidence suggesting a major role for direct efferent cholinergic stimulation of meal-induced pancreatic secretion. Intrapancreatic postganglionic cholinergic neurons would be activated by central input during the cephalic phase and by vagovagal reflexes initiated by gastric and intestinal phase stimulation. ACh released by these neurons would act directly on muscarinic receptors on acinar cells (and presumably ductal cells) to elicit secretion (Solomon, 1987).

3.2.2. Sympathetic Adrenergic Control. Although initially any effect of adrenergic stimulation on pancreatic secretion was thought to be a consequence of alterations in pancreatic blood flow (Barlow *et al.*, 1974), sympathetic innervation of acinar cells has also been implicated in controlling exocrine function (Pearson *et al.*, 1984). An adrenergic neural influence has been shown to elicit amylase release from acinar cells (Pearson *et al.*, 1984; Singh and Pearson, 1984). Furthermore, adrenergic fibers (Ahren *et al.*, 1981) and noradrenergic cell bodies (Holst, 1993) have been observed within the pancreas in rats. This innervation may also influence fluid secretion, since isoprenaline can stimulate a bicarbonate rich fluid from the rat pancreas (Linguard and Young, 1983).

3.2.3. Non-Cholinergic Non-Adrenergic Control. Evidence suggests that peptidergic nerves exhibit control of pancreatic exocrine secretion (Pearson *et al.*, 1981, 1982; Adeghate *et al.*, 1996, 1997). Thus, although most studies point to the existence or an important cholinergic regulation of pancreatic secretion, it has been known that some of the effects of electrical nerve stimulation, such as fluid and bicarbonate responses in cats and pigs (Brown *et al.*, 1967; Holst *et al.*, 1979) and some of the effects of sham-feeding in humans (Defilippi *et al.*, 1982) are atropine-resistant. From studies with adrenergic blockade it was found also to be nonadrenergic (Holst *et al.*, 1979). Many peptides have been localized in the pancreas (Holst, 1993), but the predominating one is vasoactive intestinal polypeptide (VIP), which is found in many intrapancreatic ganglionic cells and in a diffuse network of fibers in relation to islets, acini, ducts and blood vessels (Holst *et al.*, 1984). These neurons are mostly intrinsic to the pancreas. VIP is a full agonist for pancreatic secretion of fluid and bicarbonate in cats and a partial agonist in dogs and humans (Konturek *et al.*, 1976; Makhlouf *et al.*, 1978). On a molar basis, it is less potent than secretin (Konturek *et al.*, 1976) and it is released from pancreatic nerves upon stimulation of the vagus nerves (Holst *et al.*, 1984) and also by cholinergic agonists acting on nicotinic receptors, suggesting that intrinsic VIPergic neurons may be activated by cholinergic preganglionic parasympathetic fibers.

3.3. Interaction between Secretagogues

An important aspect of the physiology of the exocrine pancreas is the interaction between different secretagogues. It seems rather clear that both CCK and ACh released from intrapancreatic nerves may account for most of the enzyme secretion observed after normal meals. Although secretin may have small effects on pancreatic enzyme secretion from acinar cells (Jensen and Gardner, 1981), a possible augmentation by secretin of the pancreatic secretion of enzymes stimulated by CCK-8 has been observed only in dogs (Beglinger *et al.*, 1984). On the other hand, it has been recognized that secretin in a physiological dose range can not stimulate secretion of fluid and bicarbonate comparable to that after endogenous stimulation by food (Kim *et al.*, 1979). In humans, dogs, and rats, the action of secretin is augmented by CCK-8 or caerulein to produce the amount of pancreatic bicarbonate much greater than the sum of the respective secretions produced by the two individual hormones acting alone (Greenberg *et al.*, 1979). The great inhibition of volume and bicarbonate output estimulated by exogenous CCK-8 and secretin observed after the administration of CCK receptors antagonists support the potentiatory effect of CCK (Cantor *et al.*, 1991). Atropine was found to abolish the potentiatory effect of secretin and CCK on exocrine pancreatic secretion in humans (You and Chey, 1988), suggesting that this phenomenon is under the influence of cholinergic tone. Clearly, neural mechanisms are acting in conjunction with the hormonal ones to elicit the secretion of bicarbonate and water, and electrical stimulation of the vagus nerves has been demonstrated to potentiate the bicarbonate response to exogenous secretin in anaesthetized pigs and cats (Brown *et al.*, 1967; Holst *et al.*, 1979). In the latter species, however, the potentiatory effect was not blocked by atropine nor adrenergic blockers (Brown *et al.*, 1967), indicating that some other neurotransmitter is involved.

4. STIMULUS-SECRETION COUPLING IN THE EXOCRINE PANCREAS

This refers to the stages following receptor occupancy that lead up to secretion of enzymes, salts and fluid. In the exocrine pancreas there are several mechanisms by which secretagogues can elicit pancreatic secretion (Schulz and Stolze, 1980; Jensen and Gardner, 1981; Singh *et al.*, 1992). The action of CCK and ACh on acinar cells is associated with the hydrolysis of phosphatidylinositol 4,5-bisphosphate (PIP2) resulting in the formation of inositol 1,4,5-trisphosphate (IP3), inositol 1,3,4,5-tetrakisphosphate (IP4) and diacylglycerol (DG) (Nishizuka, 1988; Berridge, 1993). This elevation of IP3 is then thought to stimulate the release of calcium from the stores in the rough endoplasmic reticulum (RER) (Berridge and Irvine, 1989). It is also believed that the secretagogue-induced release of calcium in one part of the RER leads to a second calcium-induced calcium release at other sites in the RER. Thus, IP3 opens up a calcium channel in one compartment of the RER. This, in turn, releases calcium, which subsequently acts on a separate caffeine-sensitive calcium channel in another entirely distinct compartment of the RER (Berridge and Irvine, 1989). It seems that the major calcium release from intracellular stores in pancreatic acinar cells occurs through this channel. On the other hand, calcium availability is further enhanced by raised IP4 levels, which are thought to trigger uptake of calcium from the extracellular space (Berridge and Irvine, 1989). The elevated cytosolic calcium, together with DG (diacylglycerol) from the hydrolysis of PIP2 are then associated with the activation of their respective protein kinases: calmodulin, and the calcium-phospholipid-dependent protein kinase C, to phosphorylate zymogen granular membrane regulatory proteins resulting in migration of the granules, docking, fusion with the luminal membrane and subsequently, exocytosis and secretion (Burnham and Williams, 1982; Schulz *et al.*, 1986; Nishizuka, 1988). Another pathway in the exocrine pancreas involves the changes in intracellular adenosine 3′5′-cyclic monophosphate (cAMP) levels following stimulation with either VIP, secretin or norepinephrine (Schulz and Stolze, 1980; Jensen and Gardner, 1981; Pearson *et al.*, 1981, 1982; Pearson *et al.*, 1984). This mechanism mediates the secretion of bicarbonate from ductal cells, and enzyme from acinar cells (Jensen and Gardner, 1981). The elevation of cAMP levels by secretin and VIP results from activation of adenylate cyclase in acinar cells plasma membranes (Long and Gardner, 1977). All known effects for cAMP are mediated by cAMP-activated protein kinases (e.g protein kinase A), and are initiated by the binding of cAMP to the regulatory subunit of the kinase (Jensen and Gardner, 1978).

5. INTERACTIONS BETWEEN THE ENDOCRINE AND EXOCRINE PANCREAS: EFFECTS OF THE ISLET HORMONES ON THE EXOCRINE FUNCTION

5.1. Glucagon (Glu)

There are a number of *in vivo* studies on the effects of Glu on exocrine pancreatic secretion in different species: dog (Nakajima and Magee, 1970; Konturek *et al.*, 1974; Singer *et al.*, 1978), rat (Shaw and Health, 1973), cat (Konturek *et al.*, 1974), and man (Dyck *et al.*, 1970; Fontana *et al.*, 1975). Although little effect is seen in the resting pancreas (Williams and Goldfine, 1993), exogenous Glu has been demonstrated to inhibit exocrine pancreatic secretion induced by a meal (Shaw and Health, 1973; Konturek *et al*, 1974) and intestinal perfusion with either amino acids (DiMagno *et al.*, 1973; Konturek *et al.*, 1974) or hydrochloric acid (Konturek *et al.*, 1974). No studies have examined the effect of Glu on secretin or CCK release. In contrast, many reports show significant inhibition of pancreatic secretion stimulated by exogenous secretin, CCK or both (Dyck *et al.*, 1970; Nakajima and Magee, 1970; Konturek *et al.*, 1974), and dose-response studies (Konturek *et al.*, 1974) suggest that the inhibition of CCK-evoked pancreatic enzyme secretion by Glu is competitive in nature. The inhibitory effect of Glu has been reported to be characterized by reductions in volume and bicarbonate output (Dyck *et al.*, 1970; Nakajima and Magee, 1970; Konturek *et al.*, 1974) and also enzyme secretion (Dyck *et al.*, 1970; Konturek *et al.*, 1974). However, the inhibition of enzyme output is usually greater (Dyck *et al.*, 1970) and, in fact, some investigators could not find any inhibitory effect on bicarbonate secretion.

In vitro effects of Glu are not so clear-cut. A stimulatory effect of Glu alone on the perfused dog pancreas has been reported (Iwatsuki *et al.*, 1976). Also, Glu evoked a small increase in protein output in the perfused rat pancreas when given alone (Sommer and Kasper, 1981). In the latter study, however, Glu inhibited the volume and protein output in response to CCK and secretin stimulation. The inhibitory effect on secretin- and CCK-induced secretion was confirmed in the *in vitro* vascularly perfused cat pancreas (Wizemann *et al.*, 1974). One study on perfused dog pancreas showed no effect of a bolus infusion of Glu on secretin or CCK stimulation when given individually, but potentiation of the response to combined high concentrations of the two secretagogues (Murphy and McGeeney, 1974).

Concerning studies with pancreatic lobules, acini, or single acinar cells *in vitro*, high concentrations of Glu were reported to inhibit basal enzyme secretion by one group (Adler, 1977), but most studies have shown that Glu alone stimulates pancreatic enzyme release in different species including the mouse (Singh, 1980a), guinea pig (Pandol *et al.*, 1983) and rat (Adeghate *et al.*, 1997; Juma *et al.*, 1997a; Singh and Adeghate, 1998; Singh *et al.*, 1998). The three latter studies, performed recently in our laboratory, show that, compared to ACh, CCK-8 or EFS, Glu has small secretagogue effect on rat pancreatic segments, and, more interestingly, that in contrast to the classical secretagogues, which evoked a dose-dependent response, exogenous application of Glu alone resulted in differential effects on amylase secretion: very low concentrations initiated a marked increase in amylase output above basal levels, but the response decreased gradually with higher concentrations (Adeghate *et al.*, 1997). These results support previous findings in which Glu had been reported to exert either an inhibitory or a stimulatory effect on pancreatic secretion, and are in close agreement that these differences do not depend only on the species of the animals or on whether it is an *in vivo* or an *in vitro* effect, but also on the concentration of the peptide used. Thus, it is tempting to stress the importance of investigating proper dose-dependent secretory effects of Glu.

Apart from the above mentioned stimulatory effect of Glu alone, the peptide has been demonstrated to markedly potentiate the response to different secretagogues such as ACh, carbachol and CCK-8 (Singh, 1980b; Pandol *et al.*, 1983; Adeghate *et al.*, 1997; Juma *et al.*, 1997a). Recent experiments (Singh and Adeghate, 1998) have been focused on the interaction between Glu and nerve-mediated amylase secretion, using electrical field stimulation (EFS)—a technique which has been employed previously by other workers in the *in vitro* pancreas to study the effect of endogenous neurotransmitters on pancreatic secretion (Pearson *et al.*, 1984). The results show that combining Glu with EFS results also in a significantly enhanced amylase response in rat pancreatic segments compared to the effects of either EFS or Glu alone.

The mechanism associated with the Glu-evoked secretory response and with the potentiation is not fully understood. In *in vitro* studies using pancreatic slices, Glu had no effect on adenylate cyclase activity (Deschodt-Lanckman *et al.*, 1975). More recently, it has been demonstrated that natural Glu increases cAMP levels and amylase release from guinea pig isolated acini (Pandol *et al.*, 1983) and rat pancreatic segments (Singh, 1983), whereas synthetic Glu has no effect. In close agreement with these results, natural porcine Glu has demonstrated to evoke small secretory effects in rat isolated pancreatic acini and acinar cells (Juma *et al.*, 1997a), indicating that the islet hormone may have direct effects on pancreatic acinar cells to elicit amylase secretion. However, and in contrast to the results obtained with rat pancreatic segments (Juma *et al.*, 1997a), Glu failed to potentiate the secretory response of either ACh or CCK-8 in the isolated pancreatic acini preparation (Juma *et al.*, 1997a). Thus, intact and viable islets seem to be involved with the synergic effect of Glu and the classical secretagogues ACh and CCK.

The results showing the potentiatory effects of Glu on the secretory response to calcium-dependent agonists suggest the involvement of this intracellular mediator in the signal transduction pathway. Certainly, direct application of Glu alone elicits a small gradual rise in intracellular calcium $[Ca^{2+}]_i$ concentration (Adeghate *et al.*, 1997; Juma *et al.*, 1997a; Singh and Adeghate, 1998; Singh *et al.*, 1998). Moreover, when glucagon is combined with either ACh (Juma *et al.*, 1997a) or CCK-8 (Singh and Adeghate, 1998; Singh *et al.*, 1998), there is an enhancement of $[Ca^{2+}]_i$ concentration as compared to the secretagogues alone. In previous studies it was shown that Glu has a small but insignificant effect on calcium efflux and uptake (Singh, 1980b). Taken together, these observations indicate that the changes in $[Ca^{2+}]_i$ concentrations evoked by the combined application of the classical secretagogues and Glu cannot account for the enhancement by the latter of the secretory response. If calcium is the only intracellular mediator involved, then Glu should not augment the secretagogues-induced amylase secretion, especially when supramaximal doses of them are used. It is now well established (Jensen and Gardner, 1981; Singh *et al.*, 1992) that maximally effective doses of two secretagogues acting via the same intracellular mediator should not potentiate one another. In an attempt to ascertain the intracellular messenger involved in the potentiating actions of Glu, the same group examined the effect of genistein, an inhibitor of tyrosine kinase, on the secretory responses and $[Ca^{2+}]_i$ metabolism evoked by the combined presence of Glu and ACh (Juma *et al.*, 1997a; Singh and Adeghate, 1998) or CCK-8 (Singh *et al.*, 1998). The results showed that when genistein was combined with Glu and ACh or CCK-8, there is a marked attenuation not only of the enhanced secretory response, but also in the enhanced rise in $[Ca^{2+}]_i$. The authors conclude that tyrosine kinase activation may be associated with the potentiation of the enzyme secretion, and with the augmented increase in $[Ca^{2+}]_i$ evoked by Glu and either ACh or CCK-8, which, in turn, is in close agreement that tyrosine kinase can modulate calcium entry into pancreatic acinar cells (Yule *et al.*, 1994).

5.2. Insulin (Ins)

The effects of Ins on the exocrine pancreatic function are difficult to study *in vivo*. Traditionally, control by hormones has been demonstrated by removal of the presumed endocrine gland with restoration of function achieved by replacement of the missing hormone. It is not possible to surgically remove islets without removing the exocrine pancreas, but the supply of Ins is greatly reduced both in naturally occurring diabetes, and following treatment with specific β-cell toxins (the interactions between the islet hormones and the exocrine pancreas in streptozotocin (STZ)-induced diabetes are extensively studied in other section of this chapter. On the other hand, exogenous application of Ins in *in vivo* experiments evokes hypoglycemia and reflex vagal discharge with the resulting stimulation of the pancreas (Couture *et al.*, 1972), so this approach is not very useful. However, despite the few *in vivo* studies focusing on the interactions of Ins with the exocrine pancreas in healthy subjects, several lines of evidence have demonstrated that the islet hormone has a positive influence upon the exocrine portion of the gland (Youngs, 1972). Thus, Adler and Kern (1975) examined the effect of endogenously released Ins by infusing glucose to rats. After three days in these conditions, i.e. high Ins levels with high plasma glucose, the authors found an increase of pancreatic amylase content of about 25 %. Total protein synthesis by pancreatic lobules from the hyperinsulinemic rats was increased by 40 %. In addition, treatment with glibenclamide, a drug which is a potent secretagogue for Ins, elevated circulating Ins levels and increased total protein synthesis by 25 %. An effect of changes in the secretion of other hormones cannot be elucidated from this study, but it supports the idea that Ins may have long-term effects on the regulation of the biosynthesis of pancreatic digestive enzymes. In a more recent study (Lee *et al.*, 1990) it was shown that immunoneutralization of circulating free Ins with a specific rabbit anti-Ins serum in conscious rats prepared with pancreatic duct cannulas profoundly inhibited the pancreatic secretion stimulated by a meal, which suggests that Ins may exert short-term effects on the exocrine pancreas by potentiating the secretory response to gut hormones and neurotransmitters.

In vitro secretory effects of Ins have been examined more extensively. Although saturable Ins receptors on pancreatic acinar cells prepared from mouse, rat or guinea pig pancreas have been demonstrated and characterizated (Sankaran *et al.*, 1981; Mossner *et al.*, 1984), in most studies Ins alone has shown to have either a small stimulatory or no effect on enzyme pancreatic secretion. In contrast, the islet hormone can markedly potentiate the secretory responses to ACh and CCK-8 (Kanno *et al.*, 1976; Kanno and Saito 1976, Saito *et al.*, 1980a,b; Singh, 1985; Juma *et al.*, 1997a; Singh and Adeghate, 1998; Singh *et al.*, 1998). A potentiatory effect of Ins on nerve-mediated exocrine pancreatic secretion has been demonstrated recently as well (Singh and Adeghate, 1998; Singh *et al.*, 1998). The potentiation by Ins seems to require the presence of intact islets, since the potentiation has been confirmed in the isolated perfused pancreas (Kanno and Saito, 1976; Saito *et al.*, 1980a,b) and also in rat pancreatic segments (Singh, 1985; Juma *et al.*, 1997a; Singh *et al.*, 1998), but not in isolated acini or acinar cells (Juma *et al.*, 1997a; Williams and Goldfine, 1993; Singh *et al.*, 1998).

Direct arterial perfusion of a rabbit anti-Ins serum in totally isolated rat perfused pancreas preparation and without recirculation of portal venous effluent produced an attenuation of pancreatic secretion stimulated by secretin and CCK (Lee *et al.*, 1990a). In the same preparation, studies of endogenous Ins release have been carried out (Saito *et al.*, 1980a). When glucose was infused to stimulate Ins release, an enhanced secretory response to CCK, in terms of volume, protein content and amylase output, was observed. Since these results were found without the islet hormone entering into the systemic circulation, the authors concluded that the effects of endogenously released Ins on the exocrine pancreatic function are, at least in part, mediated by its local action on the exocrine portion of the gland.

The precise biochemical mechanisms for the slight effect of Ins alone or for the synergic effect with the classical secretagogues is still unresolved. Farese *et al.* (1981) used collagenase-digested rat pancreas to prepare a mixture of acini, islets and ductal fragments, and reported that Ins stimulated the release of amylase and induced phosphatydilinositol turnover. In a study performed on mouse pancreatic segments, Ins alone failed to influence ^{45}Ca fractional efflux, but in the continuing presence of the islet hormone, the ACh-evoked ^{45}Ca outflow was significantly attenuated, suggesting that Ins may inhibit the mechanism of calcium extrusion, which in turn would result in an elevation in cytoplasmic free calcium levels (Singh, 1985). Certainly, more recent studies employing the fluorescent probe fura-2 to load pancreatic acinar cells (Juma *et al.*, 1997a; Singh *et al.* 1998) demonstrated that, compared to much larger responses to either ACh or CCK-8, Ins alone elicits a small increase in $[Ca^{2+}]_i$. On the other hand, combining Ins with either ACh or CCK-8 resulted in a potentiation of the classical secretagogues-induced rise in $[Ca^{2+}]_i$ level. In the latter studies, genistein (an inhibitor of tyrosine kinase) was also employed. Simultaneous application of genistein with Ins and either ACh or CCK-8 resulted in an attenuation of the enhanced rise in $[Ca^{2+}]_i$ as well as in the augmented secretory response (Juma *et al.*, 1997a; Singh *et al.*, 1998). This suggests that activation of the intracellular mediator tyrosine kinase (Hunter and Cooper, 1985; Juma *et al.*, 1997a) is involved in the observed potentiating effect, and it is also in agreement with the modulation of calcium entry into pancreatic acinar cells by tyrosine kinase (Yule *et al.*, 1994). However, there are some suggestions that Ins may have an additional mode of action to enhance the effects of ACh and CCK-8. First, in the above studies, genistein failed to completely abolish the potentiation. Second, it is well known that two secretagogues acting via the same intracellular mediator do not potentiate one another (Jensen and Gardner 1981; Singh *et al.*, 1992). This is further supported by the findings of Singh (1983, 1985) in mouse pancreatic segments. In the latter, ACh decreased and Ins increased cAMP levels, but their combination produced a sustained elevation in cAMP concentration. In addition, in this and previous studies it has been shown that derivatives of cAMP potentiate cholinergic-evoked pancreatic enzyme secretion in a number of species including the rat (Singh, 1979), mouse (Singh, 1985), and guinea pig (Gardner and Jackson, 1977). It thus appears that several intracellular mediators, including calcium, cAMP and tyrosine kinase are associated with the interaction of Ins with the classical secretagogues.

5.3. Somatostatin (Soma)

The physiological role of endogenous Soma in the regulation of exocrine pancreatic function is not fully understood. Although the effects of exogenous Soma on exocrine pancreatic secretion have been investigated in multiple studies, major discrepancies between observations are apparent, most likely due to different experimental conditions, species, molecular forms of Soma, or dosages used. The *in vivo* studies on the effects of Soma on the basal secretion of water and bicarbonate have yielded conflicting results not only between but also within species. In rats, continuous intravenous infusion of Soma had little or no effect in the anaesthetized animal. However, in the conscious animal, it caused marked inhibition of volume and bicarbonate outputs (Chariot *et al.*, 1978; Singh, 1986). On the other hand, bolus injection of high doses of Soma was reported to induce a transient increase rather than decrease in pancreatic volume secretion in rats (Chariot *et al.*, 1978). In dogs, infusion of Soma strongly inhibited unstimulated water and bicarbonate secretion (Susini *et al.*, 1978). In humans, several investigators observed an inhibition of unstimulated pancreatic secretion by pharmacological doses of Soma, although the degree of inhibition varied greatly. When a physiological dose of Soma-28 was used, no inhibition of human pancreatic volume and bicarbonate secretion could be observed (Layer *et al.*, 1991). The analogue octreotide showed no effect on human basal water and bicarbonate outputs (Kohler *et al.*, 1986). The inhibitory effects of Soma on unstimulated enzyme secretion are more pronounced and less controversial: significant reduction in pancreatic enzyme secretion has been observed in both anaesthetized and conscious rats (Chariot *et al.*, 1978; Singh, 1986), dogs (Susini *et al.*, 1978) and humans in response to both Soma-14 and synthetic analogues (Kohler *et al.*, 1986; Layer *et al.*, 1987).

The available data seem to indicate that exogenous Soma administration inhibits basal pancreatic enzyme output, whereas its effects on unstimulated water and bicarbonate secretion are less well defined. No uniform effects of Soma on secretin-stimulated water and bicarbonate secretion have been established so far even within the same species: in rats, cats and rabbits, Soma had no effect (Chariot *et al.*, 1978; Singh, 1986) whereas in man, both no effect and moderate, dose-dependent inhibition in response to intravenous Soma have been reported (Dollinger *et al.*, 1976; Gullo *et al.*, 1987; Gyr *et al.*, 1987). In dogs, Soma has shown to cause a significant dose-dependent inhibition (Susini *et al.*, 1978; Konturek *et al.*, 1981a). In the latter species, the suppression of secretin-stimulated bicarbonate secretion showed the characteristics of competitive inhibition, which suggests a competition for common receptor sites between Soma and secretin (Konturek *et al.*, 1981a). In fact, it has been postulated, but not proven, that when given alone, Soma may be a weak agonist by occupying secretin binding sites, but when these are administered together, Soma inhibits the effects of secretin. This may be related to the fact that there is partial similarity in the amino acid sequences of Soma and secretin, and would explain the above mentioned stimulation of basal pancreatic volume secretion in rats after injection of a bolus of Soma (Chariot *et al.*, 1978). The inhibitory effect of Soma on endogenously stimulated water and bicarbonate secretion appears to be greater than on exogenous stimulation. After duodenal perfusion with acid or feeding a meal in dogs, pancreatic water and bicarbonate secretion is strongly inhibited by Soma whereas no effect has been observed in rats (Boden *et al.*, 1975; Chariot *et al.*, 1978; Konturek *et al.*, 1981a,b, 1985). In man, a significant reduction of water and bicarbonate secretion has been observed (Hanssen *et al.*, 1977).

Concerning the effects of Soma on the stimulated enzyme secretion, the *in vivo* studies show a marked inhibition (50–90%) of exogenously-stimulated enzyme output in all species studied (rats, cats, rabbits, dogs and humans) when Soma-14, -28 or synthetic analogues were used, not only after pharmacological doses but also in response to physiological doses (Boden *et al.*, 1975; Dollinger *et al.*, 1976; Hanssen *et al.*, 1977; Chariot *et al.*, 1978; Susini *et al.*, 1978; Susini *et al.*, 1980; Konturek *et al.*, 1981a,b; Kontureket *al.*,

1985; Singh, 1986; Gyr *et al.*, 1987). Inhibition of enzyme output was independent of the mode of exogenous stimulation, that is, stimulation by CCK, cholinergic agonists, or electrical stimulation of the vagus nerve (Singh and Adeghate, 1998; Holst *et al.*, 1984). Pancreatic enzyme secretion in response to endogenous stimulation by duodenal perfusion with nutrients, oral nutrients or feeding a meal is strongly reduced in dogs (Konturek *et al.*, 1981a,b, 1985) and humans (Johansson *et al.*, 1981) even at physiological doses.

The evidence indicates that even physiological increases in Soma plasma concentrations markedly inhibit stimulated pancreatic enzyme output. This suggests that Soma may be involved in the regulation of the pancreatic secretory response to a meal. Volume and bicarbonate secretion after endogenous stimulation is markedly decreased in response to Soma, but the effect of Soma on exogenously stimulated volume output is weaker and probably observed at pharmacological doses.

Early *in vitro* studies generally failed to show direct inhibitory effects by Soma on the basal or stimulated exocrine pancreatic secretion, either on isolated perfused or on isolated lobules or acinar preparations, independent of species or mode of stimulation—hormonal, cholinergic, neural—(Robberecht *et al.*, 1975; Singh, 1986; Layer *et al.*, 1989). Thus, in an isolated perfused stomach-pancreas preparation, high concentrations of Soma had no effect on exocrine pancreatic function, while they at the same time strongly inhibited gastric function. Receptors for Soma have been demonstrated on both intact pancreatic acinar cells and subcellular organelles of guinea pig, mouse and rat pancreas (Esteve *et al.*, 1984; Sakamoto *et al.*, 1984; Srikant and Patel, 1985). Given that the *in vivo* inhibitory effect of Soma on secretion could not be demonstrated *in vitro*, it was suggested that the above receptors may serve functions other than direct inhibition of exocrine pancreatic secretion. This idea was further supported by the findings of Esteve *et al.* (1983), who showed that high concentrations of Soma-28 stimulated, rather than inhibited, amylase release from guinea pig acini, this effect being mediated by the CCK receptor. All these observations seemed to indicate that Soma could act via indirect mechanisms to inhibit pancreatic enzyme secretion, as shown in the *in vivo* studies.

More recently, however, a number of reports have appeared supporting the concept that Soma acts directly on acinar cells. In the isolated rat pancreatic segments, Soma has been shown either to attenuate or stimulate amylase secretion depending on the concentration: high concentrations have inhibitory effects whereas physiological doses have stimulatory effects (Adeghate *et al.* 1997). A small but detectable increase in amylase output from dispersed pancreatic acinar cells by intermediate doses of Soma has been also observed (Juma *et al.*, 1997a). In previous studies it has been reported that Soma can elicit differential effects on exocrine pancreatic secretion, comprising of either a stimulatory or an inhibiting response or none depending on whether it is an *in vivo* or an *in vitro* study. The results of Adeghate *et al.* (1997) and Juma *et al.* (1997a) suggest not only that Soma may have direct effects on pancreatic acinar cells, but also that the response depends to a great extent on the concentration of the islet hormone. In addition to the effects of Soma alone, the same group examined the interactions between Soma and the classical pancreatic secretagogues, finding that Soma at low concentrations can potentiate the secretory effects of ACh and CCK-8 in rat pancreatic segments (Juma *et al.*, 1997a; Singh and Adeghate, 1998; Singh *et al.*, 1998). In contrast, there was no enhancement of the secretory response in isolated pancreatic acinar cells, which corroborates the evidence that intact and viable islets are associated with the interactions. Furthermore, Soma inhibited the nerve-mediated secretory response (Singh and Adeghate, 1998).

The site and mechanism of the direct action of Soma in the exocrine pancreas are not well defined. The earliest studies revolved around the effects of the peptide on cAMP-dependent pathways. Certainly, Soma has been shown to inhibit intracellular cAMP production in numerous cell types, including pituitary and gastric cells, and to block the effect of substances known to act though increases in cAMP (Gespach *et al.*, 1980; Batzri, 1981). In the exocrine pancreas, conflicting results have been obtained. Thus, an *in vitro* inhibitory effect of Soma was observed on secretin- and VIP-stimulated cAMP production, but it was not accompanied by an inhibition of secretion (Esteve *et al.*, 1983). In fragments of rat pancreas, Soma stimulated amylase release and evoked an increase in the tissue level of cAMP, which, on the other hand, is not a usual effect of the islet hormone (Robberecht *et al.*, 1975). Another possibility is the regulation of $[Ca^{2+}]_i$ turnover. This seems to occur in rat pancreatic acinar cells, Soma alone could not induce an increase in $[Ca^{2+}]_i$ concentration and did not enhance the ACh-induced rise in calcium compared to the response to ACh alone. In contrast, Soma potentiated the increase in $[Ca^{2+}]_i$ evoked by ACh and CCK-8 (Juma *et al.*, 1997a; Singh and Adeghate, 1998; Singh *et al.*, 1998). In these studies, a parallel potentiatory effect by Soma on the amylase output induced by both secretagogues was evident. When the inhibitor of tyrosine kinase genistein was combined with either ACh or CCK-8, there was a marked attenuation of the enhanced secretory responses. The rise in $[Ca^{2+}]_i$ level evoked by the combined presence of CCK-8 (but not ACh) and Soma was also reduced. These findings suggest that cellular calcium is involved in the interaction between Soma and CCK-8, in close agreement with those reported in other tissues (Juma *et al.*, 1997b; Singh *et al.*, 1998). The concept that the islet peptide may affect the CCK-8 and ACh-induced secretory responses via the tyrosine kinase pathway is supported by the observations of Hierowski *et al.* (1985), who showed that in undifferentiated pancreatic cancer cells, binding of Soma to their membrane receptors activates the dephosphorylation of a membrane protein whose phosphorylation is produced by an epidermal growth factor (EGF) receptor-coupled tyrosine kinase. Furthermore, the interaction of Soma with the classical secretagogues might occur at different stages throughout the stimulus-secretion coupling process, since it has been reported that in the exocrine pancreatic tissue, Soma binding to its receptors in the plasma membrane is inhibited by CCK or cholinergic agonists but also by phorbol esters, which can directly activate protein kinase C (Sakamoto *et al.*, 1984; Fearn and King, 1985). The presence of saturable binding of Soma has also been demonstrated in rat pancreatic cytosol, where the peptide was found to stimulate the activity of a phosphoprotein phosphatase (Reyl-Desmars and Lewin, 1982).

When considering the effects of Soma on the exocrine pancreas, we have to take into account several important points. Major sites of endogenous Soma release are the pancreas itself, namely the D-cells of the islets, whose final product is Soma-14, and the proximal gastrointestinal epithelial cells, which secrete Soma-28 (Susini *et al.*, 1980). Several reports indicate that the increase in plasma Soma-like immunoreactivity in response to a meal (Schusdziarra *et al.*, 1979, 1980; Zyznar *et al.*, 1981) is due exclusively to release of Soma-28 from the gut (Shoelson *et al.*, 1982). The close interaction between the islets and the exocrine tissue suggests that despite low concentration in systemic blood, Soma-14 released from the pancreas might act on the acinar tissue (where high local plasma concentrations could be achieved) via the insuloacinar portal system. Part of the effects of Soma on the exocrine pancreatic function might be exerted in this manner, and most of them are likely direct, since Soma-14 appears to bind more avidly to exocrine pancreas than Soma-28 (Srikant and Patel, 1985). On the other hand, Soma-28 released postprandially from the gut seems to have a systemic action and its effects on the exocrine pancreas may be exerted, at least in part, through simultaneous interaction with several intermediary mechanisms. Most of these support the inhibitory action of Soma observed in the *in vivo* studies, and include: a) Effects on the release of hormonal secretagogues. Somatostatin has been shown to strongly inhibit the release of secretin during duodenal acidification in different animal species (Boden *et al.*, 1975, Hanssen *et al.*, 1977). An inhibitory action of Soma on CCK release has also been reported (Schleggel *et al.*, 1977). b) Changes in the pancreatic circulation. Certainly, Soma appears to decrease splanchnic flow in general, diminishing especially the blood flow to the stomach and pancreas (Thulim *et al.*, 1979). c) Interaction with neural mechanisms. There is evidence that Soma decreases the release of ACh from both extrinsic (vagal) and intrinsic autonomic efferent nerve fibers (Chariot *et al.*, 1978; Kowal *et al.*, 1989). Also, pancreatic VIP release induced by vagal stimulation is inhibited by Soma. Other studies have shown that Soma can enhance the firing rate of pancreatic cholinergic nerves (Nakabayashi *et al.*, 1989) and that blockade of axonal neural transmission by tetrodotoxin completely abolishes the effect of Soma in the rat isolated perfused pancreas (Mulvihill *et al.*, 1990). Li and Owyang (1992) could not find an effect of Soma on pancreatic enzyme secretion evoked by direct vagal stimulation in rats, but found that the peptide blocked pancreatic secretion stimulated by 2-deoxy-D-glucose, a central vagal stimulant, concluding that Soma acts preferentially at proximal sites along the vagal pathway. d) The last possibility is that Soma affects the release of a secondary factor produced by the pancreas itself. Thus, Soma has been demonstrated to inhibit the release of pancreatic polypeptide (PP) (Taylor, 1989) which in turn inhibits pancreatic exocrine secretion (Greenberg *et al.*, 1978; Louie *et al.* 1985). On the other hand, Soma inhibits the secretory activity of α and β cells (Owyang, 1993) and high affinity of radiolabelled Soma to specific islet receptors has been found (Garry *et al.*, 1989). The closer association of α and D cells suggest that Soma may regulate α cell function through local mechanisms, and would act on the β cells through a systemic action. Thus, binding studies indicate that β cells predominantly express a Soma-28 preferring receptor. In addition, it was demonstrated that in freshly isolated islets, insulin release was blocked by Soma-28 but not Soma-14, whereas the latter did inhibit Glu secretion (Thermos *et al.*, 1990). The inhibition of Ins and Glu release, which have potentiatory effects on the secretion elicited by classical secretagogues (Juma *et al.*, 1997a; Singh and Adeghate, 1998; Singh *et al.*, 1998), explain some of the effects of Soma on the exocrine pancreas.

6. DIABETES AND MORPHOLOGY OF NORMAL AND DIABETIC PANCREAS

Diabetes mellitus is a chronic metabolic disease which is associated with hyperglycaemia. This hyperglycaemia is a result of defective insulin secretion and or function. The prolonged hyperglycaemia accompanying diabetes causes tissue damage, which results in degenerative complications in many organs including the kidney, heart, muscles, eye and many other organs. These degenerative complications arise because of the damage inflicted on the blood vessels. When blood vessels are unable to perform their regular function, the tissue they serve becomes impaired. In one study, it was shown that diabetic retinopathy is the commonest cause of blindness in industrial countries (Kahn and Hiller, 1974).

There is a selective damage to the β cells in streptozotocin(STZ)-induced experimental diabetes. Macroscopically the pancreas appears normal. In order to be able to detect the morphological abnormality in diabetic pancreas, immunohistochemistry of the islet cells is needed. This can be done using a variety of techniques. In our laboratory, the two-step sandwich method is used with reproducible results (Adeghate and Donáth, 1991; Adeghate, 1997). Our study showed the following results:

- β cells: Insulin(Ins)-positive cells are the most numerous cell type in the normal pancreas. They account for about 60–70% of the total cell population in a given islet of Langerhans (Adeghate and Donáth, 1990a; Adeghate, 1997). In experimental diabetes, however, this number decreases significantly due to selective damage of pancreatic β cells by STZ.
- α cells: Glucagon(Glu)-containing cells constitute about 28% of the total number of endocrine cells in a normal pancreatic islet (Adeghate and Donáth, 1990; Adeghate, 1997). They are located in the periphery of the islets in normal animals. However, in STZ-diabetic animals many Glu-positive cells are seen scattered within the central portion of the islets as well. The percentage of Glu-positive cells increases significantly in STZ-diabetic animals when compared to controls.
- δ cells: Somatostatin(Soma)-immunopositive cells are found mainly on the outer part of the islets of normal pancreas with a percentage distribution of about 6% (Adeghate and Donáth, 1990; Adeghate, 1997). There is no significant increase in the number of Soma-positive cells in diabetic rats. Results from our morphological studies show marked changes in the pattern of distribution of Ins-, Glu-, and Soma-positive cells in pancreatic tissue of diabetic animals compared to that of normal controls. The distribution of pancreatic islet cells in the islets follows a specific pattern which is disrupted in diabetes. Ins-positive cells become dispersed all over the islets of Langerhans while Glu-containing cells occupy areas in the central portion of the pancreas, a feature not seen in normal pancreas. The number of Ins-positive cells decreases in diabetic rats compared with control animals. There is controversy regarding the changes in the number of α cells in STZ diabetes. It has been demonstrated that the number of pancreatic α cells does not change significantly in STZ-treated rats (Papaccio *et al.*, 1992). Pons and Aoki (1995), however, reported a remarkable increase in the number of α cells in diabetic rats.
- Neurotransmitters: There seem to have no changes in the distribution of acetylcholinesterase-positive and CCK-immunopositive nerves in normal and diabetic pancreas. These nerves are distributed in the interacinar, periacinar, intrainsular and in periductal regions. CCK-reactive nerves are commonly varicose nerves located in the wall of blood vessels.

7. STIMULUS-SECRETION COUPLING IN DIABETIC PANCREAS

In pancreatic acini and segments of diabetic rats, either Insulin (Ins), Glucagon (Glu) or somatostatin (Soma) elicited a small decrease in amylase secretion (Singh *et al.*, 1998). Moreover, amylase content of diabetic rats is markedly reduced after induction of diabetes with streptozotocin and *in vivo* administration of Ins can result in the increased synthesis of pancreatic amylase (Youngs, 1972; Adler and Kern, 1975). Exogenous Ins is believed to act as a trophic factor to maintain the tissue level of amylase in acini of diabetic animals (Malaisse-Lagae *et al.*, 1975). Similarly, in pancreatic segments of diabetic animals either Ins, Glu or Soma failed to potentiate the secretory effects of either ACh, CCK-8 or nerve-stimulation (Singh *et al.*, 1998). In addition, the islet hormones had no significant effect on $[Ca^{2+}]_i$ concentration in fura-2 loaded pancreatic acini of diabetic pancreas and, moreover, they failed to potentiate the increase in cellular calcium elicited by either ACh (Juma *et al.*, 1997a) or CCK-8 (Juma *et al.*, 1997b; Singh *et al.*, 1998). The main conclusions of our work are summarized in the schematic model in Figure 1. Taken

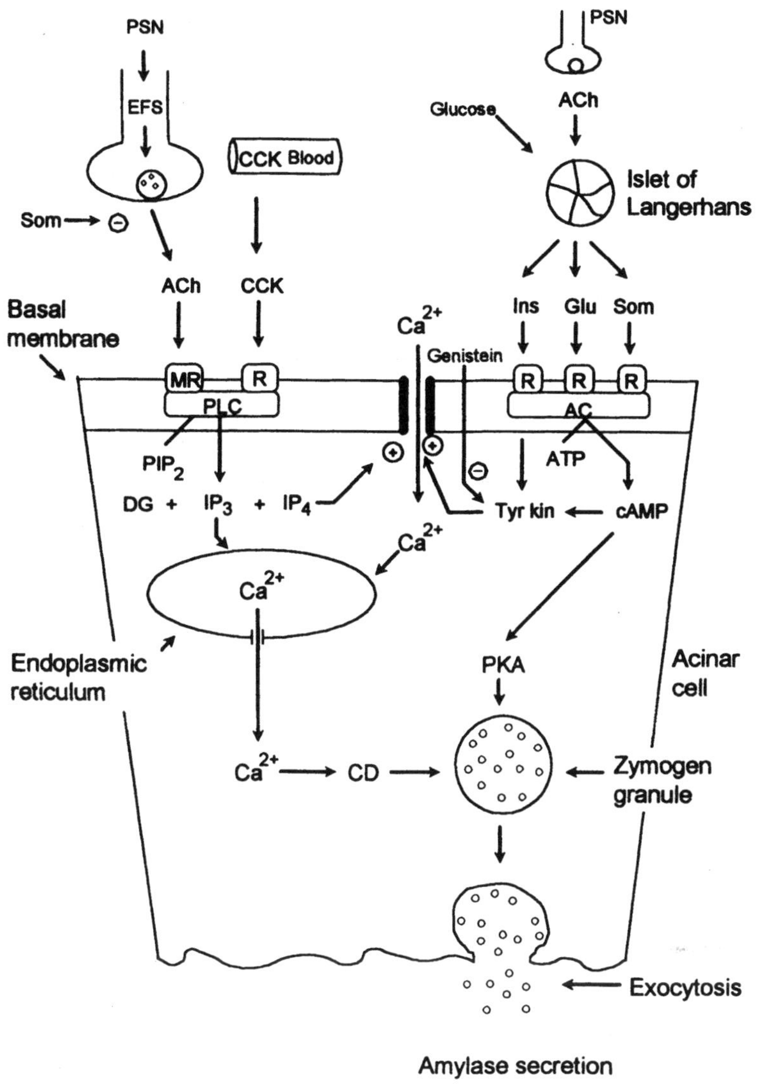

Figure 1. A schematic model of release of endogenous islet hormones insulin (Ins), glucagon (Glu) and somatostatin (Som) and stimulus-secretion coupling in exocrine pancreatic acinar cell during interaction of the islet hormones and nerve-mediated (EFS) and secretagogue-evoked amylase secretion. In the model it is postulated that both acetylcholine (ACh) and cholecystokinin (CCK) can mobilize cellular calcium from its store and its influx into acinar cell. The calcium stimulates calmodulin (CD) which, in turn, phosphorylates zymogen granular membrane to enhance enzyme secretion. On the other hand, the islet hormones activate their respective receptors leading to an elevation in cAMP and tyrosine kinase (Tyr kin). cAMP stimulates protein kinase A (PKA) which in turn, like calmodulin, phosphorylates the zymogen granular membrane proteins resulting in exocytosis and enzyme secretion. Tyrosine kinase stimulates calcium influx. Genistein inhibits tyrosine kinase leading to a decrease in calcium entry into acinar cell. When the animal is rendered diabetic, the calcium influx induced by tyrosine kinase (as well as calcium release from the endoplasmic reticulum) seems to be impaired. Thus, during reduced cellular calcium and cAMP as in diabetes, the islet hormones would have no potentiating effect on either nerve- or secretagogue-evoked amylase secretion. Finally, somatostatin (Som) seems to control acetylcholine release elicited by electrical field stimulation (EFS). (PNS = parasympathetic nerves; MR = muscarinic receptor; PLC = phospholipase C; PIP_2 = phosphatidylinositol bisphosphate; DG = diacylglycerol; IP_3 = inositol trisphosphate; IP_4 = inositol tetrakisphosphate; AC = adenilate cyclase; ATP = adenosine trisphosphate).

together, these results suggest that the potentiatory action of the islet hormones on nerve-mediated and secretagogue-evoked secretory responses seem to involve the second cellular messenger calcium as well as other mediators such as cAMP and tyrosine kinase. However, during diabetes there seems to be a derangement of the calcium signalling mechanism, either its release from intracellular stores or its entry into acinar cells, and this in turn is associated with low secretory capacity following interaction of the islet hormones with the classical secretagogues. This is an interesting area which warrants further investigation.

8. CONCLUDING REMARKS

Both the exocrine and endocrine portions of the pancreas respond to a multiplicity of extracellular signals and a variety of nutrients. The extracellular signals exert their effects via different signal transduction pathways to elicit secretion. Earlier morphological studies of the pancreas have demonstrated the dispersal of the islets throughout the exocrine pancreas, lying in close contact with acinar tissues since no rigid connective tissue capsule surrounds the islets. Morphological evidence coupled with physiological studies have indicated that the islet hormones can regulate the exocrine pancreas. Much of this regulation is locally mediated via the islet-acinar-portal vascular system. There is much evidence for a role of insulin on the regulation of the biosynthesis of pancreatic digestive enzymes and the potentiation of nerve-mediated and secretagogue-evoked secretory responses. Evidence is also available for a role of both glucagon and somatostatin in the regulation of exocrine function. The precise mechanism of the potentiation between the islet hormones and secretagogues is not yet fully established. However, studies in our laboratory have demonstrated that this complex interaction may involve several intracellular mediators including Ca^{2+}, cyclic AMP and tyrosine kinase working in tandem to elicit enzyme secretion. However, when the animals are rendered diabetic the islet hormones fail to interact with nerve stimulation and the classical pancreatic secretagogues resulting in pancreatic insufficiency. The precise mechanism of this reduced digestive enzyme secretion during diabetes is still unknown. Future directions of work are focused on the cellular mechanismms underlying pancreatic insufficiency during diabetes with the aim of developing new drugs to alleviate the problem.

ACKNOWLEDGMENTS

This work was supported by grants from the British Council, the Wellcome Trust, and the Spanish Ministry of Education and Culture.

REFERENCES

Adeghate, E., and Donáth, T. (1990). Distribution of neuropeptide Y and vasoactive intestinal polypeptide immunoreactive nerves in normal and transplanted pancreatic tissue. *Peptides* **11**, 1087–1092.

Adeghate, E., and Donáth, T. (1991). Morphometric and immunohistochemical study on the endocrine cells of pancreatic tissue transplants. *Exp. Clin. Endocrinol.* **98**, 193–199.

Adeghate, E. (1997). Immunohistochemical localisation of pancreatic hormones, neuropeptides and cytoskeletal proteins in the pancreas of the camel. *J. Morphology* **231**, 185–193.

Adeghate, E., Donáth, T., Pallot, D. J., and Parvez, H. (1995). Vasoactive intestinal polypeptidergic innervation of the canine pancreatic acinar cells. *Biogenic Amines* **11**, 255–260.

Adeghate, E., Ember, Zs., Donáth, T., Pallot, D. J., and Singh, J. (1996). mmunohistochemical identification and effects of pancreastatin, leucine-enkephalin, and galanin in the porcine pancreas. *Peptides* **17**, 503–509.

Adeghate, E., Singh, J., Pallot, D. J., and Juma, L. M. (1997). Immunohistochemical studies and secretory effects of neuropeptides and regulatory peptides in the isolated rat pancreas. *Biogenic Amines* **13**, 441–459.

Adler, G., and Kern, H. F. (1975). Regulation of exocrine pancreatic secretory process by insulin *in vivo*. *Horm. Metab. Res.* **7**, 290–296.

Adler, G. (1977). Effect of glucagon on the secretory process in the rat exocrine pancreas. *Cell Tissue Res.* **182**, 193–204.

Ahren, B., Ericson, L. E., Lundquist, Y., and Sundler, F. (1981). Adrenergic inervation of pancreatic islets and modulation of insulin secretion by the sympatho-adrenal system. *Cell Tissue Res.* **216**, 15–30.

Barlow, T. E., Greenwell, J. R., Harper, A. A., and Scratcherd, T. (1974). The influence of the splanchnic nerves on the external secretion, blood flow and electrical conductance of the cat pancreas. *J. Physiol.* **236**, 421–433.

Batzri, S. (1981). Direct action of somatostatin on dispersed mucosal cells from guinea pig stomach. *Biochim. Biophys. Acta* **677**, 521–524.

Beglinger, C., Grossman, M. I., and Solomon, T. E. (1984). Interaction between stimulants of exocrine pancreatic secretion in dog. *Am. J. Physiol.* **246**, G173–179.

Berridge, M., and Irvine, R. F. (1989). Inositol phosphate and cell signalling. *Nature* **341**, 197–204.

Berridge, M. J. (1993). Inositol trisphosphate and calcium signalling. *Nature* **361**, 315–325.

Boden, G., Sivitz, M. C., and Owen, O. E. (1975). Somatostatin suppresses secretin and pancreatic exocrine secretion. *Science* **190**, 163–165.

Bommelaer, G., Rozental, G., Bernier, C., Vaysse, N., and Ribet, A. (1981). Action of secretagogues on amylase release from the dog pancreatic acini. *Digestion* **21**, 248–254.

Brown, J. C., Harper, A. A., and Scratcherd, T. (1967). Potentiaton of secretin stimulation of the pancreas. *J. Physiol.* **190**, 519–530.

Burnham, D. B., and Williams, J. A. (1982). Effects of carbachol, CCK, and insulin on protein phosphorilation in isolated pancreatic acini. *J. Biol. Chem.* **257**, 10523–10528.

Cantor, P., Olsen, O., Gertz, B. J., Gjoerup, Y., and Worning, H. (1991). Inhibition of CCK-stimulated pancreaticobiliary output in man by the CCK receptor antagonist MK-329. *Scand. J. Gastroenterol.* **26**, 627–637.

Chariot, J., Roze, C., Vaille, C., and Debray, C. (1978). Effects of somatostatin on the external secretion of the pancreas of the rat. *Gastroenterology* **75**, 832–837.

Chey, W. Y., Kim, M. S., Lee, K. Y., and Chang, T-M. (1979). Effect of rabbit antisecretin serum on postprandial pancreatic secretion in dogs. *Gastroenterology* **77**, 1268–1275.

Coupland, R. E. (1958). The innervation of pancreas of the rat, cat and rabbit as revealed by the cholinesterase technique. *J. Anat.* **92**, 143–149.

Couture, Y., Dunnigan, J., and Morisset, J. (1972). Stimulation of pancreatic amylase secretion and protein synthesis by insulin. *Scand. J. Gastroenterol.* **7**, 257–263.

Defilippi, C., Solomon, T. E., and Valenzuela, J. E. (1982). Pancreatic secretory response to sham-feeding in humans. *Digestion* **23**, 217–223.

Deschodt-Lanckman, M., Robberecht, P., DeNut, P., Labbrie, F., and Christophe, J. (1975). *In vitro* interactions of gastrointestinal hormones on cyclic AMP levels and amylase output. *Gastroenterology* **68**, 318–325.

DiMagno, E. P., Go, V. L. W., and Summerskill, W. H. J. (1973). Intraluminal and postabsorptive affects of amino acids on pancreatic enzyme secretion. *J. Lab. Clin. Med.* **82**, 241–248.

Dollinger, H. C., Raptis, S., and Pfeiffer, E. F. (1976). Effects of somatostatin on exocrine and endocrine pancreatic function stimulated by intestinal hormones in man. *Horm. Metab. Res.* **8**, 74–78.

Dyck, W. P., Exter, E. C., Lasater, J. M., and Hightower, N. C. (1970). Influence of glucagon on pancreatic exocrine secretion in man. *Gastroenterology* **58**, 523–539.

Eisenberg, M. M., and Orahood, R. C. (1971). Vagal stimulation of the exocrine pancreas. *Ann. Surg.* **173**, 462–466.

Esteve, J. P., Susini, C., Vaysse, N., Antoniotti, H., Wunsch, E., Berthon, G., and Ribet, A. (1984). Binding of somatostatin to pancreatic acinar cells. *Am. J. Physiol.* **247**, G62–69.

Esteve, J. P., Vaysse, N., Susini, C., Kunsch, J. M., Fourmy, D., Pradayrol, L., Wunsch, E., Moroder, L., and Ribet, A. (1983). Bimodal regulation of pancreatic exocrine function *in vitro* by somatostatin 28. *Am. J. Physiol.* **245**, G208–216.

Farese, R. V., Larson, R. E., and Sabir, M. A. (1981). Insulin and its secretagogues activates calcium dependent phosphatydilinositol breakdown and amylase secretion in rat pancreas *in vitro*. *Diabetes* **30**, 396–401.

Fearn, J. C., and King, A. C. (1985). EGF receptor affinity is regulated by intracellular calcium and protein kinase C. *Cell* **40**, 991–1000.

Fontana, G., Costa, P. L., Tassari, R., and Labo, G. (1975). Effect of glucagon on pure human exocrine pancreatic secretion. *Am. J. Gastroenterol.* **63**, 490–494.

Fried, M., Erlacher, U., Schwizer, W., Lochner, C., Koerler, J., Beglinger, C., Jansen, J. B., Lamers, C. B., Harder, F., Bischof-Delaloye, A., Stalder, G. A., and Rovati, L. (1991). Role of CCK in the regulation of gastric emptying and pancreatic enzyme secretion in humans. *Gastroenterology* **101**, 503–511.

Gardner, J. D., and Jackson, M. J. (1977). Regulation of amylase release from dispersed pancreatic acinar cells. *J. Physiol.* **270**, 439–454.

Garry, D. J., Garry, M. G., Williams, J. A., Mahoney, W. C., and Sorenson, R. L. (1989). Effects of islet hormones on amylase secretion and localization of somatostatin binding sites. *Am. J. Physiol.* **256**, G897–904.

Gespach, C., Dupont, C., Bataille, D., and Rosselin, G. (1980). Selective inhibition by somatostatin of cAMP production in rat gastric glands: demonstration of a direct effect on the parietal cell function. *FEBS Lett.* **114**, 247–252.

Godfrey, D. A., and Matschinsky, M. (1975). Enzymes of the cholinergic system in islets of Langerhans. *J. Histochem. Cytochem.* **23**, 645–651.

Gorelick, F. S., and Jamieson, J. D. (1987). Structure-function relationship of the pancreas. *In* "Physiology of the Gastrointestinal Tract" (L. R. Johnson, Ed.), pp. 1089–1108. Raven Press, New York.

Greenberg, G. R., Domschke, S., Domschke, W., Rosch, W., and Bloom, S. R. (1979). Effect of low dose secretin and caerulein on pure pancreatic bicarbonate secretion and plasma secretin in man. *Acta Hepato-Gastroenterol.* **26**, 478–481.

Greenberg, G. R., McCloy, R. F., Adrian, T. E., Chadwick, V. S., Baron, J. H., and Bloom, S. R. (1978). Inhibition of pancreas and gallbladder by PP. *Lancet* **2**, 1280–1282.

Gullo, L., Priori, P., Scarpignato, C., Baldoni, F., Mattioli, G., and Barbara, L. (1987). Effect of somatostatin-14 on pure human pancreatic secretion. *Dig. Dis. Sci.* **32**, 1065–1070.

Gyr, K., Beglinger, C., Köhler, E., Trautzl, U., Keller, U., and Bloom, S. R. (1987). Circulating somatostatin, a physiological regulator of pancreatic function? *J. Clin. Invest.* **79**, 1595–1600.

Hanssen, L. E., Hanssen, F., and Myren, J. (1977). Inhibition of secretin release and pancreatic bicarbonate secretion by somatostatin infusion in man. *Scand. J. Gastroenterol.* **12**, 391–394.

Hierowski, M. T., Liebow, C., Sapin, K., and Schally, A. V. (1985). Stimulation by somatostatin of dephosphorilation of membrane proteins in pancreatic cancer MIA PaCa-2 cell line. *FEBS Lett.* **179**, 252–256.

Holst, J. J. (1993). Neural regulation of pancreatic exocrine function. *In* "The Exocrine Pancreas: Biology, Pathobiology and Diseases" (V. L. W. Go, F. P. Brooks, E. P. DiMagno, J. D. Gardner, E. Lebenthal, and G. A. Scheele, Eds.), pp. 381–402. Raven Press, New York.

Holst, J. J., Fahrenkrug, J., Knuhtsen, S., Jensen, S. L., Seier-Poulsen, S., and Nielsen, O. V. (1984). VIP in the pig pancreas. Role of VIPergic nerves in the control of fluid and bicarbonate secretion. *Regul. Pept.* **8**, 245–259.

Holst, J. J., Schaffalitzky de Muckadell, O. B., and Fahrenkrug, J. (1979). Nervous control of pancreatic exocrine secretion in pigs. *Acta Physiol. Scand.* **105**, 33–51.

Hunter, T., and Cooper, J. A. (1985). Protein tyrosine kinase. *Annu. Rev. Physiol.* **54**, 897–93.

Iwatsuki, K., Ono, H., and Hashimoto, K. (1976). Effects of glucagon on pancreatic secretion in the dog. *Clin. Exp. Pharmacol. Physiol.* **3**, 59–65.

Jensen, R. T., and Gardner, J. D. (1978). Cyclic nucleotide-dependent protein kinase activity in acinar cells from guinea pig pancreas. *Gastroenterology* **75**, 806–817.

Jensen, R. T., and Gardner, J. D. (1981). Identification and characterization of receptors for secretagogues on pancreatic acinar cells. *Fed. Proc.* **40**, 2486–2496.

Johansson, C., Kollberg, B., Effendic, S., and Uvnas-Wallensten, K. (1981). Effects of graded doses of somatostatin on gallbladder emptying and pancreatic enzyme output after oral glucose in man. *Digestion* **22**, 24–31.

Juma, L. M., Singh, J., Pallot, D. J., Salido, G. M., and Adeghate, E. (1997a). Interactions of islet hormones with acetylcholine in the isolated rat pancreas. *Peptides* **18** (9) 1415–1422.

Juma, L. M. O., Singh, J., Salido, G. M., Pariente, J. A., Tapia, J. A., and Adeghate, E. (1997b). Mechanism of interaction between islet hormones and cholecystokinin-octapeptide in isolated rat pancreatic segments and acinar cells. *J. Physiol.* **501P**, 156–157P.

Kahn, H. A., and Hiller, R. (1974). Blindness caused by diabetic retinopathy. *Am. J. Ophtalmol.* **78**, 58–65.

Kaminski, D. L., Ruwart, M. J., and Willman, V. L. (1975). The effect of electrical vagal stimulation on canine pancreatic exocrine function. *Surgery* **77**, 545–552.

Kanno, T., and Saito, A. (1976). The potentiating influence of insulin on pancreozymin-induced hyperpolarization and amylase release in the pancreatic acinar cell. *J. Physiol. (Lond)* **261**, 505–521.

Kanno, T., Ueda, N., and Saito, A. (1976). Insulo-acinar axis: a possible role of insulin potentiating the effects of pancreozymin in the pancreatic acinar cell. *In* "Endocrine Gut and Pancreas" (T. Fujita, Ed.), pp. 335–345. Elsevier, Amsterdam.

Kim, M. S., Lee, K. Y., and Chey, W. Y. (1979). Plasma secretin concentrations in fasting and postprandial states in dog. *Am. J. Physiol.* **236**, E539–544.

Köhler, E., Beglinger, C., Dettwiler, S., Whitehouse, Y., and Gyr, K. (1986). Effect of a new somatostatin analogue on pancreatic function in healthy volunteers. *Pancreas* **1**, 154–159.

Konturek, S. J., Cieszkowski, M., Bilski, J., Konturek, J., Bielanski, W., and Schally, A. V. (1985). Effects of cyclic hexapeptide analogue of somatostatin on pancreatic secretion in dogs. *Proc. Soc. Exp. Biol. Med.* **178**, 68–72.

Konturek, S. J., Pucher, A., and Radecki, T. (1976). Comparison of VIP and secretin in stimulation of pancreatic secretion. *J. Physiol.* **255**, 497–509.

Konturek, S. J., Tasler, J., Cieszkowski, M., Jaworek, J., Arimura, A., and Schally, A. V. (1981a). Studies on the inhibition of pancreatic secretion by luminal somatostatin. *Am. J. Physiol.* **241**, G109–115.

Konturek, S. J., Tasler, J., Jaworec, J., Pawlik, W., Walus, K. M., Schusdziarra, V., Meyers, C. A., Coy, D. H., and Schally, A. V. (1981b). Gastrointestinal secretory, motor, circulatory and metabolic effects of prosomatostatin. *Proc. Natl. Acad. Sci. USA* **78**, 1967–1971.

Konturek, S. J., Tasler, V., and Obtulowicz, W. (1974). Characteristics of inhibition of pancreatic secretion by glucagon. *Digestion* **10**, 138–149.

Kowal, V., Wiley, J. W., and Owyang, G. (1989). Differential action of somatostatin on peptide-induced release of acetylcholine. *Am. J. Physiol.* **257**, G221–225.

Larose, L., Poirier, G. G., Dumont, Y., Fregeau, C., Blanchard, L., and Morisset, J. (1983). Modulation of rat pancreatic amylase secretion and muscarinic receptor populations by chronic bethanechol treatment. *Eur. J. Pharmacol.* **95**, 215–223.

Layer, P., Ohe, M., Müller, M. K., and Beglinger, C. (1991). Effects of somatostatin on the exocrine pancreas. *Scand. J. Gastroenterol.* **26**, 129–136.

Layer, P., Ohe, M., Wollny, C., Beglinger, C., and Goebell, H. (1989). Human interdigestive pancreatic secretion: response to physiological doses of somatostatin-28. *Digestion* **43**, 157.

Lee, K. Y., Lu, Z., Ren, X. S., Chang, T-M., and Chey, W. Y. (1990b). An important role of endogenous insulin on exocrine pancreatic secretion in rats. *Am. J. Physiol.* **258**, G268–274.

Li, Y., and Owyang, C. (1992). Somatostatin inhibits pancreatic enzyme secretion by a central vagal site of action. *Gastroenterology* **102**, A276.

Linguard, J. M., and Young, J. A. (1983). α-adrenergic control of exocrine secretion by perfused rat pancreas *in vitro*. *Am. J. Physiol.* **245**, G690–696.

Long, B. W., and Gardner, J. D. (1977). Effects of CCK on adenylte cyclase activity in dispersed pancreatic acinar cells. *Gastroenterology* **73**, 1008–1014.

Louie, D. S., Williams, J. A., and Owyang, C. (1985). Action of PP on rat pancreatic secretion *in vivo* and *in vitro*. *Am. J. Physiol.* **249**, G489–495.

Makhlouf, F. M., Yau, W. M., Zfas, A. M., Said, S. I., and Bodanszky, M. (1978). Comparative effects of synthetic and natural VIP on pancreatic and biliary secretion and on glucose and insulin blood levels in the dog. *Scand. J. Gastroenterol.* **13**, 759–765.

Malaisse-Lagae, F., Ravazzola, M., Robberecht, P., Vandermeers, A., Malaisse, W. J., and Orci, L. (1975). Exocrine pancreas: Evidence for topographic partition of secretory function. *Science* **190**, 795–797.

Mossner, J., Logsdon, C. D., Goldfine, I. D., and Williams, J. A. (1984). Regulation of pancreatic acinar cell insulin receptors by insulin. *Am. J. Physiol.* **247**, G155–160.

Mulvihill, S. J., Bunnett, N. W., Goto, Y., and Debas, H. T. (1990). Somatostatin inhibits pancreatic exocrine secretion via a neural mechanism. *Metabolism* **39**, 143–148.

Murphy, J. J., and McGeeney, K. F. (1974). The effects of glucagon on the exocrine secretion of the perfused canine pancreas. *Ir. J. Med. Sci.* **143**, 37–41.

Nakabayashi, H., Niijima, A., Kurat, Y., Liang, Z-Y., Usukura, N., and Takeada, R. (1989). Pancreatic vagal nerve is repetitive to somatostatin in rats. *Am. J. Physiol.* **253**, R200–203.

Nakajima, S. D., and Magee, D. F. (1970). Inhibition of exocrine pancreatic secretion by glucagon and D-glucose given intravenously. *Can. J. Physiol. Pharmacol.* **48**, 299–305.

Nishizuka, Y. (1988). The molecular heterogenity of protein knase C and its implications for cellular regulation. *Nature* **334**, 661–665.

Owyang, C. (1993). Endocrine changes in pancreatic insufficiency. *In* "The Exocrine Pancreas: Biology, Pathobiology and Diseases" (V. L. W. Go, F. P. Brooks, E. P. DiMagno, J. D. Gardner, E. Lebenthal, and G. A. Scheele, Eds.), pp. 803–813. Raven Press, New York.

Pandol, S. J., Sutliff, V. E., Jones, S. W., Charlton, C. G., O'Donohue. T. L., Gardner, J. D., Jensen, R. T. (1983). Action of glucagon on pancreatic acini: Due to contamination by previously undescribed secretagogues. *Am. J. Physiol.* **245**, G703–710.

Papaccio, G., and Mezzogiorno, M. (1992). Morphological aspect of glucagon and somatostatin islet cells in diabetic bio breeding and low-dose streptozotocin-treated Wistar rats. *Pancreas* **4**, 289–294.

Pearson, G. T., Davison, J. S., Collins, R. C., and Petersen, O. H. (1981). Control of enzyme secretion by noncholinergic nonadrenergic nerves in guinea pig pancreas. *Nature* **290**, 259–261.

Pearson, G. T., Singh, J., and Petersen, O. H. (1984). Adrenergic nervous control of cAMP-mediated amylase secretion in rat pancreas. *Am. J. Physiol.* **246**, G563–573.

Pearson, G. T., Singh, J., Daoud, M. S., Davison, J. S., and Petersen, O. H. (1982). Control of pancreatic cyclic nucleotide levels and amylase secretion by noncholinergic nonadrenergic nerves. A study employing electrical field stimulation of guinea pig segments. *J. Biol. Chem.* **256**, 11025–11031.

Pons, P., and Aoki, A. (1995). Differential proliferation of somatostatin and glucagon cells in rat pancreatic islets submitted to different stimuli. *Ann. Anat.* **177**, 221–227.

Reyl-Desmars, F., and Lewin, M. J. M. (1982). Evidence for an intracellular somatostatin receptor in pancreas: a comparative study with reference to gastric mucosa. *Biochem. Biophys. Res. Commun.* **109**, 1324–1331.

Richens, C. A. (1945). The innervation of the pancreas. *J. Comp. Neurol.* **82**, 223–236.

Robberecht, P., Deschodt-Lanckman, M., de Neef, P., and Christophe, J. (1975). Effects of somatostatin os exocrine pancreatic function. Interaction with secretin. *Biochem. Biophys. Res. Commun.* **67**, 315–323.

Rosenberg, I. R., Zambrano, V. J., Janowitz, H. D., and Rudick, J. (1976). parasympathetic innervation and pancreatic secretion: the role of gastric antrum. *Ann. Surg.* **183**, 247–251.

Saito, A., Williams, J. A., and Kanno, T. (1980a). Potentiation of cholecystokinin-induced exocrine secretion by both exogenous and endogenous insulin in isolated and perfused rat pancreas. *J. Clin. Invest.* **65**, 777–782.

Saito, A., Williams, J. A., and Kanno, T. (1980b). Potentiation by insulin of the acetylcholine-induced secretory response of the perfused rat pancreas. *Biomed. Res.* **1**, 101–103.

Sakamoto, C., Goldfine, Y. D., and Williams, J. A. (1984). The somatostatin receptor on isolated acinar cell plasma membranes. *J. Biol. Chem.* **259**, 9623–9627.

Sankaran, H., Iwamoto, Y., Korc, M., Williams, J. A., and Goldfine, I. D. (1981). Insulin action in pancreatic acini from streptozotocin-treated rats. II. Binding of ^{125}I insulin to receptors. *Am. J. Physiol.* **240**, G63–68.

Schleggel, V., Raptis, S., Harvey, R. F., Oliver, J. M., and Pfeiffer, E. F. (1977). Inhibition of cholecystokinin-pancreozymin release by somatostatin. *Lancet* **2**, 166–167.

Schulz, I., and Stolze, H. H. (1980). The exocrine pancreas. the roles of secretagogues, cyclic nucleotides and calcium on enzyme secretion. *Annu. Rev. Physiol.* **42**, 127–156.

Schulz, I., Streb, H., Bayerdorffer, E., and Immanura, K. (1986). Intracellular messengers in sitmulus-secretion coupling of pancreatic acinar cells. *J. Cardiovascular Studies* **8**, suppl, 591–596.

Schusdziarra, V., Harris, V., Arimura, A., and Unger, R. H. (1979). Evidence for a role of splanchnic somatostatin in the homeostasis of ingested nutrients. *Endocrinology* **104**, 1705–1708.

Schusdziarra, V., Zyznar, E., and Rouiller, D. (1980). Splanchnic somatostatin: a hormonal regulator of nutrient homeostasis. *Science* **207**, 530–532.

Schwarzendrube, J., Niederau, M., Luthen, R., and Niederau, C. (1991). Effects of CCK receptor blockade on pancreatic and biliary function in healthy volunteers. *Gastroenterology* **100**, 1683–1690.

Shaw, H. M., and Health, T. J. (1973). The effect of glucagon on the formation of pancreatic juice and bile in the rat. *Can. J. Physiol. Pharmacol.* **51**, 1–5.

Shoelson, S., Polonsky, K., Dochsrty, H., Jaspan, J., and Rubenstein, A. (1982). Evidence for the presence for somatostatin 28 in plasma. *Diabetes* **47**, 4–7.

Singer, M. V., Tiscornia, O. M., DeOliverio, J. P. M., Demol, O., Levesque, D., and Sarles, H. (1978). Effect of glucagon on canine exocrine pancreatic secretion stimulated by a test meal. *Can. J. Physiol. Pharmacol.* **56**, 1–6.

Singh, J., and Adeghate, E. (1998). Effects of islet hormones on nerve-mediated and acetylcholine-evoked secretory responses in the isolated pancreas of normal and diabetic rats. *Int. J. Molec. Med.* **1**:627–634.

Singh, J., and Pearson, G. T. (1984). Effects of nerve stimulation on enzyme secretion from the *in vitro* rat pancreas and 3H release after incubation with catecholamines. *Naunyn-Schmiedeberg's Arch. Pharmacol.* **327**, 228–233.

Singh, J. (1983). Effects of amino acids, glucagon, insulin and acetylcholine on cyclic nucleotide metabolism and amylase secretion in isolated mouse pancreatic segments. *Biochem. Pharmacol.* **23**, 2017–2023.

Singh, J. (1985). Mechanism of action of insulin on acetylcholine-evoked amylase secretion in the mouse pancreas. *J. Physiol.* **358**, 469–482.

Singh, J., Juma, L. M. O., Adeghate, E., Pariente, J. A., and Salido, G. M. (1998). Interactive adaptation of the endocrine and exocrine pancreas: Effects of islet hormones and secretagogues. *Adaptation Biol. Med.* **2** (in press).

Singh, J., Lennard, R., Salido, G. M., Wisdom, D., Render, C. L., Pozo, M. J., Pariente, J. A., and Camello, P. J. (1992). Interaction between secretin and cholecystokinin octapeptide in the exocrine rat pancreas *in vivo* and *in vitro*. *Exp. Physiol.* **77**, 191–204.

Singh, M. (1979). Calcium and cyclic nucleotide interaction in secretion of amylase from rat pancreas *in vitro*. *J. Physiol.* **296**, 159–176.

Singh, M. (1980a). Effect of glucagon on digestive enzyme synthesis, transport and secretion in mouse pancreatic acinar cells. *J. Physiol. (Lond.)* **306**, 307–322.

Singh, M. (1980b). Amylase release from dissociated mouse pancreatic acinar cells stimulated by glucagon: effect of membrane stabilizers. *J. Physiol.* **309**, 81–91.

Singh, M. (1986). Effects of somatostatin on amylase secretion from *in vivo* and *in vitro* rat pancreas. *Dig. Dis. Sci.* **31**, 506–512.

Solomon, T. E. (1987). Control of exocrine pancreatic secretion. *In* "Physiology of the Gastrointestinal Tract" (L. R. Johnson, Ed.), pp. 1173–1207. Raven Press, New York.

Sommer, H., and Kasper, H. (1981). Effect of acetylcholine, gastrin and glucagon alone and in combination with secretin and cholecystokinin on the secretion of the isolated perfused rat pancreas. *Res. Exp. Med.* **179**, 239–247.

Srikant, C.B., and Patel, Y. C. (1985). Somatostatin receptors. *In* "Somatostatin" (Y. C. Patel, and G. S. Tannenbaum, Eds.), pp. 291–304, Plenum, New York.

Susini, C., Esteve, J. P., Bommelaer, G., Vaysse, N., and Ribet, A. (1978). Inhibition of exocrine pancreatic secretion by somatostatin in dogs. *Digestion* **18**, 384–393.

Susini, C., Esteve, J. P., Vaysse, N., Pradayrol, L., and Ribet, A. (1980). Somatostatin-28: Effects on exocrine pancreatic secretion in conscious dogs. *Gastroenterology* **79**, 720–724.

Taylor, I. L. (1989). Pancreatic polypeptide family: pancreatic polypeptide, neuropeptide Y, and peptide YY. *In* "Handbook of Physiology. The Gastrointestinal System II" (G. M. Makhlouf, and B. B. Rauner, Eds.), pp. 475–543. American Physiological Society, Maryland.

Thermos, K., Meglasson, M. D., Nelson, J., Lounsbury, K. M., and Reisine, T. (1990). Pancreatic β cell somatostatin receptors. *Am. J. Physiol.* **259**, E216–224.

Thulin, L., Tyden, G., Sammegard, H., Muhrbeck, O., and Efendic, S. (1979). Treatment of bleeding esophageal varices with somatostatin. *Acta Chir. Scand.* **145**, 395–398.

Williams, J. A., and Goldfine, I. D. (1993). The insulin-acinar relationship. *In* "The Exocrine Pancreas: Biology, Pathobiology and Diseases" (V. L. W. Go, F. P. Brooks, E. P. DiMagno, J. D. Gardner, E. Lebenthal, and G. A. Scheele, Eds.), pp. 789–802. Raven Press, New York.

Williams, J. A., Korc, M., and Dormer, R. L. (1978). Action of secretagogues on a new preparation of functionally intact isolated pancreatic acini. *Am. J. Physiol.* **235**, E517–524.

Wizemann, V., Weppler, P., and Mahrt, R. (1974). Effect of glucagon and insulin on the isolated exocrine pancreas. *Digestion* **11**, 432–435.

You, C. H., and Chey, W. Y. (1988). Atropine abolishes potentiation effect of secretin and CCK-8 on exocrine pancreatic secretion in humans. *Pancreas* **3**, 99–103.

Youngs, G. (1972). Hormonal control of pancreatic endocrine and exocrine secretion. *Gut* **13**, 154–161.

Yule, D. I., Kim, E. T., and Williams, J. A. (1994). Tyrosine kinase inhibitors attenuate capacitative calcium influx in rat pancreatic acinar cells. *Biochem. Biophys. Res. Commun.* **202**, 1697–1704.

Zyznar, E. S., Pietri, A. O., Harris, V., and Unger, R. H. (1981). Evidence for the hormonal status of somatostatin in man. *Diabetes* **30**, 883–886.

14

NEUROENDOCRINE REGULATION OF IMMUNE FUNCTION IN FISH

Yuwaraj K. Narnaware and Norman Y. S. Woo

Biology Department
The Chinese University of Hong Kong
Shatin, New Territories
Hong Kong, China

1. INTRODUCTION

Immune defense mechanisms of fish have not been studied as extensively as those of mammals but they appear to share a number of structural and functional characteristics important in the humoral, cell-mediated and non-specific aspects of the immune response (Ellis, 1982; Iwama and Nakanishi, 1996). The immune system of fish is influenced not only by environmental factors such as stress (Schreck, 1996; Narnaware *et al.*, 1994a; Narnaware and Baker, 1996), temperature (Bly and Clem, 1992), pollution (Pulsford *et al.*, 1995), environmental salinity (Narnaware *et al.*, 1998b) but also by endogenous factors such as steroid hormones (Wang and Belosevic, 1994; 1995; Narnaware and Baker, 1996), neurotransmitters (Flory, 1990; Bayne and Levy, 1991a, b; Narnaware *et al.*, 1994a), products of the sympathetic nervous system (Flory, 1990) and cytokines and interferons (Francis and Ellis, 1994; Tahir and Secombes, 1996). Thus stimulation of fish immune responses through environmental and/or hormonal manipulation is of prime importance in protecting fish against diseases. In this chapter, we shall review the existing literature on the modulation of fish immune responses by peptide hormones, neurotransmitters, cytokine-like factors, and interferons. The relationship between the neuroendocrine and the immune system of fish will be discussed.

2. NEURAL REGULATION OF IMMUNE FUNCTION IN MAMMALS

In mammals, it has been established that the brain influences immune functions by releasing the neuropeptide, corticotropin-releasing-factor (CRF) (Jain *et al.*, 1991). CRF acts as an endogenous neurotransmitter in the brain and primarily regulates the synthesis

Neural Regulation in the Vertebrate Endocrine System, edited by Prasada Rao and Peter, Kluwer Academic / Plenum Publishers, New York, 1999.

and release of pro-opiomelanocortin (POMC)-derived peptides from the pituitary gland (Harbuz and Lightman, 1992). Endogenous CRF functions as a stress neurotransmitter in the central nervous system (CNS) to produce effects on behavior (Sutton *et al.*, 1982) and activates the autonomic nervous system (ANS) by enhancing catecholamine levels (Brown *et al.*, 1982). CRF (Hollander *et al.*, 1983), CRF mRNA (Thompson *et al.*, 1987), and CRF binding sites have been identified in mammalian lymphoid tissues (Dave *et al.*, 1985). CRF stimulates the production of ACTH and β-endorphin by leukocytes (Smith *et al.*, 1986) and these products have been implicated to mediate CRF action on superoxide generation (Sharp *et al.*, 1985).

The administration of CRF to rats and mice causes a decrease in splenic (Jain *et al.*, 1991; Irwin *et al.*, 1988), peritoneal and blood natural killer (NK) cell activity (Strausbaugh and Irwin, 1992) and suppresses *in vitro* antibody production (Leu and Singh, 1993). These effects are mediated through CRF receptors on splenocytes (Dave *et al.*, 1985), suggesting that CRF communicates with the immune system through CRF receptors. In the brain, CRF is produced not only in response to stress, but also in response to infections, which can induce interleukin-1 (IL-1) production by brain immune cells, astrocytes and microglia (Bredar *et al.*, 1988). It is this centrally-produced IL-1 which stimulates CRF production and release. Thus, cytokines will depress NK cell activity indirectly, by an effect on CRF production (Saperstein *et al.*, 1992).

3. REGULATION OF FISH IMMUNITY

3.1. Central Nervous System

The ability of the CNS to influence a variety of immune processes has been established in mammals. However, few studies address the effects of the peripheral and central nervous systems on immune function of lower vertebrates. In fish, CRF-like neuropeptides are synthesized in the CNS (Mancera and Fernandez-Llebrez, 1995; Baker *et al.*, 1996) and it is the centrally-produced CRF that has been shown to stimulate synthesis of POMC-derived peptides (Baker, 1991; Baker *et al.*, 1996). Regulation of piscine immune system by POMC-derived peptides has been reported (Bayne and Levy, 1991a, b; Harris and Bird, 1997) but the effect of CRF on fish immune responses has not been fully established. Neurotransmitters also modulate various facets of piscine immune function (Schreck, 1996) and some of these modulatory effects are outlined below.

3.1.1. Catecholamines. Catecholamines are neurotransmitters which also possess immunomodulatory functions in fish. The spleen of coho salmon (*Oncorhynchus kisutch*) contains a rich adrenergic innervation and lesioning of adrenergic nerve terminals with the neurotoxin, 6-hydroxydopamine resulted in stimulation of antibody secretion cells in immunized fish (Flory, 1989). Subsequent studies by the same author on the rainbow trout (*Oncorhynchus mykiss*) have established that both adrenergic and cholinergic agents can influence directly the *in vitro* antibody response of splenic leukocytes (Flory, 1990). In experiments involving the culture of splenic leukocytes, the β-adrenergic receptor agonist, isoprenaline reduced, while the α-adrenergic receptor agonist, phenylephrine and the cholinergic agent agonist, carbachol increased the number of cells secreting antibody to trinitrophenyl-lipopolysaccharide (TNP-LPS) (Flory, 1990).

Experiments have been carried out to elucidate the modulatory effect of catecholamines on fish macrophage function. Rainbow trout (*Oncorhynchus mykiss*) injected with noradrenaline at 0.25–2.5 mg/fish exhibited significant decreases in macrophage phagocytosis and circulating lymphocyte count (Narnaware *et al.*, 1994a). In silver sea bream (*Sparus sarba*) injected with adrenaline at 50 μg/fish, a significant depression of these phagocytic activities occurred together with lowered lymphocyte count (Narnaware *et al.*, 1997b). The suppression of macrophage phagocytosis caused by noradrenaline could be prevented by phentolamine, suggesting that macrophage suppression by adrenergic agents was mediated through α-adrenergic receptors present on macrophages (Narnaware and Baker, 1996).

Interestingly, both the decrease in macrophage phagocytosis and circulating lymphocyte count caused by saline injection stress or noradrenaline could be prevented by prior cortisol and phentolamine administration (Narnaware and Baker, 1996). However, when cortisol and noradrenaline were administered together, phagocytosis was significantly suppressed as if noradrenaline were injected alone, indicating that cortisol did not antagonize the effect of exogenous noradrenaline (Narnaware and Baker, 1996) but may antagonize its release. The reduction in splenic weight of fish produced by noradrenaline administration was reversed by phentolamine injection (Narnaware and Baker, 1996). This suggested that if macrophage phagocytosis was depressed by noradrenaline, it might be due to the emigration of active macrophages out of the spleen, leaving only 'inert' macrophages behind. Phentolamine may have prevented macrophage emigration or 'retained' them in those organs for maintaining phagocytic functions. These data suggest that neurotransmitters affect macrophage phagocytosis through cell trafficking.

Whether neurotransmitters could exert their effects directly on macrophages was further elucidated using *in vitro* studies. Thus rainbow trout macrophages stimulated by phorbol myristate acetate (PMA) and incubated with the α-adrenergic agonist, phenylephrine, exhibited enhanced respiratory burst activity as measured by chemiluminescence but a decrease in burst activity was registered when the cytochrome C method was used. The α-adrenergic antagonist, phentolamine however, reduced the influence of phenylephrine and suggested that the change in respiratory burst activity was mediated through α-receptors (Bayne and Levy, 1991a, b). In other studies, addition of epinephrine, phenylephrine or carbachol enhanced the ability of lipopolysaccharide (LPS)-, concanavalin A (Con A)- and phytohemagglutinin (PHA)-stimulated pronephric leukocytes (mainly macrophages and neutrophils) to produce H_2O_2 and O_2 (Bayne and Levy 1991a, b). On the other hand, the β-adrenergic agonist, isoprenaline depressed these functions (Flory and Bayne, 1991).

In other experiments, opposite responses were found when macrophage phagocytosis was determined. Incubation of splenic or pronephric macrophages with 10 μM phenylephrine or isoprenaline resulted in depression of both percent phagocytosis and phagocytic index (Narnaware *et al.*, 1994a). It seems that adrenergic stimulation did not necessarily alter phagocytic and respiratory burst activity in the same direction.

3.1.2. Serotonin. Recent evidence shows that the neurotransmitter, serotonin (5-HT) exerts immunosuppressive effects on LPS- and PHA-stimulated proliferation of rainbow trout peripheral blood lymphocytes (Ferriere *et al.*, 1996). Furthermore, these authors also showed that the serotonin receptor agonist, 8-OH-DPAT inhibited lymphocyte proliferation. On the other hand, addition of the serotonin receptor antagonist, spiperone, reversed these inhibitory actions of serotonin suggesting that inhibition of lymphocyte proliferation was a receptor-mediated phenomenon and was mediated through serotonin receptors present on lymphocytes. However, the effect of serotonin on macrophage phagocytosis and circulating lymphocyte numbers remains to be investigated.

3.2. Regulation of Fish Immunity by Pituitary Hormones

The hormonal regulation of macrophage function has been the subject of intense investigations. Macrophage function is an important element in resistance to infectious diseases in fish. They are the principal phagocytic cells and play an integral role in the non-specific immune response (MacArthur *et al.*, 1984). Several factors, including environmental stress (Narnaware *et al.*, 1994a), pollution (Pulsford *et al.*, 1995), hormones, neurotransmitters (Sakai *et al.*, 1996a, b, c; Narnaware *et al.*, 1997a, b; 1998a) and nutrition (Blazer, 1991; Narnaware *et al.*, 1998b) are known to affect macrophage function.

Apart from the above factors, numerous pituitary hormones possess regulatory functions on the piscine immune system. These hormones include growth hormone (GH) (Sakai *et al.*, 1996 a, b; Narnaware *et al.*, 1997a), prolactin (PRL) (Sakai *et al.*, 1996b, c; Narnaware *et al.*, 1997b; 1998a) and adrenocorticotrophic hormone (ACTH) (Bayne and Levy, 1991a, b).

3.2.1. Growth Hormone (GH). GH is a polypeptide hormone produced by the vertebrate pituitary and shares a common molecular structure with PRL (Rand-Weaver and Kawauchi, 1993). Besides being actively involved in promoting growth, GH also participates in regulating metabolism, facilitating sea water adaptation (Sakamoto *et al.*, 1993; 1997), and stimulating gonadal steroidogenesis (Singh *et al.*, 1988).

As far as immunoregulatory effects of GH are concerned, this hormone is a 'macrophage activating factor' in fish (Calduch-Giner *et al.*, 1997). Its immunomodulatory role was evident when silver sea bream (*Sparus sarba*) were injected with exogenous GH at 1μg/g body weight for 7 days. In this fish, GH administration stimulated both pronephric and splenic macrophages by increasing their percent phagocytosis (i.e., % macrophages engulfing any number of yeast cells) and phagocytic index (i.e., mean number of yeast cells engulfed per macrophage in a population of 200 macrophages) (Narnaware *et al.*, 1997a). In the rainbow trout (*Oncorhynchus mykiss*) injected with GH, macrophage respiratory burst activity was markedly stimulated (Sakai *et al.*, 1995). Exogenous GH not only stimulated macrophage functions but also significantly enhanced other leukocyte functions such as hemolytic activity of serum (Sakai *et al.*, 1996a), non-specific cytotoxicity of blood, kidney and spleen leukocytes (Kajita *et al.*, 1992) and mitogenic responses of lymphocytes (Sakai *et al.*, 1996b). In a recent study, rainbow trout (*Oncorhynchus mykiss*) injected with GH exhibited protection against an experimental challenge of *Vibrio anguillarum* (Sakai *et al.*, 1997). This suggests that apart from exerting a modulatory effect on the function of macrophage and other immune cells, this hormone is also involved in protecting fish against infection. However, the protective role of GH against other fish pathogens has not been systematically investigated.

Macrophage function is also affected by environmental salinity. Thus, a direct transfer of rainbow trout from fresh water to sea water resulted in a significant increase in macrophage phagocytic activity and their lysozyme concentration leaving non-specific cytotoxicity unchanged (Marc *et al.*, 1995). It is not known, however, whether this increase in macrophage function in sea water is related to changes in GH levels.

From the above experiments, it cannot be ascertained whether GH can stimulate immune cells directly. To tackle this problem, *in vitro* experiments were carried out in which macrophages were isolated from head-kidney and incubated with physiological concentrations of GH. Rainbow trout (*Oncorhynchus mykiss*) macrophages incubated *in vitro* with GH at concentrations ranging from 10–100 ng/ml displayed augmented macrophage phagocytosis and superoxide production (Sakai *et al.*, 1996b). Furthermore, when macrophages isolated from the head-kidney and spleen of silver sea bream (*Sparus sarba*) were incubated with 100 ng/ml GH for only 6 hours, macrophage phagocytosis was stimulated (Narnaware *et al.*, 1997b). Thus, GH can exert direct stimulatory effects on macrophages within a relatively short duration and as low as 10 ng/ml concentration could be effective in stimulating macrophage function.

3.2.2. Prolactin (PRL). PRL is another polypeptide hormone that is produced by the anterior pituitary and possesses stimulatory effects on the immune system (Sakai *et al.*, 1996b; Narnaware *et al.*, 1998a). Among its many effects in fish (Hirano, 1986), its osmoregulatory role is well established (Sakamoto *et al.*, 1997). PRL is believed to act by controlling sodium permeability across the branchial epithelium, thereby preventing excessive Na^+ loss in fresh water (Ogasawara *et al.*, 1989). In addition to its hypernatremic effect, its hypercalcemic action is also important (Kaneko and Hirano, 1993).

Immunomodulatory effects of PRL in fish are not as well studied as its osmoregulatory effects. Silver seabream (*Sparus sarba*) injected with PRL at 6 μg/g body weight for 7 days exhibited a significant enhancement of macrophage phagocytosis (Narnaware *et al.*, 1998a). From this experiment alone it was not clear whether PRL could act directly on the macrophages. This was confirmed when rainbow trout head-kidney macrophages were incubated with PRL at concentrations between 10–100 ng/ml overnight. *In vitro* incubation of macrophages with PRL stimulated not only macrophage phagocytosis but also their superoxide anion production (Sakai *et al.*, 1996a). In another experiment, silver sea bream pronephric and spleen macrophages were incubated *in vitro* with PRL at 100 ng/ml for 6 hours and macrophage phagocytic activity was clearly stimulated (Narnaware *et al.*, 1997b). These *in vitro* experiments show that PRL may exert its immunostimulatory effect directly. Mitogenic activity of chum salmon (*Oncorhynchus keta*) leukocytes was enhanced when PRL was added to the cultured leukocytes at 10–100 ng/ml (Sakai *et al.*, 1996c).

Few studies have provided evidence to illustrate the mechanism of immunostimulation by both GH and PRL. Recently, Calduch-Giner *et al.* (1995) detected the presence of GH receptors on monocytes and lymphocytes of head-kidney of gilthead sea bream (*Sparus aurata*). PRL receptors were also detected in tilapia (*Oreochromis mossambicus*) kidney (Sandra *et al.*, 1995). These studies suggest that immunoregulation by GH and PRL is probably mediated through receptors present on the leukocytes. However, there is as yet no report of the presence of GH and PRL receptors in other lymphoid organs of fish such as the spleen and thymus.

3.2.3. Somatolactin (SL). SL is a putative pituitary hormone related structurally to both GH and PRL (Rand-Weaver and Kawauchi, 1993). To date, the presence of SL has only been reported in several fish species but not in other vertebrates. Physiological functions of SL is still not fully understood although it has been implicated in calcium regulation, acid-base balance, background adaptation, stress and reproduction (Kakizawa *et al.*, 1993). Exposure of rainbow trout (*Oncorhynchus mykiss*) to high calcium environments reduced the activity of SL-producing cells located in the pars intermedia. Conversely, increased activity of SL cells was observed in low-calcium environments suggesting hypercalcemic action as one possible function of SL (Kaneko and Hirano, 1993; Kakizawa *et al.*, 1995).

Since the immunostimulatory effect of GH and PRL has been well documented, studies were conducted to elucidate whether SL has similar modulatory effect on the fish immune system (Sakai *et al.*, 1996b). Apparently, SL had no effect on either macrophage phagocytosis or their extracellular and intracellular superoxide production when trout macrophages were incubated overnight with SL at concentrations ranging from 10–100 ng/ml. It is possible that the functions of SL may be restricted to that of osmoregulation in fish.

3.2.4. Adrenocorticotrophic Hormone (ACTH). It is known that lymphocytes can synthesize products of the POMC gene (Smith *et al.*, 1986). One of the product of POMC precursor is ACTH. This hormone when added to rainbow trout macrophages pre-incubated with phorbol myristate-acetate (PMA) at 5–10 μg/ml enhanced the apparent hydrogen peroxide and superoxide anion production as measured by both the chemiluminescent and cytochrome C methods (Bayne and Levy, 1991a, b).

3.2.5. Melanocyte-Concentrating-Hormone (MCH) and Melanocyte-Stimulating Hormone (α-MSH). In teleost fishes, MCH and α-MSH are two antagonistic hormones which are produced by hypothalamic neurons and are involved in physiological color change (Baker, 1991, 1994), stress (Bird and Baker, 1992; Suzuki *et al.*, 1995), and the activation of hypothalamo-pituitary-interrenal axis (HPI) (Green and Baker, 1991; Baker and Bird, 1992). The neurohypophyseal lobe of the pituitary is the main projection target for MCH fibers (Bird *et al.*, 1989) where this neuropeptide is stored (Barber *et al.*, 1987). In the trout, the abundance of immunoreactive MCH, its synthesis in the brain, and its release from the brain and pituitary depend upon the color of the background (Baker, 1991; Narnaware *et al.*, 1994b). Moreover, changes in the pituitary content of MCH in response to background color have also been observed (Bird *et al.*, 1989). However, the involvement of these two hormones in the regulation of fish immune system has not been amply studied. Evidence is emerging to reveal the possibility that mitogen-induced proliferation of blood lymphocytes is influenced by background coloration (Harris and Bird, 1997). Moreover, application of MCH and α-MSH to lymphocytes *in vitro* modulated their proliferation (Harris and Bird, 1997). Fish adapted to a black background had lower lymphocyte numbers and higher plasma cortisol level than their white background-adapted counterparts. Macrophage phagocytosis was however unchanged in fish adapted to both backgrounds (Narnaware and Baker, unpublished observation), suggesting that α-MSH released upon black background adaptation, may have depressed lymphocyte function.

3.3. Regulation of Fish Immunity by Steroid Hormones

3.3.1. Estradiol. Estradiol appears to have a differential effect on fish macrophages. Goldfish (*Carassius auratus*) macrophages when incubated with estradiol *in vitro* displayed inhibited chemotaxis and phagocytosis, but the respiratory burst activity was not affected (Wang and Belosevic, 1995). Estradiol also suppresses other immune parameters such as mitogen-induced proliferation of lymphocytes *in vitro* and further decreases disease resistance of goldfish against an experimental challenge of hemoflagellate, *Trypanosoma danilewskyi* (Wang and Belosevic, 1994).

3.3.2. Testosterone. Slater and Schreck (1993) have demonstrated that testosterone is nearly as immunosuppressive as cortisol when salmonid lymphocytes were incubated *in vitro* with this androgen. In other experiments, incubation of lymphocytes with physiological levels of testosterone and cortisol, produced additive effects on suppression of antibody forming cells (Maule and Schreck, 1990; Slater *et al.*, 1995a, b). This immunosuppression was a receptor-regulated process (Maule and Schreck, 1990; Slater *et al.*, 1995a, b). Recently, Suzuki *et al.* (1997) have demonstrated that IgM changes in rainbow trout became reversed in response to a significant increase in plasma testosterone and estradiol in females and testosterone and 11-keto-testosterone in males. An association between the decline in IgM and immunosuppression was indicated by the increased susceptibility of these fish to fungal diseases. However, direct experimentation to evaluate the role of testosterone on disease resistance in fish has not been carried out.

3.3.3. Glucocorticoids. Glucocorticoids are the most studied hormones as far as effect on the immune system is concerned. In fish, as in mammals, numerous examples can be cited regarding the suppressive effects of cortisol on the immune system (Schreck, 1996). Cortisol released upon stress-induced activation of the hypothalamo-pituitary-interrenal (HPI) axis suppresses lymphocyte proliferation (Maule *et al.*, 1987; 1989), circulating lymphocyte numbers (Pickering *et al.*, 1987; Narnaware *et al.*, 1994a), the number of antibody producing cells (Maule *et al.*, 1987; 1989), and antibody production (Wechesler *et al.*, 1986). The overall result is increased susceptibility to fungal and bacterial infections (Maule *et al.*, 1989), and high mortality in immunized fish (Houghton and Matthews, 1986). Coho salmon (*Oncorhynchus kisutch*) when implanted with cortisol exhibited reduced number of spleen plaque forming cells (PFC), splenic lymphocytes, circulating leukocytes, and increased mortality when fish were exposed to *Vibrio anguillarum* (Maule *et al.*, 1989).

Glucocorticoids either injected *in vivo* or added to a culture of macrophages *in vitro* suppressed macrophage phagocytic activity. Thus, injection of dexamethasone to rainbow trout resulted in suppression of macrophage phagocytosis (Narnaware *et al.*, 1994a). Cortisol also reduced the number of monocytes/macrophages and neutrophils at the site of inflammation (MacArthur *et al.*, 1984). Glucocorticoids are not only immunosuppressive but can also prevent stress-induced effects on the immune cells. In rainbow trout, the stress of saline injection alone caused a significant depression in macrophage phagocytosis and reduction in the numbers of circulating lymphocytes but cortisol administration following saline injection abolished these effects of stress (Narnaware *et al.*, 1996). The increase in circulatory cortisol may have acted on the HPI axis through negative feedback, preventing the release of other hormones and neurotransmitters such as catecholamines, which acts to prevent its suppressive action on macrophage and lymphocyte numbers (Narnaware and Baker, 1996).

A direct depressive effect of cortisol on macrophage phagocytic activity was obtained by incubating macrophages with varying concentrations of cortisol. For instance, when rainbow trout pronephric and splenic macrophages were incubated with 80 ng/ml cortisol, significant depression in macrophage phagocytic activity was observed. In the dab, (*Limanda limanda*) this depression however, was only significant when kidney and splenic macrophages were incubated with 320 ng/ml of cortisol (Pulsford *et al.*, 1995). Lymphocytes from the same fish showed significant depression in proliferative rates when incubated with cortisol at 160–320 ng/ml (Pulsford *et al.*, 1995). *In vitro* incubation of PMA-stimulated pronephric phagocytic cells of striped bass (*Morone saxatilis*) with hydrocortisone acetate at concentrations between 28–280 μg/ml significantly suppressed macrophage superoxide production as revealed by the chemiluminescent method (Stave and Roberson, 1985). In another study by Wang and Belosevic (1995) on goldfish, cortisol suppressed macrophage phagocytosis and chemotaxis at 0.5–0.1 μM, leaving respiratory burst activity unchanged. From these studies, it was concluded that glucocorticoids not only depressed macrophage phagocytosis but also exerted direct effects on piscine immune function.

The main question that remains to be answered is the mechanism of immunosuppression and stimulation. In response to GH and PRL, it was observed that the number of macrophages which engulfed any yeast cells (i.e., active) was increased in the spleen and pronephros, thus increasing macrophage percent phagocytosis and phagocytic index significantly (Narnaware *et al.*, 1997a; 1998a). However, cortisol and adrenergic agent administration caused an increase in the number of macrophages that did not engulf any foreign particles (i.e., inert). It is possible that in response to cortisol and adrenergic agents, these macrophages could have emigrated to other parts of the fish body such as intestine, blood or general body stroma, leaving only inert macrophages in those organs which resulted in an overall depression of macrophage phagocytic activity. It is also possible that these agents may have changed the adhesiveness of macrophages to glass slides, a phenomenon which would influence the results of the phagocytic assays. Relocation of macrophages to other parts of the body in response to stress and cortisol injection has been shown to occur in the rainbow trout (*Oncorhynchus mykiss*) (Narnaware and Baker, 1996), dab (*Limanda limanda*) (Pulsford *et al.*, 1995) and Chinook salmon (*Oncorhynchus tshawytscha*) (Maule *et al.*, 1989).

In mammals, the rapid change in leukocyte count due to stress or physical exertion is believed to result from an alteration in the adhesive interaction between the white blood cells and various tissue stroma (Ottaway and Husband, 1994). Loss of adhesiveness leads to the release of leukocytes into the circulation while enhanced adhesiveness causes their entrapment in the lymphatic or other tissues, thus removing them from the circulation. Several traffic-directing surface molecules have been identified on leukocytes (i.e. selectin and integrin) and reciprocal molecules occur on capillary endothelial cells. The abundance of such surface molecules changes rapidly in response to various hormones, neurotransmitters, and cytokines (Ottaway and Husband, 1994). Similar changes in adhesiveness and leukocyte trafficking may also explain the hormone-induced changes in macrophage phagocytic activity and lymphocyte numbers in fish.

3.4. Cytokine-Like Factors Released by Fish Immune System

Fish immune cells have been observed to release cytokine-like factors upon *in vivo* injection of mitogen-induced lymphocyte supernatant or when cells were incubated with supernatant *in vitro*. Addition of supernatant to goldfish macrophages not only stimulated phagocytosis and respiratory burst activity but also increased nitric oxide production (Neumann *et al.*, 1995). Similarly, salmon head-kidney and blood leukocytes when exposed to outer membrane protein (OMPs) isolated from *Aeromonas salmonicida* produced macrophage activating factors (MAFs) when assessed using respiratory burst activity (Francis and Ellis, 1994). When macrophages of rainbow trout (Jang *et al.*, 1995; Novoa *et al.*, 1996), goldfish (Neumann and Belosevic, 1996) or dab (Tahir and Secombes, 1996) were incubated with leukocyte-derived MAFs or tumor necrosis-factor (TNF), significantly enhanced respiratory burst occurred. These and other evidence point to the existence of MAFs and TNF in leukocyte supernatant. Elevation of respiratory burst activity by MAFs and TNF could be partially inhibited by prior treatment of the target macrophages with anti-MAFs and anti-TNF receptor monoclonal antibodies (Jang *et al.*, 1995). This strongly supports the notion that fish leukocytes secrete a variety of potent macrophage activating factors.

3.5. Bi-Directional Communication between the Endocrine and Immune Systems in Fish

In mammals, both the neuroendocrine and immune systems communicate with each other by sharing a common set of hormones, neurotransmitters and receptors. Most of these neuroendocrine hormones released under physiological conditions regulate a variety of immune parameters. Cytokines and interferons released by antigen-stimulated immune cells exert effects on the neuroendocrine center to release hormones and neurotransmitters that in turn regulate the immune system, thus completing bi-directional communication between the two systems (Blalock, 1989; 1994).

In fish, only limited hormones and neurotransmitters have been shown to regulate the immune system (Flory and Bayne, 1991; Narnaware *et al.*, 1994a; Sakai *et al.*, 1996a, b, c). Cytokine and MAF-like factors are known to be released by leukocytes but their effect on the neuroendocrine system remains to be investigated. The available study is that of Balm *et al.* (1995) who treated tilapia (*Oreochromis mossambicus*) pituitary lobe and head-kidney *in vitro* with murine interleukin-1 and LPS from *Escherichia coli*. LPS, but not interleukin-1, stimulated epidermal thickness and the interrenal as judged by the increase in cortisol production in the superfused kidney. At the same time, release of pituitary ACTH and α-MSH were inhibited (Balm *et al.*, 1995). The results from this study support the notion that under physiological conditions there may be a reciprocal communication between the two systems in fish.

4. CONCLUDING REMARKS

In mammals, there is now overwhelming evidence that cytokines, peptide hormones and neurotransmitters, as well as their receptors are endogenous to the brain, and the endocrine and immune systems. In fish, emerging evidence suggests that peptide hormones and neurotransmitters exert modulatory effects on the cells and organs of the immune system. Regarding this, only a limited number of peptide hormones have been studied. There is also a scarcity of studies on peptide hormone receptors and their relationship with immune cells and lymphoid organs. Cytokine-like factors and interferons are released from fish immune cells in response to mitogens and lipopolysaccharides but their effects on the immune system has not yet been fully investigated. Furthermore, unlike their mammalian counterparts, the protective role of most of these peptide hormones against infection in fish has not been investigated. Studies on the regulation of the piscine immune response by peptide hormones and neurotransmitters are of importance in assessing resistance of fish against infections. Neurohormonal enhancement of immune function is thus a promising approach for combating infectious diseases in cultured fish species.

ACKNOWLEDGMENT

This work was supported by Earmarked Research Grants (CUHK 353/95M, CUHK 4135/98M) awarded to NYSW by the Research Grants Council of Hong Kong.

REFERENCES

Baker, B. I. (1991). Melanin-concentrating-hormone (MCH): A general vertebrate neuropeptide. *Intl. Rev. Cytol.* **126,** 1–26.

Baker, B. I. (1994). Melanin-concentrating-hormone updated: Functional consideration. *Trends Endocrinol. Metab.* **5,** 120–126.

Baker, B. I., and Bird, D. J. (1992). The biosynthesis of melanin-concentrating-hormone in trout kept under different conditions of back-ground colour and stress, as determined by an *in vitro* method. *J. Neuroendocrinol.* **4,** 673–679.

Baker, B. I., Bird, D. J., and Buckingham, J. C. (1996). In the trout, CRH and AVT synergize ACTH release. *Regul. Peptides* **67,** 207–210.

Balm, P. H. M., Van-Lieshout, E., Lokate, J., and Wendelaar-Bonga, S. E. (1995). Bacterial lipopolysaccharide (LPS) and interleukin-1 (IL-1) exert multiple physiological effects in the tilapia, *Oreochromis mossambicus* (Teleostei). *J. Comp. Physiol.* **165B,** 85–92.

Barber, L., Baker, B. I., Penny, J. C., and Eberle, A. N. (1987). Melanin-concentrating- hormone inhibits the release of alpha-MSH from pituitary gland. *Gen. Comp. Endocrinol.* **65,** 79–86.

Bayne, C. J., and Levy, S. (1991a). The respiratory burst of rainbow trout, *Oncorhynchus mykiss* (Walbaum), phagocytes is modulated by sympathetic neurotransmitters and the 'neuro-peptide' ACTH. *J. Fish Biol.* **38,** 805–811.

Bayne, C. J., and Levy, S. (1991b). Modulation of the oxidative burst in trout myeloid cells by ACTH and catecholamines: Mechanism of action. *J. Leukocyte Biol.* **50,** 554–556.

Bird, D. J., and Baker, B. I. (1992). An immunological study of the secretory activity of neurones producing melanin-concentrating-hormone and pro-opiomelanocortin-like products in the brain of the trout and carp. *Gen. Comp. Endocrinol.* **28,** 245–251.

Bird, D. J., Baker, B. I., and Kawauchi, H. (1989). Immunocytochemical demonstration of melanin-concentrating-hormone and proopiomelanocortin-like products in the brain of the trout and carp. *Gen. Comp. Endocrinol.* **74,** 442–450.

Blalock, J. E. (1989). A molecular basis for bi-directional communication between the immune and neuroendocrine systems. *Physiol. Rev.* **69,** 1–32.

Blalock, J. E. (1994). The syntax of immune-neuroendocrine communication. *Immunol. Today* **15,** 504–511.

Blazer, V. S. (1991). Piscine macrophage function and nutritional influences: A review. *J. Aquat. Anim. Health* **3,** 77–86.

Bly, J. E., and Clem, L. W. (1992). Temperature and teleost immune functions. *Fish Shellfish Immunol.* **2,** 159–171.

Breder, C. D., Dinarello, C. A., and Saperstein, C. B. (1988). Interleukin-1 immunoreactive innervation of the human hypothalamus. *Science* **240,** 321–324.

Brown, M. R., Fisher, L. A., Spiess, J., Rivier, J., and Vale, W. (1982). Corticotrophin releasing factor: Actions on the sympathetic nervous system and metabolism. *Endocrinology* **111,** 928–931.

Calduch-Giner, J. A., Sitja-Bobadilla, A., Alvarez-Pellitero, P., and Perez-Sanchez, J. (1995). Evidence for a direct action of GH on haemopoietic cells of a marine fish, the gilthead sea bream (*Sparus aurata*). *J. Endocrinol.* **146,** 459–467.

Calduch-Giner, J. A., Sitja-Bobadilla, A., Alvarez-Pellitero, P., and Perez-Sanchez, J. (1997). Growth hormone as an *in vitro* phagocyte-activating-factor in the gilthead sea bream (*Sparus aurata*). *Cell Tissue Res.* **287,** 535–540.

Dave, J. R., Eiden, L. E., and Eskay, R. L. (1985). Corticotropin-releasing factor binding to peripheral tissue and activation of the adenylate cyclase-adenosine 3',5'-monophosphate system. *Endocrinology* **116,** 2152–2159.

Ellis, A. E. (1982). Differences between the immune mechanisms of fish and higher vertebrates. *In* "Microbial Diseases of Fish" (R. J. Roberts, Ed.), pp.1–29. Academic Press, London.

Ferriere, F., Khan, N. A., Troutaud, D., and Deschaux, P. (1996). Serotonin modulation of lymphocyte proliferation via 5-HT1A receptors in rainbow trout (*Oncorhynchus mykiss*). *Dev. Comp. Immunol.* **20,** 273–283.

Flory, C. M. (1989). Autonomic innervation of the spleen of the coho salmon, *Oncorhynchus kisutch*: a histochemical demonstration and preliminary assessment of its immunoregulatory role. *Brain Behav. Immun. 3,* 331–344.

Flory, C. M. (1990). Phylogeny of neuroimmunoregulation: Effects of adrenergic and cholinergic agents on the *in vitro* antibody response of splenic leukocytes from the rainbow trout, *Oncorhynchus mykiss. Dev. Comp. Immunol.* **14,** 283–294.

Flory, C. M., and Bayne, C. J. (1991). The influence of adrenergic and cholinergic agents on the chemiluminescence and mitogenic responses of leukocytes from the rainbow trout, *Oncorhynchus mykiss*. *Dev. Comp. Immunol.* **15,** 135–142.

Francis, C. H., and Ellis, A. E. (1994). Production of a lymphokine (macrophage activating factor) by salmon (*Salmo salar*) leukocytes stimulated with outer membrane protein antigens of *Aeromonas salmonicida*. *Fish Shellfish Immunol.* **4,** 489–497.

Green, J. A., and Baker, B. I. (1991). Influence of repeated stress on the release of melanin- concentrating-hormone in the rainbow trout. *J. Endocrinol.* **128,** 261–266.

Harbuz, M. S., and Lightman, S. L. (1992). Stress and the hypothalamo-pituitary-adrenal axis: acute, chronic and immunological activation. *J. Endocrinol.* **134,** 327–339.

Harris, J., and Bird, D. J. (1997). The effects of alpha-MSH and MCH on the proliferation of rainbow trout (*Oncorhynchus mykiss*) lymphocytes. Abstract presented at XIII International Congress of Comparative Endocrinology, (November 16–21, 1997, Yokohama, Japan).

Hirano, T. (1986). The spectrum of prolactin actions in teleosts. *In* "Comparative Endocrinology: Developments and Direction" (C. L. Ralph, Ed.), pp. 53–74. Alan Liss, New York.

Hollander, C. S., Audhya, T., Russo, M., Passarelli, J., and Schlesinger, D. (1983). Levels of corticotropin releasing factor-like immunoreactivity in mammalian hypothalamic and extra-hypothalamic brain tissue as determined with a monoclonal antibody to the bovine material. *Endocrinology* **112,** 2206–2208.

Houghton, G., and Matthews, R. A. (1986). Immunosuppression of carp (*Cyprinus carpio* L.) to ichthyophthiriasis using the corticosteroid trimcinolone acetonide. *Vet. Immunol. Immunopathol.* **12,** 413–419.

Irwin, M., Haugher, R. L., Brown, M. R., and Britton, K. T. (1988). Corticotropin releasing factor activates the autonomic nervous system and reduces natural cytotoxicity. *Am. J. Physiol.* **255,** R744-R747.

Iwama, G. K., and Nakanishi, T. (1996). "The Fish Immune System: Organism, Pathogen, and Environment." Academic Press, San Diego.

Jain, R., Zwickler, D., Hollander, C. S., Brand, H., Saperstein, A., Hutchinson, B., Brown, C., and Audhya, T. (1991). Corticotropin-releasing-factor modulates the immune response to stress in rat. *Endocrinology* **129,** 329–336.

Jang, S. I., Hardie, L. J., and Secombes, C. J. (1995). Elevation of rainbow trout, *Oncorhynchus mykiss* macrophage respiratory burst activity with macrophage-derived supernatants. *J. Leukocyte Biol.* **57,** 943–947.

Kajita, Y., Sakai, M., Kobayashi, M., and Kawauchi, H. (1992). Enhancement of non-specific cytotoxic activity of lymphocytes in rainbow trout, *Oncorhynchus mykiss* injected with growth hormone. *Fish Shellfish Immunol.* **2,** 155–157.

Kakizawa, S., Kaneko, T., Hasegawa, S., and Hirano, T. (1993). Activation of somatolactin cells in the pituitary of the rainbow trout *Oncorhynchus mykiss* by low environmental calcium. *Gen. Comp. Endocrinol.* **91,** 298–306.

Kakizawa, S., Kaneko, T., Hasegawa, S., and Hirano, T. (1995). Effects of feeding, fasting, background, acute stress, and exhaustive exercise on the plasma somatolactin concentrations in rainbow trout. *Gen. Comp. Endocrinol.* **98,** 137–146.

Kaneko, T., and Hirano, T. (1993). Role of prolactin and somatolactin in calcium regulation in fish. *J. Exp. Biol.* **184,** 31–45.

Leu, S. J. C., and Singh, V. K. (1993). Suppression of *in vitro* antibody production by corticotropin-releasing factor neurohormone. *J. Neuroimmunol.* **45,** 23–30.

MacArthur, J. I., Fletcher, T. C., Pirie, B. J. S., Davidson, R. J. L., and Thomson, A. W. (1984). Peritoneal inflammatory cells in plaice, *Pleuronectes platessa*: Effect of stress and endotoxin. *J. Fish Biol.* **25,** 69–81.

Mancera, J. M., and Fernandez-Llebrez, P. (1995). Localization of corticotropin-releasing factor immunoreactivity in the brain of the teleost *Sparus aurata*. *Cell Tissue Res.* **281,** 569–572.

Marc, A. M., Quentel, C., Severe, A., Le-Bail, P. Y., and Boeuf, G. (1995). Changes in some endocrinological and non-specific immune parameters during sea water exposure in the brown trout. *J. Fish Biol.* **46,** 1065–1081.

Matera, L. (1996). Endocrine, paracrine and autocrine actions of prolactin on immune cells. *Life Sci.* **59,** 599–614.

Maule, A. G., and Schreck, C. B. (1990). The glucocorticoid receptors in leukocytes and gill of juvenile coho salmon (*Oncorhynchus kisutch). Gen. Comp. Endocrinol.* **77,** 448- 455.

Maule, A. G., Schreck, C. B., and Kaattari, S. L. (1987). Changes in the immune system of the coho salmon (*Oncorhynchus kisutch*) during the parr-smolt transformation and after implantation of cortisol. *Can. J. Fish. Aquat. Sci.* **44,** 161–166.

Maule, A. G., Tripp, R. A., and Kaattari, S. L. (1989). Stress alters immune function and disease resistance in Chinook salmon (*Oncorhynchus tshawytscha*). *J. Endocrinol.* **120,** 135–142.

Narnaware, Y. K., and Baker, B. I. (1996). Evidence that cortisol may protect against the immediate effects of stress on circulating leukocytes in the rainbow trout. *Gen. Comp. Endocrinol.* **103,** 359–366.

Narnaware, Y. K., Baker, B. I., and Tomlinson, M. G. (1994a). The effect of various stresses, corticosteroids and adrenergic agents on phagocytosis in the rainbow trout, *Oncorhynchus mykiss*. *Fish Physiol. Biochem.* **13,** 31–40.

Narnaware, Y. K., Suzuki, M., Baker, B. I., Levy, A., and Lightman, S. (1994b). A second group of MCH neurones in the trout brain. Abstract presented at The Annual Meeting of the British Neuroendocrine Group", September 12–13, 1994. Manchester, England.

Narnaware, Y. K., Kelly, S. P., and Woo, N. Y. S. (1997a). Effect of injected growth hormone on phagocytosis in silver sea bream (*Sparus sarba*) adapted to hyper- and hypo-osmotic salinities. *Fish Shellfish Immunol.* **7,** 515–517.

Narnaware, Y. K., Woo, N. Y. S., Kelly, S. P., and Lo, K. M. (1997b). Neuroendocrine regulation of immunity in silver sea bream (*Sparus sarba*). Satellite Symposium on Neural Control of the Endocrine System, November 26–28, 1997, p. 5. (Abstract), Nagpur University, Nagpur, India.

Narnaware, Y. K., Kelly, S. P., and Woo, N. Y. S. (1998a). Stimulation of macrophage phagocytosis and lymphocyte count by exogenous prolactin administration in silver sea bream (*Sparus sarba*) adapted to hyper- and hypo-osmotic salinities. *Vet. Immunol. Immunopathol.* **61,** 387–391.

Narnaware, Y. K., Kelly, S. P., and Woo, N. Y. S. (1998b). Effect of salinity and ration size on macrophage phagocytosis in juvenile sea bream (*Mylio macrocephalus*). *J. Appl. Ichthyol.* (In press).

Neumann, N. F., and Belosevic, M. (1996). Deactivation of primed respiratory burst response of goldfish macrophages by leukocyte-derived macrophage activating factor(s). *Dev. Comp. Immunol.* **20,** 427–439.

Neumann, N. F., Fagan, D., and Belosevic, M. (1995). Macrophage activating factor(s) secreted by mitogen stimulated goldfish kidney leukocytes synergize with bacterial lipopolysaccharide to induce nitric oxide production in teleost macrophages. *Dev. Comp. Immunol.* **19,** 473–482.

Novoa, B., Figueras, A., Ashton, I., and Secombes, C. J. (1996). *In vitro* studies on the regulation of rainbow trout (*Oncorhychus mykiss*) macrophage respiratory burst activity. *Dev. Comp. Immunol.* **20,** 207–216.

Ogasawara, T., Hirano, T., Akiyama, K., Arai, S., and Tagawa, M. (1989). Changes in plasma prolactin and growth hormone concentrations during fresh water adaptation of juvenile chum salmon (*Oncorhynchus keta*) reared in sea water for a prolonged period. *Fish Physiol. Biochem.* **7,** 309–313.

Ottaway, C. A., and Husband, A. J. (1994). The influence of neuroendocrine pathways on lymphocyte migration. *Immunol. Today* **6,** 97–116.

Pickering, A. D., Pottinger, T. G., and Sumpter, J. P. (1987). On use of dexamethasone to block the pituitary-interrenal axis in the brown trout, *Salmo trutta* L. *Gen. Comp. Endocrinol.* **65,** 346–353.

Pulsford, A. L., Crampe, M., Langston, A., and Glynn, P. J. (1995). Modulatory effects of disease, stress, copper, TBT and vitamin E on the immune response of flatfish. *Fish Shellfish Immunol.* **5,** 631–643.

Rand-Weaver, M., and Kawauchi, H. (1993). Growth hormone, prolactin and somatolactin: A structural overview. *In* "Molecular Biology Frontiers" (P. W. Hochachka and T. P. Mommsen, Eds.) pp-39–56. Elsevier, Amsterdam.

Sakai, M., Kobayashi, M., and Kawauchi, H. (1995). Enhancement of chemiluminescence response of phagocytic cells from rainbow trout, *Oncorhynchus mykiss*, injected with growth hormone. *Fish Shellfish Immunol.* **5,** 375–379.

Sakai, M., Kobayashi, M., and Kawauchi, H. (1996a). Increased in haemolytic activity of serum from rainbow trout , *Oncorhychus mykiss* injected with exogenous growth hormone. *Fish Shellfish Immunol.* **6,** 615–617.

Sakai, M., Kobayashi, M., and Kawauchi, H. (1996b). *In vitro* activation of fish phagocytic cells by GH, prolactin and somatolactin. *J. Endocrinol.* **151,** 113–118.

Sakai, M., Kobayashi, M., and Kawauchi, H. (1996c). Mitogenic effect of growth hormone and prolactin on chum salmon, *Oncorhynchus keta* leucocytes *in vitro*. *Vet. Immunol. Immunopathol.* **53,** 185–189.

Sakai, M., Kobayashi, M., and Kawauchi, H. (1997). Immunostimulatory effects of growth hormone: *In vivo* administration of GH in rainbow trout enhances resistance to *Vibrio anguillarum* infection. *Vet. Immunol. Immunopathol.* **57,** 147–152.

Sakamoto, T., McCormick, S. D., and Hirano, T. (1993) Osmoregulatory actions of growth hormone and its mode of action in salmonids: A review. *Fish Physiol. Biochem.* **11,** 155–164.

Sakamoto, T., Sheperd, B. S., Madsen, S. S., Nishioka, R. S., Siharath, K., Richman, N. H. III, Bern, H. A., and Grau, E. G. (1997). Osmoregulatory actions of growth hormone and prolactin in an advanced teleost. *Gen. Comp. Endocrinol.* **106,** 95–101.

Sandra, O., Sohm, F., De-Luze, A., Prunet, P., Edery, M., and Kelly, P. A. (1995). Expression cloning of a cDNA encoding a fish prolactin receptor. *Proc. Natl. Acad. Sci USA* **92,** 6037–6041.

Saperstein, A., Band, H., Audhya, T., Nabriski, D., Hutchinson, B., Rosenzweig, S., and Hollander, C. S. (1992). Interleukin-1-beta mediates stress-induced immunosuppression via corticotropin-releasing-factor. *Endocrinology* **130,** 152–158.

Schreck, C. B. (1996). Immunomodulation: Endogenous factors. *In* "The Fish Immune System: Organism, Pathogen, and Environment", (G. K. Iwama and T. Nakanishi, Eds.) pp 311–337. Academic Press, San Diego.

Sharp, B. M., Keane, W. F., Suh, H. J., Tsukuyama, D., and Peterson, P. K. (1985). Opioid peptides rapidly stimulate superoxide production by human polymorphonuclear leukocytes and macrophages. *Endocrinology* **117,** 793–801.

Singh, H., Griffith, R. W., Takahashi, K., Kawauchi, H., Thomas, P., and Stegeman, J. J. (1988). Regulation of gonadal steroidogenesis in *Fundulus heteroclitus* by recombinant salmon growth hormone and purified salmon prolactin. *Gen. Comp. Endocrinol.* **72,** 144–153.

Slater, C. H., and Schreck, C. B. (1993). Testosterone alters the immune response of Chinook salmon (*Oncorhynchus tshawytscha*). *Gen. Comp. Endocrinol.* **89,** 871–874.

Slater, C. H., Fitzpatrick, M. S., and Schreck, C. B. (1995a). Androgen and immunocompetence in salmonids: Specific binding in and reduced immunocompetence of salmonid lymphocytes exposed to natural and synthetic androgens. *Aquaculture* **136,** 363–370.

Slater, C. H., Fitzpatrick, M. S., and Schreck, C. B. (1995b). Characterization of an androgen receptor in salmonid lymphocytes: Possible link to androgen and immunosuppression. *Gen. Comp. Endocrinol.* **100,** 218–225.

Smith, E. M., Morill, A. C., Meyer, W. J., and Blalock, J. E. (1986). Corticotropin releasing factor induction of leukocyte-derived immunoreactive ACTH and endorphin. *Nature* **321,** 881.

Stave, J. W., and Roberson, B. S. (1985). Hydrocortisone suppresses the chemiluminescent response of striped bass phagocytes. *Dev. Comp. Immunol.* **9,** 77–84.

Strausbaugh, H., and Irwin, M. (1992). Central corticotropin releasing hormone reduces cellular immunity. *Brain Behav. Immun.* **6,** 11–17.

Sutton, R. E., Kobb, G. F., Moal, L. M., Rivier, J., and Vale, W. (1982). Corticotropin releasing factor produces behavioural activation in rats. *Nature* **297,** 331–333.

Suzuki, M., Narnaware, Y. K., Baker, B. I., and Levy, A. (1995). Influence of environmental colour and diurnal phase on MCH gene expression in the trout. *J. Neuroendocrinol.* **7,** 319–328.

Suzuki, Y., Otaka, T., Sato, S., Hou, Y. Y., and Aida, K. (1997). Reproduction related immunoglobulin changes in rainbow trout. *Fish Physiol Biochem.* **17,** 415–421.

Tahir, A., and Secombes, C. J. (1996). Modulation of dab (*Limanda limanda*) macrophage respiratory burst activity. *Fish Shellfish Immunol.* **6,** 135–146.

Thompson, R. C., Seasholtz, A. F., Douglas, J. O., and Herbert, E. (1987). The rat corticotropin-releasing hormone gene. *Ann. N. Y. Acad. Sci.* **512,** 1–11.

Wang, R., and Belosevic, M. (1994). Estradiol increases susceptibility of goldfish to *Trypanosoma danilewskyi*. *Dev. Comp. Immunol.* **18,** 377–387.

Wang, R., and Belosevic, M. (1995). The *in vitro* effects of estradiol and cortisol on the function of a long-term goldfish macrophage cell line. *Dev. Comp. Immunol.* **19,** 327- 336.

Wechsler, S. J., McAllister, P. E., Hetrick, F. M., and Henderson, D.P. (1986). Effect of exogenous corticosteroids on circulating virus and neutralizing antibodies in striped bass (*Morone saxatilis*) infected with infectious pancreatic necrosis virus. *Vet. Immunol. Immunopathol.* **12,** 305–311.

INDEX

GPSR Compliance

The European Union's (EU) General Product Safety Regulation (GPSR) is a set of rules that requires consumer products to be safe and our obligations to ensure this.

If you have any concerns about our products, you can contact us on ProductSafety@springernature.com

In case Publisher is established outside the EU, the EU authorized representative is:

Springer Nature Customer Service Center GmbH
Europaplatz 3
69115 Heidelberg, Germany

FSC
www.fsc.org
MIX
Papier | Fördert gute Waldnutzung
FSC® C083411

Kolibri 360 GmbH
Ferdinand-Jühlke-Straße 7
99095 Erfurt, Deutschland
produktsicherheit@kolibri360.de